T0348745

Gene Expression at the Beginning of Animal Development

ADVANCES IN DEVELOPMENTAL BIOLOGY AND BIOCHEMISTRY

Volume 12

Series Editor

Paul M. Wassarman
Mount Sinai School of Medicine
Mount Sinai Medical Center
New York, USA

GENE EXPRESSION AT THE BEGINNING OF ANIMAL DEVELOPMENT

Editor

Melvin L. DePamphilis

National Institute of Child Health and Human Development
National Institutes of Health
Bethesda, MD,
USA

2002

ELSEVIER
Amsterdam – London – New York – Oxford – Paris – Tokyo – Boston – San Diego –
San Francisco – Singapore – Sydney

Elsevier
Radarweg 29, PO Box 211, 1000 AE Amsterdam, The Netherlands
The Boulevard, Langford Lane, Kidlington, Oxford OX5 1GB, UK

First edition 2002

Library of Congress Cataloging-in-Publication Data

Gene expression at the beginning of animal
development / editor, Melvin L. DePamphilis.—
1st ed.
    p.   cm.—(Advances in developmental biology
    and biochemistry; v. 12)
Includes bibliographical references and index.
ISBN 0-444-51048-6
1. Developmental genetics.   2. Genetic regulation.
3. Embryology.   I. DePamphilis, Melvin L.
II. Series.
QH453.R44   2002
571.8′51—dc21   2002069762

British Library Cataloguing in Publication Data

Gene expression at the beginning of animal
development.—(Advances in developmental
biology and biochemistry; v. 12)
1. Genetic regulation.   2. Gene expression.
3. Embryology.   I. DePamphilis, Melvin L.
572.8′651
ISBN 0-444-51048-6

Transferred to digital print, 2007
Printed and bound by CPI Antony Rowe, Eastbourne

**This book is dedicated to the memory of
Alan P. Wolffe, Ph.D. (1959–2001)**

Most of us would agree that, in the end, it is neither the papers we publish nor the prizes we win that define us, but the life we lived, the lives we touched, and the memories we left behind with those who knew us. Alan Wolffe was truly admired by friends and colleagues alike for his energy, enthusiasm, brilliance, and charm. His knowledge of the chromatin field was legendary. His accomplishments in research were internationally recognized. His reputation as a scientist was first-class. His passion for science was inspirational. His willingness to help others was boundless. Quite simply, Alan was among the best and brightest of us all. We shall miss him.

Contents

Preface

"There are more things in heaven and earth, Horatio, than are dreamt of in our philosophy."

(Shakespeare's *Hamlet*, act 1, sc. 5).

Significance

The beginning of life may be a miracle to some, and a mystery to others, but it is certainly one of the most exciting and perhaps controversial fields of scientific investigation in the 21st century. To understand the molecular mechanisms that allow life to begin, is to understand more clearly what we are, where we have been, and where we are going. Among the metazoa, life begins when an egg is fertilized by a sperm. The sperm provides a genetic blueprint from the father and perhaps some critical proteins. The egg provides a genetic blueprint from the mother together with a large reservoir of mRNAs and proteins that are required for DNA replication, cell division and the onset of zygotic gene expression. All of the thousands of genes in these two mature gametes are transcriptionally silent and remain so until fertilization.

Fertilization triggers a sequence of events that transforms the sperm and egg into a replicate of their species. Soon after fertilization, the paternal and maternal genomes are replicated and the newly formed embryo (or zygote) undergoes its first mitosis and cleaves into two cells, each with a single nucleus that contains one copy of Mom's genome and one of Dad's. The newly formed embryo must then decide when to begin zygotic gene expression (i.e. transcription-dependent gene expression) and which of its thousands of genes to turn on and which to keep silent. Making the right choices results in a healthy adult animal. Making the wrong choices results in its spontaneous abortion. By the time a blastocyst has formed, the first clearly differentiated cells have appeared while the embryo itself still exists as a cluster of "embryonic stem cells" that have the potential to develop into any type of cell within the adult animal. Thus, in the case of humans, some of the most critical decisions occur during the first 7 to 10 cell divisions (the first two weeks of embryonic development) as a single fertilized egg develops into a blastocyst that can then implant into the mother's uterine wall.

These events are characteristic of most, if not all, of the metazoa, but they have been studied most extensively in mice, frogs and flies using a wide variety of techniques in molecular biology, genetics, and biochemistry. Therefore, we have focused on these three biological systems in the hope of providing the reader with a clear understanding of the current state of affairs, and the ability to identify common principles as well as critical differences that are responsible for beginning the process of animal development. Accordingly, we hope that the essays presented here will be of practical value to all those who are interested in improving fertilization in vitro, in designing novel methods of contraception, in developing preimplantation genetic diagnosis for various diseases, in cloning animals by transplanting nuclei from adult cells to an enucleated egg, and in the application of embryonic stem cells to curing genetic diseases or replacing damaged tissues. But above all, we offer these essays to those who simply have an insatiable curiosity about life and its beginnings.

Commonalities

The authors of the essays you will read have noted the following similarities and differences among the metazoa at the beginning of their development.

Spermatogenesis. Reproduction serves to maintain a species, but sexuality (meiosis and fertilization producing genetic variability in offspring) enables each species to adapt to its own environment ("survival of the fittest"). To this end, males and females have developed different strategies for germ cell development. Females produce a few large germ cells before they are borne. Development of these germ cells is regulated through meiosis (two arrest points) and through the accumulation of large amounts of RNA and proteins that will be used to carryout the initial rounds of cell cleavage. In contrast, males produce a great number of small germ cells throughout their life. Development of these germ cells involves expression of transition proteins for DNA-repair and protamines for chromatin condensation resulting in an arrest of transcription. Gene activity is regulated at the level of transcription prior to chromatin condensation and at the level of translation after chromatin condensation.

The mechanism of chromatin condensation via protamine is common to spermatogenesis among most, if not all, species. Despite the presence of multiple genes for transition proteins and for protamines, within a single species, generally only one is expressed for transition protein and one for protamine during spermatogenesis. If two proteins are expressed (e.g. protamine-1 and -2 in man, mouse, horse, and hamster), their ratio is species-specific and thus presumably of fundamental importance for the fertilization capacity of the sperm.

Oogenesis. Oocyte growth takes place while oocytes are continually arrested in meiosis, however, the timing and extent of oocyte growth varies widely among mammalian species. The latter is even more evident when comparing oocyte growth in mammals with other metazoa. Oocyte growth in mammals results in tremendous enlargement of the cell and is characterized by marked changes in oocyte ultrastructure and unusually high transcriptional and translational activity. As a result, the mature egg contributes an extensive array of macromolecules and organelles to the zygote that, to varying degrees, supports the early embryo's nutritional, synthetic, energetic, and regulatory requirements. This situation is similar to that observed in virtually all other vertebrates. However, it is noteworthy that while oocytes from fish, birds, and amphibians accumulate enormous amounts of yolk protein during oogenesis, oocytes from mammals do not. On the other hand, the extracellular coat (vitelline envelope or zona pellucida) that surrounds eggs from fish, birds, amphibians, and mammals, in each case is laid down during oocyte growth. Fully-grown oocytes undergo meiotic maturation and become unfertilized eggs at the time of ovulation.

Meiotic maturation occurs as a result of changes in intercellular communication between follicular components, as well as changes in the levels of various factors, including cyclic AMP, calcium, and steroids. In flies (*Drosophila*), the molecular signals triggering oocyte maturation have not yet been identified.

Early Embryogenesis. The actual time required for early development varies considerably among animal species. For example, in two days, a fertilized fly (*Drosophila*) egg produces the second instar larva, a fertilized frog (*Xenopus*) egg produces a swimming tadpole, but a fertilized mouse egg produces only 4-cells! The larger the maternal stockpile, the more rapid the cell cleavage cycle, and the greater number of cell cleavages that occur in the absence of transcription. Most of this difference reflects the amount of maternally inherited gene products that can support DNA replication and cell cleavage in the absence of zygotic gene expression. Therefore, following fertilization, all animals exhibit a transition from dependence on maternally inherited gene products to dependence on zygotically expressed gene products. Moreover, common to all animals is a series of events following fertilization that include chromatin remodeling, DNA replication, chromatin-mediated global repression of transcription, and acquisition of specific transcription factors and their co-activators that allow this repression to be relieved in selected genes. These changes occur through activation of a preset group of early zygotic genes, followed later by activation of a larger group of zygotic genes.

What differs dramatically is either the time required or the number of nuclear divisions required for this transition to take place, and by implication, the molecular mechanisms that regulate these events. In mammals, there appears to be a time dependent change in the status of a maternally inherited protein, because zygotic gene activation (ZGA) is independent of the number of S-phases or mitoses that occur. Thus, ZGA begins after the amount of time required to produce from 2 to 16-cells, depending on the species, regardless of whether or not cells undergo S- and M-phases of the cell division cycle. In contrast, ZGA in amphibians, fish, sea urchins, and flies appears to be determined by the ratio of nuclei to cytoplasm, suggesting that the genome titrates an inhibitor out of the cytoplasm. ZGA in these animals occurs after a preset number of nuclear divisions has occurred.

In *Xenopus*, the mid-blastula transition consists of three coincident transitions: a progressive loss of cell cycle synchrony, the acquisition of cell motility, and the onset of embryonic transcription. These transitions do not necessarily occur simultaneously in other species. However, embryonic development in most species is characterized by a period during which the zygotic or embryonic genome is transcriptionally quiescent, although the timing of the onset of transcription relative to the embryo's development differs. A potentially important factor in these differences is whether embryonic development is external, as it is in *Xenopus* and *Drosophila*, or occurs in utero, as for example in the mouse. Properties associated with external versus internal development, such as the size of the egg and its rate of development, might determine the extent of pre-patterning of the egg (mosaic versus regulative development), maternal deposition of regulatory components, and the timing of the onset of transcription. Historically, a model in which the cytoplasm-to-nucleus ratio regulates the onset of transcription has enjoyed much interest. However, it has become increasingly clear that the onset of transcription in *Xenopus* is more complicated, and a variety of molecular mechanisms have been implicated, including regulation of the extent to which chromatin is repressive towards transcription, developmental regulation of the subcellular localization of transcription factors, developmental regulation of the abundance of one of the general transcription factors, and developmental regulation of the extent to which transcription is activated over basal levels of transcription.

Drosophila uses modified cell cycles to achieve developmental goals during oogenesis and embryogenesis. The embryo is stockpiled with maternal products necessary for the early embryonic divisions and differentiation events. This is accomplished by the massive production of these proteins and mRNAs by the polyploid nurse cells, followed by their deposition into the oocytes. The use of increased ploidy to augment gene expression is not unique to the nurse cells, being used widely in *Drosophila* tissues, but also as a strategy in mammalian trophectoderm cells that contribute to the placenta. In addition to an increase in genomic ploidy, *Drosophila* employs a more specialized mechanism of amplifying specific genes needed for eggshell production to facilitate high levels and rapid gene expression by the somatic follicle cells during oogenesis.

The presence of large maternal stockpiles in animals such as *Xenopus* and *Drosophila* permits rapid divisions during early embryogenesis that occur in the absence of gene transcription and without cell growth. Consequently, the early embryonic cell cycle does not have gap phases and consists of direct oscillations between S-phase and mitosis. There are unique cell cycle regulators used solely for these S-M cycles. Transcription of the zygotic genome then requires the addition of a gap phase to the cell cycle; this occurs by the insertion of a G2 phase in response to developmental signals.

Early embryogenesis in all animals involves the establishment of a body pattern, but patterning of the anterior-posterior body axis in insects is a well characterized process that is clearly reflected by the segmented body of the larva and adult. In *Drosophila*, the transcription factors that initiate this pattern are pre-loaded into the embryo by the mother, but are prevented from acting until sufficient numbers of cells required for the segmental pattern are produced. The activation and function of these transcription factors and the transcriptional cascade that follows can be traced in some detail. Many of the transcription factors that control this process are shared by vertebrates where they control related aspects of anterior–posterior patterning. However, *Drosophila* is unique in that these transcription factors can interact in an uninterrupted cascade. This is because the first few hours of *Drosophila* development proceed within a cellular syncytium, which alleviates the need for intercellular signaling pathways.

Embryonic Stem Cells. Embryonic stem or ES cells can be made to differentiate into a wide variety of cell types in vitro, because they are pluripotent and when appropriately combined with other embryos in vivo can form an entire mouse. This ability to shuttle the mouse's life cycle between an in vitro and in vivo environment has provided the experimental basis to the genetic manipulation of the mouse's genome that has revolutionized mammalian genetics. The fact that cells with very similar properties to mouse ES cells could be isolated from human blastocysts, has acted as a stimulus to the use of ES cells to generate differentiated cells or tissues that could be used for therapeutic purposes, like treating diseases such as diabetes or Parkinson's. This has greatly motivated the use of these cells to study mammalian development in vitro. Much interest is now centered on understanding the molecular basis of pluripotency in ES cells. How do stem cells proliferate, and how can ES cells be made to differentiate, uniformly, into specific cell types or tissues? In addressing these questions, researchers in the coming years, should make major advances in understanding the molecular basis to not only mouse embryonic growth and differentiation but also of how human embryos begin development, an area that still remains in complete obscurity.

Genomic Imprinting. Genomic imprinting describes the expression of certain genes from either the maternally or paternally inherited copy, a phenomenon that is restricted to mammals in the animal kingdom. Making direct cross-species comparisons therefore seems initially difficult. However, although flies, frogs, fish and worms do not undergo genomic imprinting, the mechanisms which are utilized by mammals to imprint the genome are very similar to those used by lower organisms for silencing genes. Such mechanisms include the use of DNA methylation, modification of histones, and remodeling of higher-order chromatin structure.

The universality of these systems means that students of imprinting can learn much from all organisms. Mechanistic aspects aside, perhaps the most intriguing question to ask is why mammals have imprinting while other animals do not. The most convincing arguments relate to the fact that mammals develop *in utero* while all the other species discussed here do not. Much is made of the "battle of the sexes" hypothesis, whereby paternally expressed genes are responsible for promoting embryonic growth and maternally expressed genes suppress growth. A balance of maternal and paternal contributions may have evolved as a result of this intimate link between maternal resources and competitive paternal genes. Unfortunately, the real-life situation is not as simple as that, with many imprinted genes not fitting neatly into such a model. Perhaps when we understand the evolutionary driving force towards *in utero* development, we will discover why mammals have genomic imprinting but other species do not.

List of Contributors

Katharine L. Arney
Wellcome Trust/Cancer Research UK Institute of Cancer and Developmental Biology,
University of Cambridge, Tennis Court Road, Cambridge CB2 1QR, UK
e-mail: kla21@hermes.cam.ac.uk

Giovanni Bosco
Whitehead Institute for Biomedical Research, Nine Cambridge Center,
Cambridge, MA 02142, USA
e-mail: bosco@wi.mit.edu

Bruce H. Dietrich
Institut Curie, Dynamique Nucléaire et Plasticité du Génome, Pavillon Pasteur,
26 rue d'Ulm, 75248 Paris Cedex 05, France
e-mail: bruce.dietrich@curie.fr

Melvin L. DePamphilis
National Institute of Child Health and Human Development, Building 6, Room 416,
National Institutes of Health, Bethesda, MD 20892-2753, USA
e-mail: depamphm@mail.nih.gov

Sylvia Erhardt
Wellcome Trust/Cancer Research UK Institute of Cancer and Developmental Biology,
University of Cambridge, Tennis Court Road, Cambridge CB2 1QR, UK
e-mail: se226@mole.bio.cam.ac.uk

Luca Jovine
Department of Molecular, Cell and Developmental Biology, Mount Sinai School of
Medicine, One Gustave L. Levy Place, New York, NY 10029-6574, USA
e-mail: jovinl02@doc.mssm.edu

Kotaro J. Kaneko
National Institute of Child Health and Human Development, Building 6, Room 416,
National Institutes of Health, Bethesda, MD 20892-2753, USA
e-mail: KanekoK@mail.nih.gov

Henry M. Krause
Banting and Best Department of Medical Research, University of Toronto and
C.H. Best Institute, 112 College Street, Toronto, Ontario, Canada M5G 1L6
e-mail: h.krause@utoronto.ca

Eveline S. Litscher
Department of Molecular, Cell and Developmental Biology, Mount Sinai School of
Medicine, One Gustave L. Levy Place, New York, NY 10029-6574, USA
e-mail: eveline.litscher@mssm.edu

Andrzej Nasiadka
Abt. Entwicklungsbiologie, Institut für Biologie I, Universität Freiburg,
Hauptstrasse 1, 79104 Freiburg, Germany
e-mail: andrzej.nasiadka@biologie.uni-freiburg.de

Terry Orr-Weaver
Department of Biology, Massachusetts Institute of Technology and Whitehead Institute,
Nine Cambridge Center, Cambridge, MA 02142, USA
e-mail: weaver@wi.mit.edu

Klaus Steger
Institut für Veterinär-Anatomie, -Histologie und Embryologie der Justus-Liebig-
Universität, Frankfurter Strasse 98, 35392 Giessen, Germany
e-mail: Klaus.Steger@vetmed.uni-giessen.de

Colin L. Stewart
Cancer and Developmental Biology Laboratory, Basic Research Division,
National Cancer Institute, Frederick, MD 21702, USA
e-mail: stewartc@ncifcrf.gov

M. Azim Surani
Wellcome Trust/Cancer Research UK Institute of Cancer and Developmental Biology,
University of Cambridge, Tennis Court Road, Cambridge CB2 1QR, UK
e-mail: as10021@mole.bio.cam.ac.uk

Alex Vassilev
National Institute of Child Health and Human Development, Building 6, Room 416,
National Institutes of Health, Bethesda, MD 20892-2753, USA
e-mail: vassilev@mail.nih.gov

Gert Jan C. Veenstra
Department of Molecular Biology, Nijmegen Center for Molecular Life Sciences,
University of Nijmegen, Geert Grooteplein 26–28, 6525 GA Nijmegen, The Netherlands
e-mail: G.Veenstra@ncmls.kun.nl

Paul M. Wassarman
Department of Molecular, Cell and Developmental Biology, Mount Sinai School of
Medicine, One Gustave L. Levy Place, New York, NY 10029-6574, USA
e-mail: paul.wassarman@mssm.edu

Advances in Developmental Biology and Biochemistry, Vol. 12
M. DePamphilis (Editor)

Gene expression during mouse spermatogenesis

Klaus Steger

Institut für Veterinär-Anatomie, -Histologie und Embryologie der Justus-Liebig-Universität,
Frankfurter Strasse 98, 35392 Giessen, Germany

Summary

During spermatogenesis, somatic histones are partially replaced by testis-specific subtypes appearing together with transition proteins. In haploid spermatids, both histones and transition proteins are replaced by protamines. DNA-protamine interactions result in chromatin condensation causing cessation of transcription in elongating spermatids. This occurs at a time when many proteins need to be synthesized and assembled for the complete condensation of the chromatin, the development of the acrosome, and the formation of the flagellum. To ensure complete differentiation of round spermatids into mature spermatozoa, the precise temporal regulation of gene expression via transcriptional and translational control mechanisms is of pivotal importance. These processes involve the binding of transcription factors to the promoter sequence of the genes and protein repressors to the 3´-untranslated region and poly-A tail of the transcripts. Human male infertility has been associated with both decreased levels of protamines and increased protamine-1 to protamine-2 ratios. These findings are in line with data obtained from haploinsufficient mice carrying a mutation in either protamine-1 or protamine-2 resulting in male sterility. Despite some species-specific peculiarities, there is broad conformity between human and murine spermatogenesis recommending mouse an excellent animal model for the study of human spermatogenesis. To further unravel the mechanisms being involved in the regulation of gene expression during spermatogenesis, knockout and transgenic mice will play an important role in answering questions about the function of a specific gene and the sequences being responsible for the manner in which this gene is expressed during spermatogenesis.

Contents

1. Sequential gene expression in differentiating male germ cells

Sequential gene expression in differentiating germ cells is due to synchronized development within the seminiferous epithelium. Within the seminiferous epithelium, spermatogenesis occurs on the surface of somatic Sertoli cells. Functional Sertoli cells are required for normal spermatogenic progression resulting in the continuous production of numerous fertile spermatozoa which, in turn, is necessary to maintain Sertoli cells in their functional state. Adjacent Sertoli cells develop Sertoli–Sertoli junctional complexes which divide the seminiferous epithelium in a basal and an adluminal compartment (Dym and Fawcett, 1970). During spermatogenesis, germ cells migrate from the basal to the adluminal compartment successively passing through the following developmental stages:

- First multiplying phase (mitosis): Following mitosis of a spermatogonial stem cell, one spermatogonium is conserved as a spermatogonial stem cell, while the other spermatogonium undergoes further mitoses and, finally, differentiates into primary spermatocytes. The latter cells double their DNA content and, subsequently, enter meiosis.
- Second multiplying phase (meiosis): During the first meiotic division, one primary spermatocyte (DNA content: 4c) give rise to two secondary spermatocytes (DNA content: 2c each). During the second meiotic division, each of the two secondary spermatocytes give rise to two round spermatids (DNA content: 1c each).
- Differentiation phase (spermiogenesis): Round spermatids do not divide but differentiate into mature spermatozoa undergoing numerous morphological, biochemical, and physiological modifications. During spermiogenesis, nuclear chromatin condensation, development of the acrosome, and formation of the flagellum occur simultaneously in haploid spermatids.

In the course of spermatogenesis, germ cells are subjected to permanent proliferation and differentiation processes resulting in the appearance of various germ cell populations each representing a particular phase of germ cell development. A defined arrangement of germ cell populations is called the stage of the seminiferous epithelium. A complete series of changes in stages arranged in the logical sequence of germ cell maturation is called the cycle of the seminiferous epithelium. In mice, the seminiferous epithelial cycle is divided into twelve stages (Fig. 1).

The differentiation of spermatogonial stem cells into fertile sperm requires stringent temporal and stage-specific gene expression. This becomes evident when studying the sequential nucleoprotein expression in developing germ cells resulting in histone-to-protamine exchange in haploid spermatids (reviewed in Steger, 1999). First, part of the

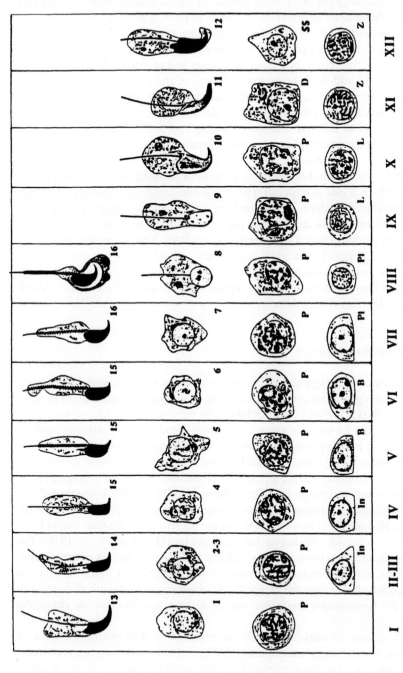

Fig. 1. The cycle of the seminiferous epithelium in the mouse (Oakberg, 1956; Russell et al., 1990). The seminiferous epithelial cycle represents twelve stages (I-XII). Due to the shape of the acrosome, spermatid differentiation is further subdivided into sixteen steps (1-16). To follow the spermatogenic progress, procede horizontally left to right until reaching the far right of a particular row. Then continue from left to right on the next row up and so forth until the stage during which sperm release is indicated. Note that somatic Sertoli cells and A-type spermatogonia both being present at all stages of the seminiferous epithelial cycle are not shown in this cycle map. Intermediate-type spermatogonia (In), B-type spermatogonia (B), primary spermatocytes at preleptotene (PL), leptotene (L), zygotene (Z), pachytene (P), and diplotene (D) phase of meiosis, secondary spermatocytes (SS).

somatic histones are replaced by testis-specific histones. Subsequently, transition protein-1 and -2 occur together with testis-specific histones. Finally, both histones and transition proteins are replaced by protamine-1 and -2. In mice, transcripts of testis-specific histone H1t are expressed in mid and late pachytene spermatocytes, while the corresponding proteins are present from mid pachytene spermatocytes to elongated spermatids (Drabent et al., 1996). Transcripts of transition proteins and protamines are expressed in spermatids of step 7-13 and 7-15, respectively. However, the corresponding proteins are synthesized, with temporal delay, in spermatids of step 12-14 and 13-16, respectively (Mali et al., 1989; Yelick et al., 1989; Alfonso and Kistler, 1993). While the stage-specific but different expression of mRNAs and corresponding proteins is due to the storage of translationally repressed transcripts representing a common phenomenon in haploid spermatids (Pentillä et al., 1995; Kleene, 1996; Cataldo et al., 1999), the sequential gene expression of nucleoproteins in differentiating germ cells is based on synchronized germ cell development which is due to a special type of cell communication, namely intercellular bridges. Intercellular bridges are formed by incomplete cytokinesis during the telophase of mitosis resulting in an open cytoplasmic continuity between germ cells of a clone. As has been demonstrated for protamine and a RNA-binding protein binding to protamine-mRNA, transcripts produced by some of the cells of a clone can move through intercellular bridges and, subsequently, will be expressed in all of the cells of the clone. Haploid spermatids thus represent functional diploid cells (Braun et al., 1989a, Caldwell and Handel, 1991; Morales et al., 1998).

2. Histone-to-protamine exchange

Histone-to-protamine exchange causes chromatin condensation followed by cessation of transcription in elongating spermatids. The extent of chromatin condensation in spermatids and spermatozoa is known to be correlated with the degree of histone-to-protamine exchange being essential for both the shape of the sperm head and the fertilizing capacity of the sperm. While in round spermatids, testis-specific histones occur together with transition proteins, in elongating spermatids, both histones and transition proteins are removed from the condensing chromatin and are replaced by protamines (reviewed in Steger, 1999).

In spermatogonia and spermatocytes, like in somatic cells, the DNA double helix is wound around nucleosomes, histone octamers consisting of two molecules of H2A, H2B, H3, and H4 each. Histone H1 is localized between the nucleosomes. Histone-bound DNA is further coiled into solenoids with six nucleosomes per turn (Finch and Klug, 1976). The highly conserved nature of histone proteins and the various possibilities of posttranslational modifications, such as acetylation, phosphorylation, methylation, and ubiquitination (Bradbury, 1992), which can alter the charges, conformation, and strength of binding to DNA reinforce the fundamental regulatory role of histone modifications on chromatin structure.

There are two classes of enzymes which are involved in determining the state of histone acetylation, histone acetyl transferases (HATs) and histone deacetylases (HDACs) (Davie, 1998). Substrates for these enzymes include amino groups of lysine residues located in the amino-terminal tails of core histones reducing the electrostatic alteration between histones and DNA. Subsequently, the basic nature of histones is somewhat neutralized which decreases their affinity for DNA within the nucleosomes facilitating histone removal (Meistrich et al., 1992). Recently, a detailed analysis of waves of histone acetylation and deacetylation during mouse spermatogenesis has been reported demonstrating a strong decrease in histone deacetylases in elongating spermatids, the stage of spermiogenesis where histone-to-protamine exchange takes place (Hazzouri et al., 2000). Although phosphorylation is not as well understood as acetylation, recent studies suggest a coordinate pattern of histone modification. In-vitro, phosphorylation of Ser-10 in histone H3 has been demonstrated to promote acetylation on nearby Lys-14 suggesting promoter-specific regulation by a kinase/acetyltransferase enzyme pair (Lo et al., 2000).

The role of transition proteins is still unclear and may, in addition, be different between transition protein-1 and -2. It has been reported that, in-vitro, transition protein-1 decreases (Singh and Rao, 1987) and transition protein-2 increases (Kundu and Rao, 1995) the melting temperature of DNA causing an decrease or increase of the DNA-histone-interactions, respectively. Since the frequency of DNA strand breaks becomes less prominent as the level of transition protein-1 increases, this protein has, by contrast, been suggested to enhance local DNA–DNA-interactions by neutralizing the negative charges of the phosphate backbone (Levesque et al., 1998). Recently, transition protein-1 has been demonstrated to stimulate resealing of DNA single strand breaks in-vitro and be involved in DNA repair processes in-vivo (Caron et al., 2001). It is assumed that transition protein-1 act as an alignment factor holding the broken DNA ends together until an as-yet unidentified ligase bridges the gap.

Protamines are the nucleoproteins of elongated spermatids and mature spermatozoa exhibiting several characteristic features:

- In mice, protamine-1 is synthesized as a mature protein of 50 amino acids, whereas protamine-2 is generated from a precursor of 106 amino acids (Yelick et al., 1987).
- Both protamine-1 and protamine-2 are highly basic proteins resulting from an unusually high content of arginine residues. The lack of conservation suggests that the overall basicity of these proteins is more important to their function than a particular amino acid sequence (Balhorn et al., 1984).
- Protamines can be modified by the addition of phosphate groups to serine residues. While serin-arginine (SR) protein-specific kinase-1 phosphorylates Ser-10 and, to a lesser extent, Ser-8 of protamine-1 (Papoutsopoulou et al., 1999), Ca^{2+}-dependent protein kinase-IV (Camk-IV) phosphorylates Ser-14 of protamine-2 (Wu et al., 2000). Protamine phosphorylation is assumed to facilitate the correct binding of protamines to DNA, while subsequent dephosphorylation increases the positive charge and attraction for DNA and thus is associated with an increase in sperm chromatin condensation (Oliva and Dixon, 1991).
- Protamines contain cystein residues which set up disulfide linkages making the sperm nucleus a highly insoluble and chemically stable structure (Bedford and Calvin, 1991).
- Protamines bind zinc through cysteine and histidine side chains and thus could be considered as zinc finger proteins representing one C2H2 motif (Reinicke and Chevaillier, 1991).

Protamines are thought to bind lengthwise within the minor groove of the DNA double helix with their central polyarginine segment crosslinking and neutralizing the phosphodiester backbone of the DNA double helix, while the amino- and carboxy-terminal residues of the protein participate in the formation of intra- and interprotamine binding involving hydrophobic and disulfide bonds. These DNA-protamine complexes of one DNA double helix fit exactly into the major grooves of a parallel DNA double helix and are packed side by side in a linear array within the sperm nucleus (Balhorn et al., 1984). In addition, Ward (1993) performed a model for the packing of the entire haploid genome into the sperm nucleus in which DNA loop domains are packed as doughnuts attached to the sperm nuclear matrix. Protamine-bound DNA is coiled into large concentric circles that collapse into a doughnut in which the DNA-protamine complexes are tightly packed together by van der Waal's forces (Fig. 2).

Protamine-DNA interactions result in chromatin condensation causing cessation of transcription in elongating spermatids. This occurs at a time when many proteins need to be synthesized and assembled for the complete condensation of the chromatin, the development of the acrosome, and the formation of the flagellum. It is evident that precise temporal regulation of gene expression via transcriptional and translational control mechanisms is of fundamental importance to ensure complete differentiation of round spermatids into mature spermatozoa.

3. Gene expression in haploid spermatids

The stage-specific but different expression of mRNAs and corresponding proteins in haploid spermatids is due to temporal uncoupling of transcription and translation (Eddy, 1998; Hecht, 1998; Steger, 1999, 2001; Braun, 2000). Post-transcriptional events, such as processing, transport, and storage of mRNAs, thus play important roles in determining when transcripts become functionally available for translation.

In mice, chromatoid bodies, electron dense material in the vicinity of the nucleus, and ribonucleoprotein (RNP) particles have been colocalized in pachytene spermatocytes and round spermatids which are known to contain translationally silent mRNAs (Biggiogera et al., 1990). Within chromatoid bodies of the rat, several mRNAs, including transition protein-2 mRNA, have been demonstrated (Söderström, 1981; Walt and Armbruster, 1984; Saunders et al., 1992; Moussa et al., 1994). In addition, in mice, several proteins which are known to be involved in the translational repression of mRNAs have been identified to be components of RNP particles and chromatoid bodies, such as Y-box proteins (Kwon et al., 1993; Tafuri et al., 1993), poly-A binding protein (PABP) (Gu et al., 1995), 48/52 kDa mRNA-binding proteins (Oko et al., 1996), as well as the zinc finger protein MOK2 (Arranz et al., 1997). Chromatoid bodies, therefore, may serve as storage organelle for translationally repressed mRNAs being attached to RNP particles (Fig. 3).

3.1. Transcriptional regulation of gene expression during spermatogenesis

3.1.1. (De)methylation of cytosines in the promoter region of the genes
Histone acetylation and phosphorylation have already been mentioned to be involved in histone-to-protamine exchange. In addition, methylation of histones is highly significant to genetic regulation (Bradbury, 1992). Addition of each methyl-group eliminates one

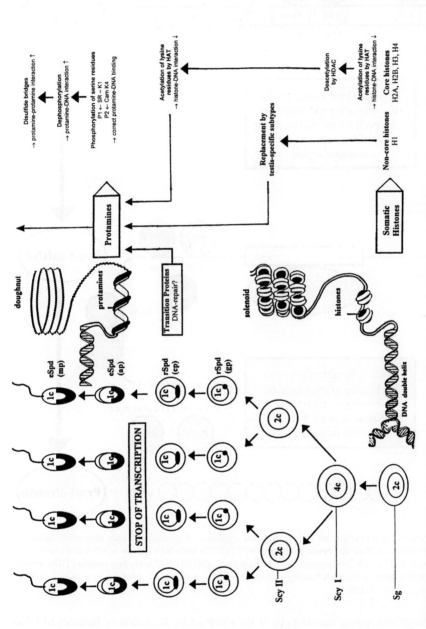

Fig. 2. Histone-to-protamine exchange during spermatogenesis. Spermatogonia (Sg), primary spermatocytes (ScyI), secondary spermatocytes (ScyII), round spermatids (rSpd) in Golgi phase (gp) and cap phase (cp), elongated spermatids (eSpd) in acrosome phase (ap) and maturation phase (mp), histone acetyltransferase (HAT), histone deacetylase (HDAC), serine–arginine protein-specific kinase-1 (SR-K1), Ca²⁺-dependent protein kinase-IV (Camk-IV), increasing (↑) and decreasing (↓).

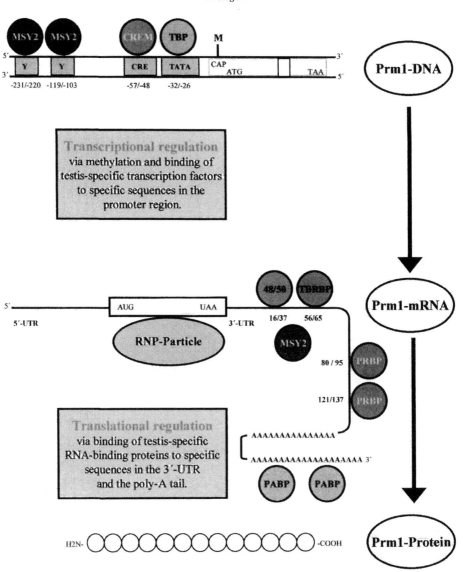

Fig. 3. Mechanisms of transcriptional and translational regulation of protamine-1 gene expression in haploid spermatids of the mouse. Note that mouse Y-box protein-2 (MSY2) binds both specific to DNA and unspecific to RNA. Methylation (M), cAMP-responsive element modulator (CREM), TATA-binding protein (TBP), ribonucleoprotein (RNP), testis/brain-RNA-binding protein (TB-RBP), protamine-1 RNA-binding protein (PRBP), poly-A binding protein (PABP).

positive charge and reduce the strength of the electrostatic attraction of histones to DNA resulting in conversion of genes from inactive to active form.

DNA methylation of specific cytosines following a guanine in the sequence CpG (p represents the phosphate group connecting C and G nucleotides) is performed by the enzyme

DNA (cytosine-5)-methyltransferase (DNA MTase). Applying Northern and Western blotting, as well as enzyme activity assays, analysis of purified germ cells from the adult testis revealed high levels of expression of both DNA MTase mRNA and protein in haploid round spermatids (Benoit and Trasler, 1994; Numata et al., 1994). In sperm, almost all available CpGs are fully methylated. The very few unmethylated CpGs occur in the promoter regions of genes, such as histones, that become active at very early embryonic steps following fertilization. This suggests that methyl groups may be removed from CpGs in 5' flanking control regions well before genes actually become active in transcription. However, demethylation by itself is not sufficient to initiate transcription. Other control steps, such as the binding or release of regulatory proteins, are necessary for transcription to take place. Especially for genes specifically expressed in germ cells, the relationship between DNA methylation and gene activity is not yet clear. While genes for transition protein-2, protamine-1, and protamine-2 are fully methylated when they are actively transcribed (Choi et al., 1997), the opposite occurs in the gene for transition protein-1 which shows demethylation in the 5' region associated with gene activity (Trasler et al., 1990). Recent data suggest that cytosine methylation may contribute to the transcriptional silencing of the testis-specific histone H1t gene in non-expressing tissues (Singal et al., 2000).

3.1.2. Protein binding to the promoter region of the genes

Various promoters have been found to have a restricted pattern of activity due to the presence of germ cell specific factors. However, general transcription factors, in addition, may be differentially regulated in germ cells (Goldberg, 1996).

The TATA-box is present in all protamine genes (Oliva and Dixon, 1991) and in the transition protein-1 gene (Heidaran et al., 1989). TATA-binding protein (TBP) has been demonstrated to be overaccumulated in the testis and to be particularly rich in round spermatids. The appearance of TBP and polymerase-II overexpressing spermatids is accompanied by an increase in whole-organ levels of total RNA. It has been demonstrated that round spermatids contain roughly 1000-fold more TBP-mRNA than somatic cells. Quantitative analyses revealed that testis-specific overaccumulation of TBP-mRNA is caused by both modest up-regulation of the somatic TBP promoter and, in addition, recruitment of at least two major and three minor testis-specific promoters (Schmidt and Schibler, 1997; Schmidt et al., 1997).

Transcriptional regulation via the adenyl-cyclase signalling pathway is mediated by cAMP-response element (CRE) nuclear factors, cAMP-response element binding (CREB) protein and cAMP-response element modulator (CREM) protein. Proteinkinase K phosphorylates and thereby activates CREB or CREM. Phosphorylation of a serine residue at position 133 (CREB) or at position 117 (CREM) by proteinkinase A endogenous to germ cells leads to activation of CRE (Lalli et al., 1996; Walker and Habener, 1996; Tamai et al., 1997). Recent data suggest a phosphorylation independent mechanism of CREM activation in the testis by the specific transcription activator of CREM, referred to as ACT (Activator of Crem in the Testis) (Fimia et al., 1999).

CRE consits of the eight nucleotide palindromic sequence 5'TGACGTCA3' (Roesler et al., 1988). It is invariably present at position −57/−48 in all protamine genes (Oliva and Dixon, 1991) and in the transition protein-1 gene (Kistler et al., 1994). The CREM gene gives rise to both full-length activator proteins and truncated repressor proteins generated by alternative exon splicing and alternative start-sites for translation (Delmas et al., 1992; Masquilier et al., 1993; Walker et al., 1994). Inducible cAMP early repressor (ICER) constitutes the only CREB protein which functions as a repressor of cAMP-induced transcription. It is generated from an alternative CREM promoter and represses the activity of its own promoter, thus constituting a negative autoregulatory feedback loop (Walker et al., 1998).

In the course of spermatogenesis, there is an abrupt switch in CREM gene expression (Foulkes et al., 1992; Lamas et al., 1996). CREMτ differs from CREM inhibitors by the coordinate insertion of two glutamine-rich domains that confer transcriptional activation function. Interestingly, CREM-mRNA repressor isoforms are expressed at low levels in spermatogonia, whereas during meiosis the CREMτ activator transcript is abundantly expressed and stabilized under the influence of FSH (Foulkes et al., 1993). However, the effect of FSH cannot be direct, since FSH-receptors are solely expressed by Sertoli cells (Kliesch et al., 1992). Recent data showing that CREM expression in rat testis is maintained despite gonadotropin deficiency further reduce the role of FSH (Behr and Weinbauer, 1999) (Figs. 4 and 5).

A protamine-1 promoter sequence of 113 nucleotides (Zambrowicz et al., 1993) and a protamine-2 promoter sequence of 859 nucleotides (Stewart et al., 1988) have been shown to be sufficient for spermatid-specific expression. A number of ubiquitous and testis-specific proteins bind within this region (Zambrowicz and Palmiter, 1994), including Tet-1, a testis-specific trans-acting nuclear protein, which recognizes the 11-mer sequence 5′TGACTTCATAA3′ at position −64. Although the first 8-mer in the Tet-1 11-mer shares homology with the CRE-box, Tet-1 was demonstrated to be distinct from known CRE nuclear factors (Tamura et al., 1992).

Analyses of the mouse protamine-2 promoter by in vitro transcription assays have identified a potential positive regulatory region from −170 to −82 (Bunick et al., 1990). Mobility shift assays from −140 to −23 have detected the binding of both ubiquitous and testis-specific proteins (Johnson et al., 1991). In addition, deletion and mutational analyses revealed two positive regulatory sequences for protamine-2 transcription at positions −84/−72 (5′ACAATCAATCAGG3′) (Yiu and Hecht, 1997) and −64/−48 (5′CCGACAAGGT-CACAG3′) containing the single core motif 5′AGGTCA3′ recognized by orphan nuclear receptors (Enmark and Gustafsson, 1996) and the half-site motif 5′GTCA3′ of the CRE (Delmas et al., 1993). Removal or alteration of one of these two sites leads to a significant reduction in the transcription of mouse protamine-2, while binding of the protamine activating factor-1 (PAF-1) to site 1 and of the Y-box protein p48/p52 to site 2 induces a more than 5-fold increase of mouse protamine-2 transcription (Yiu and Hecht, 1997).

The testis-specific protein PAF-1 reaches high levels in round spermatids at the time of protamine-2 transcription. PAF-1 is thought to be a novel orphan receptor in the nuclear receptor superfamily of ligand-activated transcription factors. In addition, TAK1 (Hirose et al., 1995a), Tr2-11 (CH Lee et al., 1996), and germ cell nuclear factor/retinoid receptor-related testis-associated receptor (GCNF/RTR) (Chen et al., 1994; Hirose et al., 1995b; YL Zhang et al., 1998) are orphan receptors that have been demonstrated in testis of mouse. Like PAF-1, GCNF/RTR is expressed in round spermatids, but declines abruptly as spermatids start to elongate. However, PAF-1 and GCNF/RTR reveal different binding-sites, PAF-1 response element, PAF-RE (5′GACAGGTCA3′) (Yiu and Hecht, 1997), and

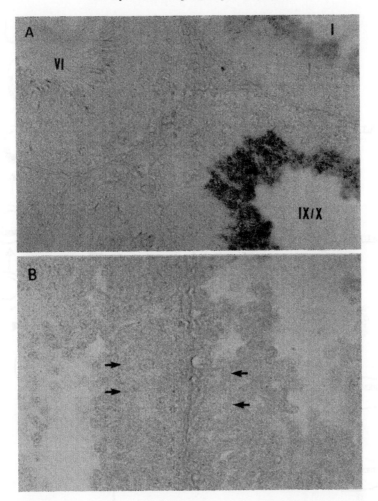

Fig. 4. In-situ hybridization on paraffin sections from testes of mice with normal spermatogenesis (A) and CREM-knockout mice (B) using digoxygenin-labeled RNA-probes against transition protein-1 mRNA. In mice with normal spermatogenesis, transcripts for transition protein-1 can be observed from round step 7 spermatids (stage VII) to elongating step 13 spermatids (stage I). Male mice lacking the gene for the transcription factor CREM exhibit spermatogenic arrest at the level of round spermatids (arrows). In addition, round spermatids are devoid of transcripts for transition protein-1 the gene of which is known to contain a CRE-box within its promoter sequence. Roman numbers indicate stages of the seminiferous epithelial cycle (see Figure 1).

GCNF/RTR response element, GCNF/RTR-RE (5'TCAAGGTCA3') (Yan et al., 1997), respectively. Both binding-sites contain the sequence 5'GTCA3' which is also present in CRE. However, mobility shift assays revealed that CREM is not the protein binding to PAF-RE or GCNF/RTR-RE.

Y-box proteins (the name refers to a promoter element conserved in MHC class II genes with the sequence 5'CTGATTGGCCAA3' containing an inverted CCAAT box) (Didier et al., 1988) comprise a family of transcription factors which are general modifiers

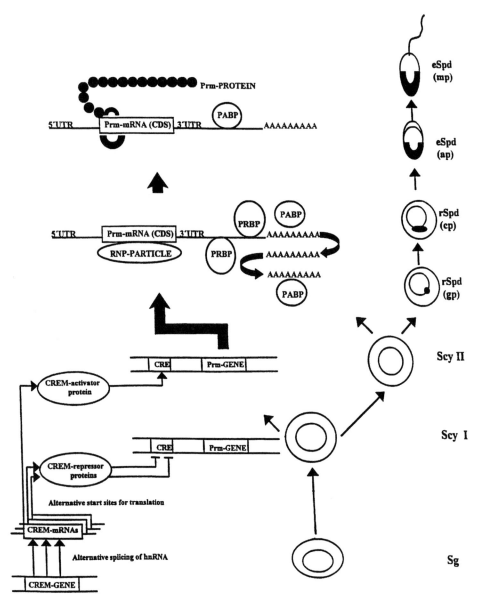

Fig. 5. Regulation of protamine gene expression via CREM. Spermatogonia (Sg), primary spermatocytes (ScyI), secondary spermatocytes (ScyII), round spermatids (rSpd) in Golgi phase (gp) and cap phase (cp), elongated spermatids (eSpd) in acrosome phase (ap) and maturation phase (mp), coding sequence (CDS), ribonucleoprotein (RNP), protamine-1 RNA-binding protein (PRBP), poly-A binding protein (PABP).

of gene activity. The consequence of their binding to promoter elements depends on the interaction with other tissue-specific regulatory proteins. Several genes, identified as functionating specifically in germ cells, have consensus Y-box sequences in their promoters, e.g. mouse protamine-1 at positions -226 and -100 (Zambrowicz et al., 1993) and mouse protamine-2 at positions -489, -178, and -72 (Nikolajczyk et al., 1995; Yiu and Hecht, 1997) (Table 1).

3.2. Translational regulation of gene expression during spermatogenesis

3.2.1. Protein binding to the 3´-UTR of the transcripts

Specific interactions of cytoplasmic protein repressors with the 3′-UTR of mRNAs play a key role in the regulation of translation, degradation, and polyadenylation of mRNAs in haploid spermatids. Although some protein repressors may, in addition, bind to the 5′-UTR to block the association of the ribosomal initiation complex with the site of initiation on mRNA or the movement of the ribosomal initiation complex along the transcript towards the initiation site, it has been demonstrated that repression of ferritin-mRNA requires, in addition, binding of the same protein repressor to the 3′-UTR (Dickey et al., 1988).

Table 1
DNA-binding proteins and corresponding DNA-binding domains

DNA-binding protein	Target DNA	DNA-binding domain	References
TATA-binding (TBP) + Polymerase-II	Tnp1, Prm1, Prm2	TATA-box $-32/-28$	Heidaran et al., 1989, Oliva and Dixon 1991, Schmidt and Schibler 1997
cAMP-responsive element modulator (CREM) protein	Tnp1, Prm1, Prm2	CRE-box $-57/-48$	Oliva and Dixon 1991, Kistler et al., 1994, Tamai et al., 1997
?	Prm1, Prm2	CAAT-box $-87/-67$	Han et al., 1997
?	Prm2	? $-170/-23$	Johnson et al., 1991
Tet-1 protein	Prm1	? $-64/-54$	Tamura et al., 1992
Y-box protein p48/p52	Prm1	Y-box $-231/-220$ and $-119/-108$	Zambrowicz et al., 1994, Yiu and Hecht 1997
Y-box protein p48/p52	Prm2	Y-box $-500/-489$, $-189/-178$, and $-83/-72$	Nikolajczyk et al., 1995, Yiu and Hecht 1997
Protamine activating factor-1 (PAF-1)	Prm2	PAF-1-responsive element (PAF-RE) $-64/-48$	Yiu and Hecht 1997
Germ cell nuclear factor / retinoid receptor-related testis-associated receptor (GCNF/RTR)	Prm2	GCNF/RTR-responsive element (GCNF/RTR-RE) $-60/-52$	Yan et al., 1997
?	Prm2	Hormone-responsive element (HRE) box $-328/-311$ and $-239/-210$	Han et al., 1997

Protein repressors that have, so far, been identified can be divided into three groups, according to their RNA-binding domain.

Repressor proteins of the first group, such as deleted in azoospermia (DAZ) protein (Reijo et al., 1995) and poly-A binding protein (PABP) (Hornstein et al., 1999), reveal a 90 amino acids ribonucleoprotein (RNP) motif, also known as RNA recognition motif (RRM) being present in one to four copies within the protein repressor. Within the RNP motif, the two most conserved sequences are known as RNP1 and RNP2 exhibiting the consensus sequences KGYGFVHF and NLYVKN, respectively (Adam et al., 1986). Remarkably, over 60% and over 30% of the amino acids of RNP1 and RNP2, respectively, are formed by basic/aromatic amino acids. However, although RNP1 and RNP2 are determinant for RNA-binding, specific RNA recognition depends instead on the interspersed variable regions (Kenan et al., 1991; Burd and Dreyfuss, 1994; Derrigo et al., 2000).

Repressor proteins of the second group, such as MSY2 (Gu et al., 1998), are members of the Y-box protein family containing a cold shock domain (CSD). In analogy to RNP1 and RNP2, the two most conserved sequences within the CSD are called RNP1-like and RNP2-like motifs exhibiting the consensus sequences NGYGFINR and DVFVHQ, respectively (Grauman and Marahiel, 1998). Again, over 35% and over 30% of the amino acids of RNP1-like and RNP2-like motifs, respectively, are formed by basic/aromatic amino acids. Binding depends on phosphorylation. Dephosphorylation may be one means to release these proteins from mRNAs thereby activating the stored transcripts (Herbert and Hecht, 1999).

Protein repressors of the third group, such as protamine-1 RNA-binding protein (PRBP) (K Lee et al., 1996), contain the 21 amino acids RNA-binding motif (RBM) GTGP-SKKAAKHKAAEVALKHL. The three basic amino acids, arginine, histidine, and lysine together comprise 12% of the amino acids of the protein but over 30% of the amino acids of the RBM. Although the three aromatic amino acids, phenylalanine, tryptophane, and tyrosine, together comprise 3% of the amino acids of the protein, the RBM is completely devoid of any aromatic amino acids. PRBP has been suggested to interact with two fragments of protamine-1 mRNA 3'-UTR which are proposed to form alternative stem-loop structures that might be stabilized by the protein binding (K Lee et al., 1996).

The best characterized protein repressors are members of the Y-box protein family. In mice, spermatogenic cells have been demonstrated to express five isoforms of Y-box proteins. Applying UV-crosslinking, the 18 kDa RNA-binding protein has been demonstrated to bind to the Y-box element in the 3'-UTR of both protamine-1 mRNA and protamine-2 mRNA (Kwon and Hecht, 1991, 1993). The 48/52 kDa Y-box protein is the murine homologue of the Xenopus 54/56 kDa mRNA-binding protein FRGY2/mRNP$_{3+4}$ (Kwon et al., 1993). While pachytene spermatocytes contain primarily the 52 kDa subunit, round spermatids contain primarily the 48 kDa subunit, suggesting that 48/52 kDa Y-box protein may have temporally distinct regulatory functions (Kwon et al., 1993). Indeed, Y-box proteins are believed to be specific DNA-binding proteins and non-specific RNA-binding proteins at the same time. Therefore, they have the potential to regulate gene expression either as a transcription factor or as a modulator repressing translation of accumulating mRNAs (Matsumoto and Wolffe, 1998). The 48/52 kDa Y-box protein has been renamed mouse Y-box protein-2 (MSY2). MSY2 has, immunohistochemically, been demonstrated in pachytene spermatocytes and round spermatids (Oko et al., 1996). Both MSY2 and MSY4 have been demonstrated to bind specifically to a site within the 5' most 37 nucleotides in the 3'-UTR of protamine-1 mRNA (Mastrangelo and Kleene, 2000). Polysome analysis demonstrates that MSY4 is associated with RNP particles being consistent with MSY4 having a role in storing repressed transcripts (Davies et al., 2000).

The 26 kDa testis/brain RNA-binding protein (TB-RBP) is known to bind to the Y-box within the 3'-UTR of protamine-1 and protamine-2 mRNA but is thought to represent no Y-box protein (Han et al., 1995; Gu et al., 1998), since binding requires both phosphorylation and dimerization mediated by a leucine zipper motif located in the carboxy-terminal end of the protein (Wu et al., 1998). TB-RBP has been shown to facilitate mRNA movement from the nucleus to the cytoplasm and through intercellular bridges (Morales et al., 1998). In addition to TB-RBP, the 48/50 kDa RNA-binding protein is also no member of the Y-box protein family. It has been demonstrated that this protein repressor binds to conserved motifs located outside the Y-box element in the 3'-UTR of both protamine-1 mRNA (5'caaguccau3') and protamine-2 mRNA (5'ccauuccau3') (Fajardo et al., 1994).

Two non-overlapping regions of protamine-1 mRNA 3'-UTR have been shown to be involved in translational repression. The first maps to the 5' most 37 nucleotides which contains a conserved nine nucleotide motif (5'caaguccau3') being also present in the 3'-UTR of protamine-2 mRNA (5'ccauuccau3') and binds the 48/50 kDa RNA-binding protein (Fajardo et al., 1994). The second lies within the 3' most 62 nucleotides and contains a 17 nucleotide sequence which is also present in protamine-2 mRNA 3'-UTR (Braun, 1990). This 62 nucleotide element contains both the polyadenylation consensus motif (PCM) 12 nucleotides upstream of the poly-A tail and a conserved 17 nucleotide motif just 5' to the PCM. A second copy of the 17 nucleotide motif is present just 5' to the 62 nucleotide element. Both copies reveal a nearly perfect direct repeat. Either repeat is able to base pair with the other forming a region of dsRNA with a similarly sized single-stranded loop (K Lee et al., 1996). The 17 nucleotide motif, in addition, is present in the 3'-UTR of protamine-2 mRNA. Here, this sequence has been denoted Z-box (Kwon and Hecht, 1991, 1993). However, RNA-binding proteins solely bind to the Y-box, but not to the Z-box.

Conserved sequences which may be involved in RNA-binding both in the 3'-UTR and in RNA-binding proteins are summarized in Tables 2 and 3, respectively.

3.2.2. Protein binding to the poly-A tail of the transcripts

Primary transcripts (hnRNAs) are cleaved within 30 nucleotides downstream of the PCM being localized within the 3'-UTR (Zhao et al., 1999). In pachytene spermatocytes and round spermatids, however, this cleavage is rapidly followed by the addition of a poly-A tail of about 180 adenine residues. Polyadenylated mRNAs are sequestered in RNP particles. Translation subsequently takes place in elongated spermatids after polyadenylated mRNAs undergo a partial poly-A shortening by deadenylation. In mice, the length of the poly-A tail on transition protein and protamine mRNAs has been reported to correlate with the translational activity of the transcripts (Kleene et al., 1984; Kleene, 1989, 1993, 1996; Cataldo et al., 1999).

Both stability and translation of the transcripts is mediated by the interaction with a poly-A binding protein (PABP). As has been demonstrated in in-vitro studies, polyadenylated mRNAs are degraded faster when PABP is absent. It is assumed that PABP migrates from the poly-A tail to AU-rich elements within the 3'-UTR leaving the poly-A tail naked and vulnerable to degradation (Bernstein and Ross, 1989; Bernstein et al., 1989).

K. Steger

Table 2
RNA-binding proteins and corresponding RNA-binding domains

RNA-binding protein	Target RNA	RNA-binding domain	References
48/52 kDa RNA-binding protein = Mouse Y-box protein-2 (MSY2)	Specific DNA-binding Non-specific RNA-binding Binding depends on phosphorylation	CSD Y-box	Kwon and Hecht, 1993, Ladomery and Sommerville, 1994, Grauman and Marahiel, 1998, Gu et al., 1998a, Matsumoto and Wolffe, 1998, Herbert and Hecht, 1999
18 kDa RNA-binding protein	Prm1-mRNA, 3'-UTR Prm2-mRNA, 3'-UTR Binding depends on phosphorylation	56/65 (Y-box) 116/129 (Y-box)	Kwon and Hecht, 1991
Testis/brain RNA-binding protein (TB-RBP)	Prm1-mRNA, 3'-UTR Prm2-mRNA, 3'-UTR Binding depends on phosphorylation and dimerization	56/65 (Y-box) 116/129 (Y-box)	Han et al., 1995 Wu et al., 1998
48/50 kDa RNA-binding protein	Prm1-mRNA, 3'-UTR Prm2-mRNA, 3'-UTR Binding depends on phosphorylation	16/37 (No Y-box) 85/104 (No Y-box)	Fajardo et al., 1994
Ferritin-receptor RNA-binding protein	Ferritin-mRNA, 5'-UTRand 3'-UTR	?	Leibold and Munro, 1988
Transferrin-receptor RNA-binding protein	Transferrin-mRNA, 5'-UTR and 3'-UTR	?	Mullner et al., 1989
Protamine-1 mRNA-binding protein (PRBP)	Prm1-mRNA, 3'-UTR Prm2-mRNA, 3'-UTR	80/95 and 121/137 (RBM) 157/173 (RBM)	K Lee et al., 1996
Spermatid perinuclear RNA-binding protein (Spnr)	Prm1-mRNA, 3'-UTR	RBM	Schumacher et al., 1995b
Testis nuclear RNA-binding protein (Tenr)	Prm1-mRNA, 3'-UTR	RBM	Schumacher et al., 1995a
Poly-A binding protein 1 (PABP1)	poly-A tail	RRM	Adam et al., 1986
Testis-specific poly-A binding protein (PABPt = PABP2)	poly-A tail	RRM	Kleene et al., 1998

Table 3
Consensus sequences of RNA-binding domains in RNA-binding proteins

RNA-binding domain in RNA-binding protein	Consensus sequence	Examples of RNA-binding proteins	References
RNA recognition motif (RRM) = ribonucleoprotein (RNP) motif	Sequence of about 90 amino acids (one or more copies) RNP1 (8 amino acids): KGFGFV*F RNP2 (6 amino acids): NLYVKN	PABP, RBM, DAZ	Adam et al., 1986 Kenan et al., 1991
Cold shock domain (CSD)	Sequence of about 90 amino acids (one or more copies) RNP1-like motif (8 amino acids): NGYGFI** RNP2-like motif (6 amino acids): DVFVH*	Y-box proteins: 18 kDa RNA-binding protein, 48/52 kDa RNA-binding protein, MSY2	Kwon and Hecht, 1991, 1993, Grauman and Marahiel, 1998, Gu et al., 1998, Matsumoto and Wolffe, 1998, Herbert and Hecht, 1999
RNA-binding motif (RBM)	Sequence of 21 amino acids (two copies) G*G*SKK*AK**AAE*AL**L	PRBP, Spnr, and Tenr	Schumacher et al., 1995a, b K Lee et al., 1996
Zinc finger motif	Various zinc finger motifs $C(X)_{2-4}C(X)_{12}H(X)_{3-5}H = C_2H_2$	MOK2	Arranz et al., 1997 Klug, 1999

In mice, there exist at least four PABPs. The best characterized PABP is the 70 kDa PABP1 (Jackson and Standart, 1990). PABP1 exhibits four RRMs (Adam et al., 1986; Kenan et al., 1991) and is associated with mRNAs containing poly-A tails ranging from 30 to 180 adenine residues confirming that PABP1 binds to both active and stored forms of the same mRNA (Kleene et al., 1984; Gu et al., 1995). PABP1-mRNA is expressed at a level being at least 10-fold higher in testis than in somatic tissues (Kleene et al., 1994).

Recently, a testis-specific isoform, PABPt or PABP2, has been suggested to be an expressed retroposon (Kleene et al., 1998; Kleene and Mastrangelo, 1999). Retroposons are generated by making a reverse transcriptase copy of a mRNA and inserting the DNA copy into genomic DNA in a germ-line cell. Characteristics are the absence of introns, a 3' A-rich terminus representing the remnant of a poly-A tail, short flanking direct repeats resulting from insertion of the retroposon into a staggered break in genomic DNA, and a chromosomal locus differing from the intron-containing progenitor gene (Weiner et al., 1986).

4. Knockout and transgenic mice

What do knockout and transgenic mice tell us about the regulation of gene expression during spermatogenesis? Creating a knockout mouse, to date, is thought to be the only way of proving an essential function of a specific gene. Surprisingly, some widely expressed genes not originally thought to be involved in spermatogenesis caused male sterility when knocked out. This defect, however, may be secondary to the health of the animals. By contrast, male mice lacking genes which are thought to play an essential role in spermatogenesis, such as acrosin (Baba et al., 1994) or testis-specific histone H1t (Lin et al., 2000), are fertile and reproduce as wild-type mice. This unexpected finding can, at least in part, be explained by redundancy, since gene expression of other H1 subtypes has been demonstrated to be enhanced during spermatogenesis creating a normal H1-to-nucleosome ratio and presumably compensating for H1t functions in H1t-deficient mice (Drabent et al., 2000).

Male mice lacking the gene for transition protein-1 exhibit elevated levels of transition protein-2 and protamine-2 precursors. Although sperm motility is severely reduced, approximately 40% of transition protein-1 deficient mice are fertile (Yu et al., 2000). Similar results have been obtained in transition protein-2 knockout mice (Adham et al., 2001).

By contrast, both protamine-1 and protamine-2 are essential for the production of structurally and functionally intact sperm. Recently, it has been demonstrated that haploinsufficiency caused by a mutation in one allele of protamine-1 or protamine-2 prevents genetic transmission of both mutant and wild-type alleles (Cho et al., 2001). Interestingly, in sperm from protamine-1 chimeras, solely protamine-1 was reduced, whereas in sperm from protamine-2 chimeras, both protamine-1 and mature protamine-2 were reduced with the reduction being greater for protamine-1 than for mature protamine-2. In addition, in both chimeras, a protamine-2 precursor form different from that analyzed in sperm from wild-type mice was present as a doublet.

Male knockout mice for Ca^{2+}/calmodulin-dependent protein kinase IV (Camk-IV), a serine/threonine protein kinase phosphorylating protamine-2, are infertile due to prolonged retention of transition protein-2 and complete absence of protamine-2 in elongated spermatids (Wu et al., 2000).

Male knockout mice lacking the genes for transcription factors TATA-binding protein-related factor (TRF) (D Zhang et al., 2001) or cAMP-responsive element modulator (CREM) (Blendy et al., 1996; Nantel et al., 1996) binding to the TATA-box and the CRE-box, respectively, known to be present in the promoter region of transition protein-1 and both protamines are infertile due to spermatogenic arrest at the level of round spermatids.

Male mice with a null allele for DAZ-like-1 (DAZL1), the mouse homologue of human deleted in azoospermia (DAZ) encoding a testis-specific RNA-binding protein, are infertile due to a meiotic entry defect followed by severe germ cell depletion (Ruggiu et al., 1997). Although DAZL1 knockout mice remained infertile when the human DAZ transgene was introduced, histological examination revealed a pronounced increase in the germ cell population with a germ cell survival to spermatocytes of the pachytene stage of meiosis (Slee et al., 1999) (Table 4).

In order to define promoter elements and transcription factors responsible for testis-specific expression, reporter genes have frequently been fused to testis-specific promoters creating transgenic mice. In a classical experiment, the 3′ most 62 nucleotide element of

Table 4
Male knockout mice lacking genes with known function in spermatogenesis

Male knockout mice	Phenotype	References
Acrosin	Normal fertility	Baba et al., 1994
cAMP-responsive element modulator (CREM)	Infertility due to spermatogenic arrest at the level of round spermatids	Blendy et al., 1996 Nantel et al., 1996
Deleted in azoospermia-like-1 (DAZL1)	Infertility due to meiotic entry defect followed by germ cell depletion	Ruggiu et al., 1997
Casein kinase II (Ck2) alpha'	Infertility due to round-headed spermatozoa (globozoospermia)	Xu et al., 1999
Protamine-1 RNA-binding protein (PRBP)	Infertility due to delayed replacement of transition proteins and subsequent failure of spermiation	Zhong et al., 1999
Testis-specific histone H1t	Normal fertility	Drabent et al., 2000 Lin et al., 2000
FSH-receptor	Reduced fertility due to a decrease in the percentage of elongated spermatids exhibiting reduced chromatin condensation	Krishnamurthy et al., 2000
Ca^{2+}/calmodulin-dependent protein kinase IV (Camk-IV)	Infertility due to delayed replacement of transition protein-2 and complete absence of protamine-2	Wu et al., 2000
Transition protein-1 and -2	Reduced fertility due to decreased sperm motility and reduced chromatin condensation	Yu et al., 2000 Adham et al., 2001
Protamine-1 and -2	Infertility due to delayed replacement of transition proteins and subsequent failure of spermiation	Cho et al., 2001
LH-receptor	Infertility due to spermatogenic arrest at the level of round spermatids	Lei et al., 2001 FP Zhang et al., 2001
TATA-binding protein-related factor (TRF)	Infertility due to spermatogenic arrest at the level of round spermatids	D Zhang et al., 2001

mouse protamine-1 mRNA 3'-UTR has been demonstrated to be sufficient for proper temporal and stage-specific translation of the human growth hormone reporter transgene (Braun et al., 1989b). It has been suggested that translational repression in round spermatids is achieved by the binding of sequence-specific RNA-binding proteins to this 62 nucleotide element followed by the assembly of mRNAs into RNP particles, whereas translation in elongating spermatids is achieved by covalent modification of the RNP complex and release of translatable protamine-1 mRNA. Subsequently, it has been proposed that the 3'-UTR does not function to target mRNAs to assemble into RNP particles but rather is a part of the timing mechanism that determines when various mRNAs within RNP particles will be translated. It is assumed that RNA-binding proteins with no sequence specificity for RNA-binding mediate RNP particle assembly, while RNA-binding proteins exhibiting specific RNA-binding are involved in timing the release of individual transcripts for translation (Schmidt et al., 1999).

The sperm-specific expression of the testis-specific isoform of angiotensin converting enzyme (ACE) has been demonstrated to be mediated by a 91 nucleotide promoter containing a CRE-like element (Howard et al., 1993). Using the protamine-1 promoter for direct transcription in transgenic mice, it has been shown that a region from -150 to -37 successfully directs spermatid-specific transcription (Peschon et al., 1987). While overexpression of protamine-1 has been demonstrated to have no negative effect on male fertility (Zambrowicz et al., 1993), premature translation of protamine-1 mRNA causes precocious chromatin condensation and arrests spermatid differentiation resulting in male infertility (Lee et al., 1995). Translational repression of protamine-1 mRNA involves binding of protamine-1 RNA-binding protein (PRBP) to the 3'-UTR of protamine-1 mRNA. Male mice carrying a targeted disruption of the Tarbp2 gene encoding PRBP are infertile due to delayed replacement of transition proteins and subsequent failure of spermiation (Zhong et al., 1999).

Multiple factor-binding regions within specific promoter sequences of different genes have been demonstrated to influence the pattern of tissue-specific demethylation followed by initiation of transcription. The single disruption of any one of these binding activities reduced, but did not abolish, transgene expression being consistent with an enhancement-like function in this promoter involving multiple bound activator proteins that interact in a combinatorial manner to synergistically promote testis-specific transcription (LP Zhang et al., 1998).

Summarized, transgenic mice answer the question which sequences are responsible for the manner in which genes are expressed, whereas knockout mice answer the question what happens when a specific gene is absent. Note that some genes, such as acrosin or testis-specific histone H1t, exhibit redundant expression and absence of these genes is followed by no prominent effect on spermatogenesis, while the expression of other genes, such as protamines, is of pivotal importance for the production of fertile sperm and absence of these genes results in male infertility. The expression of a specific gene is influenced by the gene products of previous expressed genes and influences itself the expression of subordinate genes, thus representing only one part of a complex gene cascade. Since reproduction is a main task in the life of each organism, it is evident that sequential gene expression during spermatogenesis is regulated via various control mechanisms.

5. Model system for human infertility

Mouse spermatogenesis provides a model for the study of human infertility. It is known that, in man, equal amounts of both protamine-1 and protamine-2 are required for the production of structurally and functionally intact sperm (Balhorn et al., 1987; Belokopytova et al., 1993; Steger et al., 2001). Protamine-1 was present in all mammalian spermatozoa analyzed so far (Oliva and Dixon, 1991), whereas protamine-2 only has been detected in spermatozoa of man (Ammer et al., 1986), stallion (Belaiche et al., 1987), and mouse (Bellve et al., 1988). Since protamine-2 comprises about 50% in human sperm (Oliva and Dixon, 1991) but about 70% in mouse sperm (Balhorn et al., 1984), there may be substantial differences in the formation of the DNA-protamine complex in sperm nuclei in various mammalian species. Interestingly, the protamine-2 gene is transcribed in rat, boar, ram, and bull, although the corresponding protein has, so far, not been detected in these species (Bower et al., 1987).

Since transcripts homologous to mouse protamine-2 have been detected on polysomes, it has been suggested that rat protamine-2 mRNA is also translated, but is not properly processed (Bower et al., 1987). In-vitro run-off transcription assays performed in either rat or mouse testis-derived transcription systems, in addition, revealed that the protamine-2 promoter in rat is only 30% as efficient as in mouse (Bunick et al., 1990). Therefore, the lack of protamine-2 in rat spermatozoa is caused by both lowered transcription rate and altered processing sites of the rat protamine-2 gene.

In man, histone-to-protamine exchange followed by chromatin condensation involves three characteristic features:

First, the replacement of histones is only 85% complete (Tanpaichitr et al., 1978; Gatewood et al., 1987; Prigent et al., 1996). This may be the reason for nuclear vacuoles which can frequently be observed in human spermatozoa assuming regional differences in levels of chromatin condensation (Zamboni et al., 1971; Bedford et al., 1973).

Second, the human transition protein-2 is expressed at only a very low level. This has been suggested to be due to insufficient storage of transition protein-2 mRNA probably caused by the absence of the conserved 5'GCCATCAC3' motif in the 3'-UTR. This eight nucleotide motif is present in the 3'-UTR of transition protein-2 mRNA of mouse, rat, boar, and bull, which express the gene at a high level (Schlüter et al., 1993).

Third, while protamine-2 is absent in rat but is the predominant form in mouse, human spermatids and spermatozoa exhibit an equal distribution of the two protamine variants. Interestingly, human male infertility has been associated with decreased levels of protamines (Lescoat et al., 1988; Blanchard et al., 1990; Sakkas et al., 1996), complete absence of protamine-2 (Ziyyat et al., 1999), and increased protamine-1 to protamine-2 ratios at the level of both protein (Balhorn et al., 1987; Belokopytova et al., 1993) and mRNA (Steger et al., 2001). These findings are in line with data obtained from haploinsufficient mice carrying a mutation in either protamine-1 or protamine-2 resulting in male sterility (Cho et al., 2001).

Both transition proteins and protamines contain a CRE-box within their promoter region being transcribed through binding of CREM protein (Oliva and Dixon, 1991; Kistler et al., 1994). Testicular biopsies from men with round spermatid maturation arrest were found negative for both transition protein-1 (Steger et al., 1999) and CREM (Weinbauer et al., 1998;

Steger et al., 1999). Since the histology was similar to that of CREM knockout mice (Blendy et al., 1996; Nantel et al., 1996), it has been suggested that the failure of round spermatids to progress to mature spermatids may also in man be caused by a defect of the CREM gene.

It is known that about 13% of azoospermic men reveal a mutation on the long arm of their Y-chromosome (Yq11.21–23) involving the multigene locus for azoospermia factor (AZFa-c) (Ma et al., 1993). While the gene product of AZFa is still unknown, AZFb encodes for RNA-binding motif (RBM) protein and AZFc encodes for deleted in azoospermia (DAZ) protein. Deletions within this multigene locus are followed by highly variable testicular defects ranging from Sertoli-cell-only (SCO) syndrome to spermatogenic arrest with occasional production of condensed spermatids. Specific deletion of AZFa results in SCO syndrome (Grimaldi et al., 1998), while specific deletion of AZFb causes spermatogenic arrest at the level of primary spermatocytes (Elliot et al., 1997). The most frequent finding is a deletion of the AZFc region encoding the DAZ protein. Complete absence of this protein results in severe hypospermatogenesis suggesting that the specific deletion of AZFc is not sufficient to cause a total lack of spermatogenesis (Reijo et al., 1995; Ferlin et al., 1999). Men with a deletion of the AZFc region, therefore, exhibit a phenotype similar to male knockout mice lacking the DAZ-like-1 (DAZL1) gene, the mouse homologue of the human DAZ gene (Ruggiu et al., 1997).

In summary, despite some species-specific peculiarities, the broad conformity between murine and human spermatogenesis recommends mouse a good animal model (superior to rat) for the study of both normal and impaired human spermatogenesis.

Acknowledgments

The author thanks Prof. Dr. G.F. Weinbauer, Covance Laboratories GmbH, Münster, and Dr. R. Behr, Institute of Reproductive Medicine, Universitiy of Münster, for providing the mouse testes, and Prof. Dr. M. Bergmann, Institute of Veterinary Anatomy, University of Giessen, for many valuable discussions. Apologies are given to authors whose publications have not been cited. Funding of this research program was provided by grant STE 892/ 1-3 of the Deutsche Forschungsgemeinschaft (DFG).

References

Adam, S.A., Nakagawa, T., Swanson, M.S., Woodruff, T.K., Dreyfuss, G. 1986. mRNA polyadenylate-binding protein: gene isolation and identification of a ribonucleoprotein consensus sequence. Mol. Cell. Biol. 6, 2932–2943.

Adham, P.M., Nayernia, K., Burkhardt-Gottges, E., Topaloglu, O., Dixkens, C., Holstein, A.F., Engel, W. 2001. Teratozoospermia in mice lacking the transition protein 2 (Tnp2). Mol. Hum. Reprod. 7, 513–520.

Alfonso, P.J., Kistler, W.S. 1993. Immunohistochemical localization of spermatid nuclear transition protein 2 in the testes of rats and mice. Biol. Reprod. 48, 522–529.

Ammer, H., Henschen, A., Lee, C.H. 1986. Isolation and amino acid sequence analysis of human sperm protamines P1 and P2. Biol. Chem. 367, 515–522.

Arranz, V., Harper, F., Florentin, Y., Puvion, E., Kress, M., Ernoult-Lange, M. 1997. Human and mouse MOK2 proteins are associated with nuclear ribonucleoprotein components and bind specifically to RNA and DNA through their zinc finger domains. Mol. Cell. Biol. 17, 2116–2126.

Baba, T., Azuma, S., Kashiwabara, S., Toyoda, Y. 1994. Sperm from mice carrying a targeted mutation of the acrosin gene can penetrate the oocyte zona pellucida and effect fertilization. J. Biol. Chem. 269, 31845–31849.

Balhorn, R., Reed, S., Tanphaichitr, N. 1987. Aberrant protamine 1/protamine 2 ratios in sperm of infertile human males. Experientia 44, 52–55.

Balhorn, R., Weston, S., Thomas, C., Wyrobek, A.J. 1984. DNA packaging in mouse spermatids. Synthesis of protamine variants and four transition proteins. Exp. Cell Res. 150, 298–308.

Bedford, J.M., Bent, M.J., Calvin, H.J. 1973. Variation in the structural character and stability of the nuclear chromatin in morphologically normal spermatozoa. J. Reprod. Fertil. 33, 19–29.

Bedford, J.M., Calvin, H.I. 1991. The occurrence and possible functional significance of -S-S- crosslinkins in sperm heads, with particular reference to eutherian mammals. J. Exp. Zool. 188, 137–155.

Behr, R., Weinbauer, G.F. 1999. Germ cell-specific cyclic adenosine 3'5' monophosphate response element modulator expression in rodent and primate testis is maintained despite gonadotropin deficiency. Endocrinology 140, 2746–2754.

Belaiche, D., Loir, M., Kruggle, M., Sautiere, P. 1987. Isolation and characterization of two protamines St1 and St2 from stallion spermatozoa, and amino acid sequence of the major protamine St1. Biochem. Biophys. Acta 913, 145–149.

Bellve, A.R., McKay, D.J., Renaux, B.S., Dixon, G.H. 1988. Purification and characterization of mouse protamines P1 and P2. Amino acid sequence of P2. Biochemistry 27, 2890–2897.

Belokopytova, I.A., Kostyleva, E.I., Tomilin, A.N., Vorob'ev, V.I. 1993. Human male infertility may be due to a decrease of the protamine P2 content in sperm chromatin. Mol. Reprod. Dev. 34, 53–57.

Benoit, G., Trasler, J.M. 1994. Developmental expression of DNA methyltransferase messenger ribonucleic acid, protein, and enzyme activity in the mouse testis. Biol. Reprod. 50, 1312–1319.

Bernstein, P., Peltz, S.W., Ross, J. 1989. The poly (A)-poly (A) binding protein complex is a major determinant of mRNA stability in vitro. Mol. Cell. Biol. 9, 659–670.

Bernstein, P., Ross, J. 1989. Poly(A), poly(A) binding protein and the regulation of mRNA stability. Trends Biochem. Sci. 14, 373–377.

Biggiogera, M., Fakan, S., Leser, G., Martin, T.E., Gordon, J. 1990. Immunoelectron microscopical visualization of ribonucleoproteins in the chromatoid body of mouse spermatids. Mol. Reprod. Dev. 26, 150–158.

Blanchard, Y., Lescoat, D., LeLannou, D. 1990. Anomalous distribution of nuclear basic proteins in round-headed human spermatozoa. Andrologia 22, 549–555.

Blendy, J.A., Kaestner, K.H., Weinbauer, G.F., Nieschlag, E., Schütz, G. 1996. Severe impairment of spermatogenesis in mice lacking the CREM gene. Nature 380, 162–165.

Bower, P.A. Yelick, P.C. Hecht, N.B. 1987. Both protamine P1 and 2 genes are expressed in the mouse, hamster, and rat. Biol. Reprod. 37, 479–488.

Bradbury, E.M. 1992. Reversible histone modifications and the chromosome cell cycle. BioEssays 14, 9–16.

Braun, R.E. 1990. Temporal translational regulation of the protamine 1 gene during mouse spermatogenesis. Enzyme 44, 120–128.

Braun, R.E. 2000. Temporal control of protein synthesis during spermatogenesis. Int. J. Androl. (Suppl. 2) 23, 92–94.

Braun, R.E., Behringer, R.R., Peschon, J.J., Brinster, R.L., Palmiter, R.D. 1989a. Genetically haploid spermatids are phenotypically diploid. Nature 337, 373–376.

Braun, R.E., Peschon, J.J., Behringer, R.R., Brinster, R.L., Palmiter, R.D. 1989b. Protamine 3' untranslated sequence regulate temporal traslational control of growth hormone in spermatids of transgenic mice. Genes Dev. 3, 793–802.

Bunick, D., Johnson, P.A., Johnson, T.R., Hecht, N.B. 1990. Transcription of the testis-specific mouse protamine 2 gene in a homologous in vitro transcription system. Proc. Natl. Acad. Sci. USA 87, 891–895.

Burd, C.G., Dreyfuss, G. 1994. Conserved structures and diversity of functions of RNA-binding proteins. Science 265, 615–621.

Caldwell, K.A., Handel, M.A. 1991. Protamine transcript sharing among postmeiotic spermatids. Proc. Natl. Acad. Sci. USA 88, 2407–2411.

Caron, N., Veilleux, S., Boissonneault, G. 2001. Stimulation of DNA repair by the spermatidal TP1 protein. Mol. Reprod. Dev. 58, 437–443.

Cataldo, L., Mastrangelo, M.A., Kleene, K.C. 1999. A quantitative sucrose gradient analysis of the translational activity of 18 mRNA species in testes from adult mice. Mol. Hum. Reprod. 5, 206–213.

Chen, F., Cooney, A.J., Wang, Y., Law, S.W., O'Malley, B.W. 1994. Cloning of a novel orphan receptor (GCNF) expressed during germ cell development. Mol. Endocrinol. 8, 1434–1444.

Cho, C., Willis, W.D., Goulding, E.H., Jung-Ha, H., Choi, Y.C., Hecht, N.B., Eddy E.M. 2001. Haploinsufficiency of protamine-1 or -2 causes infertility in mice. Nat. Genet. 28, 82–86.

Choi, Y.C., Aizawa, A., Hecht, N.B. 1997. Genomic analysis of the mouse protamine 1, protamine 2, and transition protein 2 gene cluster reveals hypermethylation in expressing cells. Mamm. Genome 8, 317–323.

Davie, J.R. 1998. Covalent modifications of histones: expression from chromatin templates. Curr. Opin. Genet. Dev. 8, 173–178.

Davies, H.G., Giogini, F., Fajardo, M.A., Braun, R.E. 2000. A sequence-specific RNA binding complex expressed in murine germ cells contains MSY2 and MSY4.

Delmas, V., Laoide, B., Masquilier, D., deGroot, R.P., Foulkes, N.S., Sassone-Corsi, P. 1992. Alternative usage of initiation codons in CREM generates regulators with opposite functions. Proc. Natl. Acad. Sci. USA 89, 4226–4230.

Delmas, V., van der Hoorn, F., Mellstrom, B., Jegou, B., Sassone-Corsi, P. 1993. Induction of CREM activator proteins in spermatids: down-stream targets and implications for haploid germ cell differentiation. Mol. Endocrinol. 7, 1502–1514.

Derrigo, M., Cestelli, A., Savettieri, G., di Liegro, I. 2000. RNA-protein interactions in the control of stability and localization of messenger RNA. Int. J. Mol. Med. 5, 111–123.

Dickey, L.F., Wang, Y.H., Shull, G.E., Wortman, I.A., Theil, E.C. 1988. The importance of the 3'-untranslated region in the translational control of ferritin mRNA. J. Biol. Chem. 263, 3071–3074.

Didier, D.K., Schiffenbauer, J., Woulfe, S.L., Zacheis, M., Schwart, B.D. 1988. Characterization of the cDNA encoding a protein binding to the major histocompatibility complex class II Y box. Proc. Natl. Acad. Sci. USA 85, 7322–7326.

Drabent, B., Bode, C., Bramlage, B., Doenecke D. 1996. Expression of the mouse testicular histone gene H1t during spermatogenesis. Histochem. Cell Biol. 106, 247–251.

Drabent, B., Saftig, P., Bode, C., Doenecke, D. 2000. Spermatogenesis proceeds normally in mice without linker histone H1t. Histochem. Cell Biol. 113, 433–442.

Dym, M., Fawcett, D.W. 1970. The blood-testis barrier in the rat and the physiological compartmentation of the seminiferous epithelium. Biol. Reprod. 3, 308–326.

Eddy, E.M. 1998. Regulation of gene expression during spermatogenesis. Cell Dev. Biol. 9, 451–457.

Elliott, D.J., Millar, M.R., Oghene, K., Ross, A., Kiesewetter, F., Pryor, J., McIntyre, M., Hargreave, T.B., Saunders, P.T.K., Vogt, P.H., Chandley, A.C., Cooke H. 1997. Expression of RBM in the nuclei of human germ cells is dependent on a critical region of the Y chromosome long arm. Proc. Natl. Acad. Sci. USA 94, 3848–3853.

Enmark, E., Gustafsson, J. 1996. Orphan nuclear receptors—the first eight years. Mol. Endocrinol. 10, 1293–1307.

Fajardo, M.A., Butner, K.A., Lee, K., Braun, R.E. 1994. Germ cell-specific proteins interact with the 3' untranslated regions of Prm-1 and Prm-2 mRNA. Dev. Biol. 166, 643–653.

Ferlin, A., Moro, E., Onisto, M., Toscano, E., Bettella, A., Foresta, C. 1999. Absence of testicular DAZ gene expression in idiopathic severe testiculopathies. Hum. Reprod. 14, 2286–2292.

Fimia, G.M., DeCesare, D., Sassone-Corsi, P. 1999. CBP-independent activation of CREM and CREB by the LIM-only protein ACT. Nature 398, 165–169.

Finch, J.T., Klug A. 1976. Solenoid model for superstructure in chromatin. Proc. Natl. Acad. Sci. USA 73, 1897–1901.

Foulkes, N.S., Mellstrom, B., Benusiglio, E., Sassone-Corsi, P. 1992. Developmental switch of CREM function during spermatogenesis: from antagonist to activator. Nature 355, 80–84.

Foulkes, N.S., Schlotter, F., Pevet, P., Sassone-Corsi, P. 1993. Pituitary hormone FSH directs the CREM functional switch during spermatogenesis. Nature 362, 264–267.

Gatewood, J.M., Cook, G.R., Balhorn, R., Bradbury, E.M., Schmid, C.W. 1987. Sequence-specific packaging of DNA in human sperm chromatin. Science 236, 962–964.

Goldberg, E. 1996. Transcriptional regulatory strategies in male germ cells. J. Androl. 17, 628–632.

Grauman, P.L., Marahiel, M.A. 1998. A superfamily of proteins that contain the cold-shock domain. Trends Biochem. Sci. 23, 286–290.

Grimaldi, P., Scarponi, C., Rossi, P., March, M.R., Fabbri, A., Isidori, A., Spera, G., Krausz, C., Geremia, R. 1998. Analysis of Yq microdeletions in infertile males by PCR and DNA hybridization techniques. Mol. Hum. Reprod. 4, 1116–1121.

Gu, W., Kwon, Y., Oko, R., Hermo, L., Hecht, N.B. 1995. Poly (A) binding protein is bound to both stored and polysomal mRNAs in the mammalian testis. Mol. Reprod. Dev. 40, 273–285.

Gu, W., Tekur, S., Reinbold, R., Eppig, J.J., Choi, Y.C., Zheng, J.Z., Murray, M.T., Hecht, N.B. 1998. Mammalian male and female germ cells express a germ cell-specific Y-box protein, MSY2. Biol. Reprod. 59, 1266–1274.

Han, J.R., Gu, W., Hecht, N.B. 1995. Testis/brain RNA-binding protein, a testicular translational regulatory RNA-binding protein, is present in the brain and binds to the 3′ untranslated regions of transported brain mRNAs. Biol. Reprod. 53, 707–717.

Hazzouri, M., Pivot-Pajot, C., Faure, A.K., Usson, Y., Pelletier, R., Sele, B., Khochbin, S., Rousseaux, S. 2000. Regulated hyperacetylation of core histones during mouse spermatogenesis: involvement of histone deacetylases. Eur. J. Cell Biol. 79, 950–960.

Hecht, N.B. 1998. Molecular mechanisms of male germ cell differentiation. BioEssays 20, 555–561.

Heidaran, M.A., Kozak, C.A., Kistler, W.S. 1989. Nucleotide sequence of the Stp-1 gene coding for rat spermatid nuclear transition protein 1 (TP1): homology with protamine P1 and assignment of the mouse Stp-1 gene to chromosome 1. Gene 75, 39–46.

Herbert, T.P., Hecht, N.B. 1999. The mouse Y-box protein, MSY2, is associated with a kinase on non-polysomal mouse testicular mRNAs. Nucleic Acids Res. 27, 1747–1753.

Hirose, T., O'Brien, D.A., Jetten, A.M. 1995a. Cloning of the gene encoding the murine orphan receptor TAK1 and cell-type-specific expression in testis. Gene 163, 239–242.

Hirose, T., O'Brien, D.A., Jetten, A.M. 1995b. RTR: a new member of the nuclear receptor superfamily that is expressed in murine testis. Gene 152, 247–251.

Hornstein, E., Git, A., Braunstein, I., Avni, D., Meyuhas, O. 1999. The expression of poly(A)-binding protein gene is translationally regulated in a growth-dependent fashion through a 5′-terminal oligopyrimidine tract motif. J. Biol. Chem. 274, 1708–1714.

Howard, T., Balogh, R., Overbeek, P., Bernstein, K.E. 1993. Sperm-specific expression of angiotensin-converting enzyme (ACE) is mediated by a 91-base-pair promoter containing a CRE-like element. Mol. Cell. Biol. 13, 18–27.

Jackson, R.J., Standart, N. 1990. Do the poly(A) tail and 3′ untranslated region control mRNA translation? Cell 62, 15–24.

Johnson, P.A., Bunick, D., Hecht, N.B. 1991. Protein binding regions in the mouse and rat protamine-2 genes. Biol. Reprod. 44, 127–134.

Kenan, D.J., Query, C.C., Keene, J.D. 1991. RNA recognition: towards identifying determinants of specificity. Trends Biochem. Sci. 16, 214–220.

Kistler, M.K., Sassone-Corsi, P., Kistler, W.S. 1994. Identification of a functional cyclic adenosine 3′,5′-monophosphate response element in the 5′ flanking region of the gene for transition protein 1 (TP1), a basic chromosomal protein of mammalian spermatids. Biol. Reprod. 51, 1322–1329.

Kleene, K.C. 1989. Poly (A) shortening accompanies the activation of translation of five mRNAs during spermiogenesis in the mouse. Development, 106, 367–373.

Kleene, K.C. 1993. Multiple controls over the efficiency of translation of the mRNAs encoding transition proteins, protamines, and the mitochondrial capsule selenoprotein in late spermatids in mice. Dev. Biol. 159, 720–731.

Kleene, K.C. 1996. Patterns of translational regulation in the mammalian testis. Mol. Reprod. Dev. 43, 268–281.

Kleene, K.C., Distal, R.J., Hecht, N.B. 1984. Translational regulation and coordinate deadenylation of a haploid mRNA during spermiogenesis in mouse. Dev. Biol. 105, 71–79.

Kleene, K.C., Mastrangelo, M.A. 1999. The promoter of the poly(A) binding protein 2 (Pabp2) retroposon is derived from the 5′-untranslated region of the Pabp1 progenitor gene. Genomics 61, 194–200.

Kleene, K.C., Mulligan, E., Steiger, D., Donohue, K., Mastrangelo, M.A. 1998. The mouse gene encoding the testis-specific isoform of poly(A) binding protein (Pabp2) is an expressed retroposon: intimations that gene expression in spermatogenic cells facilitates the creation of new genes. J. Mol. Evol. 47, 275–281.

Kleene, K.C., Wang, M.Y., Cutler, M., Hall, C., Shih, H. 1994. Developmental expression of poly(A) binding protein mRNAs during spermatogenesis in the mouse. Mol. Reprod. Dev. 39, 355–364.

Kliesch, S., Penttilä, T.L., Gromoll, J., Saunders, P.T., Nieschlag, E., Parvinen, M. 1992. FSH receptor mRNA is expressed stage-dependently during rat spermatogenesis. Mol. Cell Endocrinol. 84, 45–49.

Klug, A. 1999. Zinc finger peptides for the regulation of gene expression. J. Mol. Biol. 293, 215–218.

Krishnamurthy, H., Danilovich, N., Morales, C.R., Sairam, M.R. 2000. Qualitative and quantitative decline in spermatogenesis of the follicle-stimulating hormone receptor knockout (FORKO) mouse. Biol. Reprod. 62, 1146–1159.

Kundu, K.T., Rao, M.R.S. 1995. DNA condensation by the rat spermatidal protein TP2 shows GC-rich sequence preference and is zinc dependent. Biochemistry 34, 5143–5150.

Kwon, Y.K., Hecht, N.B. 1991. Cytoplasmic protein binding to highly conserved sequences in the 3′ untranslated region of mouse protamine 2 mRN.A. a translationally regulated transcript of male germ cells. Proc. Natl. Acad. Sci. USA 88, 3584–3588.

Kwon, Y.K., Hecht, N.B. 1993. Binding of a phosphoprotein to the 3′ untranslated region of the mouse protamine 2 mRNA temporally represses its translation. Mol. Cell. Biol. 13, 6547–6557.

Kwon, Y.K., Murray, M.T., Hecht, N.B. 1993. Proteins homologous to the Xenopus germ cell-specific RNA-binding proteins p54/p56 are temporally expressed in mouse male germ cells. Dev. Biol. 158, 99–100.

Ladomery, M., Sommerville, J. 1994. Binding of Y-box proteins to RNA: involvement of different protein domains. Nucleic Acids Res. 22, 5582–5589.

Lalli, E., Lee, J.S., Lamas, M., Tamai, K., Zazaopoulos, E., Nantel, F., Penna, L., Foulkes, N.S., Sassone-Corsi, P. 1996. The nuclear response to cAMP: role of transcription factor CREM. Philos. Trans. R. Soc. Lond. 351, 201–209.

Lamas, M., Monaco, L., Zazapopoulos, E., Lalli, E., Tamai, K., Penna, L., Mazzucchelli, C., Nantel, F., Foulkes, N.S., Sassone-Corsi, P. 1996. CREM: a master-switch in the transcriptional response to cAMP. Philos. Trans. R. Soc. Lond. B Biol. Sci. 351, 561–567.

Lee, C.H., Chang, L., Wie, L.N. 1996. Molecular cloning and characterization of a mouse nuclear orphan receptor expressed in embryos and testes. Mol. Reprod. Dev. 44, 305–314.

Lee, K., Fajardo, M.A. Braun, R.E. 1996. A testis cytoplasmic RNA-binding protein that has the properties of a translational repressor. Mol. Cell. Biol. 16, 3023–3034.

Lee, K., Haugen, H.S., Clegg, C.H., Braun, R.E. 1995. Premature translation of protamine-1 mRNA causes precocious nuclear condensation and arrests spermatid differentiation in mice. Proc. Natl. Acad. Sci. USA 92, 12451–12455.

Lei, Z.M., Mishra, S., Zou, W., Xu, B., Foltz, M., Li, X., Rao, C.V. 2001. Targeted disruption of luteinizing hormone/human chorionic gonadotropin receptor gene. Mol. Endocrinol. 15, 184–200.

Leibold, E.A., Munro, H.N. 1988. Cytoplasmic protein binds in vitro to a highly conserved sequence in the 5′ untranslated region of ferritin heavy- and light-subunit mRNA. Proc. Natl. Acad. Sci. USA 85, 2171–2175.

Lescoat, D., Colleu, D., Boujard, D., LeLannou, D. 1988. Electrophoretic characteristics of nuclear proteins from human spermatozoa. Arch. Androl. 20, 35–40.

Levesque, D., Veilleux, S., Caron, N., Biossonneault, G. 1998. Architectural DNA-binding properties of the spermatidal transition proteins 1 and 2. Biochem. Biophys. Res. Commun. 252, 602–609.

Lin, Q., Sirotkin, A., Skoultchi, A.I. 2000. Normal spermatogenesis in mice lacking the testis-specific linker histone H1t. Mol. Cell. Biol. 20, 2122–2128.

Lo, W.S., Trievel, R.C., Rojas, J.R., Duggan, L., Hsu, J.Y., Allis, C.D., Marmorstein, R., Berger, S.L. 2000. Phosphorylation od serine 10 in histone H3 is functionally linked in vitro and in vivo to Gcn5-mediated acetylation at lysine 14. Mol. Cell 5, 917–926.

Ma, K., Inglis, J.D., Sharkey, A., Bickmore, W.A., Speed, R.M., Thomson, E.J., Jobling, M., Taylor, K., Wolfe, J., Cooke, H.J., Hargraeve, T.B., Chandley, A.C. 1993. A Y chromosome gene family with RNA-binding protein homology: candidates for the azoospermia factor AZF controlling human spermatogenesis. Cell 75, 1287–1295.

Mali, P., Kaipia, A., Kangasniemi, M., Toppari, J., Sandberg, M., Hecht, N.B. Parvinen, M. 1989. Stage-specific expression of nucleoprotein mRNAs during rat and mouse spermiogenesis. Reprod. Fertil. Dev. 1, 369–382.

Masquilier, D., Foulkes, N.S., Mattei, M.G., Sassone-Corsi, P. 1993. Human CREM gene: evolutionary conservation, chromosomal localization, and inducibility of the transcript. Cell. Growth Diff. 4, 931–937.

Mastrangelo, M.A., Kleene, K.C. 2000. Developmental expression of Y-box protein 1 mRNA and alternatively spliced Y-box protein 3 mRNA in spermatogenic cells in mice. Mol. Hum. Reprod. 6, 779–788.

Matsumoto, K., Wolffe, A.P. 1998. Gene regulation by Y-box proteins: coupling control of transcription and translation. Trends Cell Biol. 8, 318–323.

Meistrich, M.L., Trostle-Weige, P.K., Lin, R., Bhatnagar, Y.M., Allis, C.D. 1992. Highly acetylated H4 is associated with histone displacement in rat spermatids. Mol. Reprod. Dev. 31, 170–181.

Morales, C.R., Wu, X.Q., Hecht, N.B. 1998. The DNA/RNA-binding protein, TB-RBP, moves from the nucleus to the cytoplasm and through intercellular bridges in male germ cells. Dev. Biol. 201, 113–123.

Moussa, F., Oko, R., Hermo, L. 1994. The immunolocalization of small nuclear ribonucleoprotein particles in testicular cells during the cycle of the seminiferous epithelium of the adult rat. Cell Tissue Res. 278, 363–378.

Mullner, E.W., Neupert, B., Kuhn, L.C. 1989. A specific mRNA binding factor regulates the iron-dependent stability of cytoplasmic transferrin receptor mRNA. Cell 28, 373–382.

Nantel, F., Monaco, L., Foulkes, N.S., Masquilier, D., LeMeur, M., Henriksen, K., Dierich, A., Parvinen, M., Sassone-Corsi, P. 1996. Spermiogenesis deficiency and germ cell apoptosis in CREM-mutant mice. Nature 380, 159–162.

Nikolajczyk, B.S., Murray, M.T., Hecht, N.B. 1995. A mouse homologue of the Xenopus germ cell-specific ribonucleic acid/deoxyribonucleic acid-binding proteins p54/p56 interacts with the protamine 2 promoter. Biol. Reprod. 52, 524–530.

Numata, M., Ono, T., Iseki, S. 1994. Expression and localization of the mRNA for DNA (cytosine-5)-methyl-transferase in mouse seminiferous tubules. J. Histochem. Cytochem. 42, 1271–1276.

Oakberg, E.F. 1956. Duration of spermatogenesis in the mouse and timing of stages of the cycle of the semi-iniferous epithelium. Am. J. Anat. 99, 507–516.

Oko, R., Korley, R., Murray, M.T., Hecht, N.B., Hermo L. 1996. Germ cell-specific DNA and RNA binding proteins p48/52 are expressed at specific stages of male germ cell development and are present in the chromatoid body. Mol. Reprod. Dev. 44, 1–13.

Oliva, R., Dixon, G.H. 1991. Vertebrate protamine genes and the histone-to-protamine replacement reaction. Prog. Nucleic Acids Res. Mol. Biol. 40, 25–94.

Papoutsopoulou, S., Nikolakaki, E., Chalepakis, G., Kruft, V., Chevaillier, P., Giannakouros, T. 1999. SR protein-specific kinase 1 is highly expressed in testis and phosphorylates protamine 1. Nucleic Acids Res. 27, 2972–2980.

Pelletier, R.M., Friend, D.S. 1983. The Sertoli cell junctional complex: structure and permeability to filipin in the neonatal and adult guinea pig. Am. J. Anat. 168, 213–228.

Penttilä, T.L., Yuan, L., Mali, P., Höög, C., Parvinen, M. 1995. Haploid gene expression: temporal onset and storage patterns of 13 novel transcripts during rat and mouse spermiogenesis. Biol. Reprod. 53, 499–510.

Peschon, J.J., Behringer, R.R., Brinster, R.L., Palmiter, R.D. 1987. Spermatid-specific expression of protamine-1 in transgenic mice. Proc. Natl. Acad. Sci. USA 84, 5316–5319.

Peschon, J.J., Behringer, R.R., Palmiter, R.D., Brinster, R.L. 1989. Expression of mouse protamine-1 genes in transgenic mice. Ann. NY Acad. Sci. 564, 186–197.

Prigent, Y., Müller, S., Dadoune, J.P. 1996. Immunoelectron microscopical distribution of histones H2B and H3 and protamines during human spermiogenesis. Mol. Hum. Reprod. 2, 929–935.

Reijo, R., Lee, T.Y., Salo, P., Alagappan, R., Brown, L.G., Rosenberg, M., Rozen, S., Jaffe, T., Straus, D., Hovatta, O., de la Chapelle, A., Silber, S., Page, D.C. 1995. Diverse spermatogenic defects in humans caused by Y chromosome deletions encompassing a novel RNA-binding protein gene. Nat. Genet. 10, 383–393.

Reinicke, K., Chevaillier, P. 1991. Human protamines are zinc-binding proteins. Mol. Androl. 3, 49–65.

Roesler, W.J., Vandenbark, G.R., Hanson, R.W. 1988. Cyclic AMP and the induction of eukaryotic gene transcription. J. Biol. Chem. 263, 9063–9066.

Ruggiu, M., Speed, R., Taggart, M., McKay, S.J., Kilanowski, F., Saunders, P., Dorin, J., Cooke, H.J. 1997. The mouse Dazla gene encodes a cytoplasmic protein essential for gametogenesis. Nature 389, 73–77.

Russell, L.D., Ettlion, R.A., Strika-Hikin, A.P., Clegg, E.D. 1990. *Histological and Histopathological Evaluation of the Testis*. Cache River Press, Clearwater.

Sakkas, D., Urner, F., Bianchi, P.G., Bizzaro, D., Wagner, I., Jaquenoud, N., Manicardi, G., Campana, A. 1996. Sperm chromatin anomalies can influence decondensation after intracytoplasmic sperm injection. Hum. Reprod. 11, 837–843.

Saunders, P.T.K., Millar, M.R., Maguire, S.M., Sharpe, R.M. 1992. Stage-specific expression of rat transition protein 2 mRNA and possible localization to the chromatoid body of step 7 spermatids by in situ hybridization using a nonradioactive riboprobe. Mol. Reprod. Dev. 33, 385–391.

Schlüter, G., Schlicker, M., Engel, W. 1993. A conserved 8 bp motif (GCYATCAY) in the 3′ UTR of transition protein 2 as a putative target for a transcript stabilizing protein factor. Biochem. Biophys. Res. Commun. 197, 110–115.

Schmidt, E.E., Hanson, E.S., Capecchi, M.R. 1999. Sequence-independent assembly of spermatid mRNAs into messenger ribonucleoprotein particles. Mol. Cell. Biol. 19, 3904–3915.

Schmidt, E.E., Ohbayashi, T., Makino, Y., Tamura, T., Schibler U. 1997. Spermatid-specific overexpression of the TATA-binding protein gene involves recruitment of two potent testis-specific promoters. J. Biol. Chem. 272, 5326–5334.

Schmidt, E.E., Schibler, U. 1997. Developmental testis-specific regulation of mRNA levels and mRNA translational efficiencies for TATA-binding protein mRNA isoforma. Dev. Biol. 184, 138–149.

Schumacher, J.M., Lee, K., Edelhoff, S., Braun, R.E. 1995a. Distribution of Tenr, an RNA-binding protein, in a lattice-like network within the spermatid nucleus in the mouse. Biol. Reprod. 52, 1274–1283.

Schumacher, J.M., Lee, K., Edelhoff, S., Braun, R.E. 1995b. Spnr, a murine RNA-binding protein that is localized to cytoplasmic microtubules. J. Cell Biol. 129, 1023–1032.

Singal, R., van Wert, J., Bashambu, M., Wolfe, S.A., Wilkerson, D.C., Grimes S.R. 2000. Testis-specific histone H1t gene is hypermethylated in nongerminal ceöös in the mouse. Biol. Reprod. 63, 1237–1244.

Singh, J., Rao, MR.S. 1987. Interaction of rat testis protein, TP with nucleic acids in vitro. J. Biol. Chem. 262, 734–740.

Slee, R., Grimes, B., Speed, R.M. Taggart, M., Maguire, S.M. Ross, A., McGill, N.I., Saunders, P.T., Cooke, H.J. 1999. A human DAZ transgene confers partial rescue of the mouse Dazl null phenotype. Proc. Natl. Acad. Sci. USA 96, 8040–8045.

Söderström, K.O. 1981. Labeling of the chromatoid body by [^{3}H] uridine in the rat pachytene spermatocytes. Exp. Cell Res. 131, 488–491.

Steger, K. 1999. Transcriptional and translational regulation of gene expression in haploid spermatids. Anat. Embryol., 199, 471–487.

Steger, K. 2001. Haploid spermatids exhibit translationally repressed mRNAs. Anat. Embryol. 203, 323–334.

Steger, K., Failing, K., Klonisch, T., Behre, H.M., Manning, M., Weidner, W., Hertle, L., Bergmann, M., Kliesch, S. 2001. Round spermatids from infertile men exhibit decreased levels of protamine-1 and 2 mRNA. Hum. Reprod. 16, 709–716.

Steger, K., Klonisch, T., Gavenis, K., Behr, R., Schaller, V., Drabent, B., Doenecke, D., Nieschlag, E., Bergmann, M., Weinbauer, G.F. 1999. Round spermatids show normal testis-specific H1t but reduced cAMP-responsive element modulator (CREM) and transition protein 1 (TP1) expression in men with round spermatid maturation arrest. J. Androl. 20, 747–754.

Tafuri, S.R., Familari, M., Wolffe, A.P. 1993. A mouse Y-box protein, MSY1, is associated with paternal mRNA in mouse spermatocytes. J. Cell Chem. 268, 12213–12220.

Tamai, K.T., Monaco, L., Nantel, F., Zazopoulos, E., Sassone-Corsi, P. 1997. Coupling signalling pathways to transcriptional control: nuclear factors responsive to cAMP. Recent Prog. Horm. Res. 52, 121–140.

Tamura, T., Makino, Y., Mikoshiba, K., Muramatsu, M. 1992. Demonstration of a testis-specific trans-acting factor Tet-1 in vitro that binds to the promoter of the mouse protamine 1 gene. J. Biol. Chem. 267, 4327–4332.

Tanphaichitr, N., Sohbon, P., Taluppeth, N., Chalermisarachai, P. 1978. Basic nuclear proteins in testicular cells and ejaculated spermatozoa in man. Exp. Cell Res. 117, 347–350.

Trasler, J.M., Hake, L.E., Johnson, P.A., Alcivar, A.A., Millette, C.F., Hecht, N.B. 1990. DNA methylation and demethylation events during meiotic prophase in the mouse testis. Mol. Cell. Biol. 10, 1828–1834.

Walker, W.H., Daniel, P.B., Habener, J.F. 1998. Inducible cAMP early repressor ICER down-regulation of CREB gene expression in Sertoli cells. Mol. Cell Endocrinol. 25, 167–178.

Walker, W.H., Habener, J.F. 1996. Role of transcription factors CREB and CREM in cAMP-regulated transcription during spermatogenesis. Trends Endocrinol. Metab. 7, 133–138.

Walker, W.H., Sanborn, B.M., Habener, J.F. 1994. An isoform of transcription factor CREM expressed during spermatogenesis lacks the phosphorylation domain and represses cAMP-induced transcription. Proc. Natl. Acad. Sci. USA 91, 12423–12427.

Walt, H., Armbruster, B.L. 1984. Actin and RNA are components of the chromatoid bodies in spermatids of the rat. Cell Tissue Res. 236, 487–490.

Ward, W.S. 1993. Deoxyribonucleic acid loop domain tertiary structure in mammalian spermatozoa. Biol. Reprod. 48, 1193–1203.

Weinbauer, G.F., Behr, R., Bergmann, M., Nieschlag, E. 1998. Testicular cAMP responsive element modulator (CREM) protein is expressed in round spermatids but is absent or reduced in men with round spermatid maturation arrest. Mol. Hum. Reprod. 4, 9–15.

Weiner, A.M., Deininger, P.L., Efstratiadis, A. 1986. Nonviral retroposons: genes, pseudogenes and transposable elements generated by reverse flow of genetic information. Ann. Rev. Biochem. 55, 631–661.

Wu, J.Y., Ribar, T.J., Cummings, D.E., Burton, K.A., McKnight, G.S., Means, A.R. 2000. Spermiogenesis and exchange of basic nuclear proteins are impaired in male germ cells lacking Camk4. Nat. Genet. 25, 448–452.

Wu, X.Q., Xu, L., Hecht, N.B. 1998. Dimerization of the testis brain RNA-binding protein (translin) is mediated through its C-terminus and is required for DNA- and RNA-binding. Nucleic Acids Res. 26, 1675–1680.

Xu, X., Toselli, P.A., Russell, L.D., Seldin, D.C. 1999. Globozoospermia in mice lacking the casein kinase II alpha' catalytic subunit. Nat. Genet. 23, 118–121.

Yan, Z.H., Medvedev, A., Hirose, T., Gotoh, H., Jetten, A.M. 1997. Characterization of the response element and DNA binding properties of the nuclear orphan receptor germ cell nuclear factor/retinoid receptor-related testis-associated receptor. J. Biol. Chem. 272, 10565–10572.

Yelick, P.C., Balhorn, R., Johnson, P.A., Corzett, M., Mazrimas, J.A., Kleene, K.C., Hecht, N.B. 1987. Mouse protamine 2 is synthesized as a precursor whereas mouse protamine 1 is not. Mol. Cell. Biol. 7, 2173–2179.

Yelick, P.C., Kwon, Y.H., Flynn, J.F., Borgorgzadeh, A., Kleene, K.C., Hecht, N.B. 1989. Mouse transition protein 1 is translationally regulated during postmeiotic stages of spermatogenesis. Mol. Reprod. Dev. 1, 193–200.

Yiu, G.K., Hecht, N.B. 1997. Novel testis-specific protein-DNA interactions activate transcription of the mouse protamine 2 gene during spermatogenesis. J. Biol. Chem. 272, 26926–26933.

Yu, Y.E., Zhang, Y., Unni, E., Shirley, C.R., Deng, J.M., Russell, L.D., Weil, M.M., Behringer, R.R., Meistrich, M.L. 2000. Abnormal spermatogenesis and reduced fertility in transition nuclear protein-1 deficient mice. Proc. Natl. Acad. Sci. USA 97, 4683–4688.

Zamboni, L., Zemjanis, R., Stefanini, M. 1971. The fine structure of monkey and human spermatozoa. Anat. Rec. 169, 129–137.

Zambrowicz, B.P., Harendza, C.J., Zimmermann, J.W., Brinster, R.L., Palmiter, R.D. 1993. Analysis of the mouse protamine 1 promoter in transgenic mice. Proc. Natl. Acad. Sci. USA 90, 5071–5075.

Zambrowicz, B.P., Palmiter, R.D. 1994. Testis-specific and ubiquitous proteins bind to functionally important regions of the mouse protamine-1 promoter. Biol. Reprod. 50, 65–72.

Zhao, J., Hyman, L., Moore C. 1999. Formation of mRNA 3' ends in eukaryotes: mechanism, regulation and interrelationships with other steps in mRNA synthesis. Microbiol. Mol. Biol. Rev. 63, 405–455.

Zhang, D., Penttila, T.L., Morris, P.L., Teichmann, M., Roeder R.G. 2001. Spermiogenesis deficiency in mice lacking the Trf2 gene. Science 292, 1153–1155.

Zhang, F.P., Poutanen, M., Wilbertz, J., Huhtaniemi, I. 2001. Normal prenatal but arrested postnatal sexual development of luteinizing hormone receptor knockout (LuRKO) mice. Mol. Endocrinol. 15, 172—183.

Zhang, L.P., Stroud, J.C., Walter, C.A., Adrian, G.S., McCarrey, J.R. 1998. A gene-specific promoter in transgenic mice directs testis-specific demethylation prior to transcriptional activation in vivo. Biol. Reprod. 59, 284–292.

Zhang, Y.L., Akmal, K.M., Tsuruta, J.K., Shang, Q., Hirose, T., Jetten, A.M., Kim, K.H., O'Brien, D.A. 1998. Expression of germ cell nuclear factor (GCNF/RTR) during spermatogenesis. Mol. Reprod. Dev. 50, 93–102.

Zhong, J., Peters, A.H., Lee, K., Braun, R.E. 1999. A double-stranded RNA-binding protein required for activation of repressed messages in mammalian germ cells. Nat. Genet. 22, 171–174.

Ziyyat, A., Lassalle, B., Testart, J., Briot, P., Amar, E., Finaz, C., Lefevre, A. 1999. Flow cytometry isolation and reverse transcriptase-polymerase chain reaction characterization of human round spermatids in infertile patients. Hum. Reprod. 14, 379–387.

Advances in Developmental Biology and Biochemistry, Vol. 12
M. DePamphilis (Editor)

Egg zona pellucida, egg vitelline envelope, and related extracellular glycoproteins

Luca Jovine, Eveline S. Litscher and Paul M. Wassarman

Department of Molecular, Cell and Developmental Biology, Mount Sinai School of Medicine,
One Gustave L. Levy Place, New York, NY 10029-6574, USA

Summary

Vertebrate eggs are surrounded by an extracellular coat that performs vital functions during development, especially during fertilization. In mammalian and non-mammalian animals the egg coat is called a *zona pellucida* (ZP) and *vitelline envelope* (VE), respectively. The ZP of mammalian eggs, from mice to humans, and VE of fish, bird, and amphibian eggs, consists of only a few glycoproteins that are related to each other and often possess several recognizable elements (e.g., a signal sequence, ZP domain, consensus furin cleavage-site, and transmembrane domain). In most instances, ZP and VE glycoproteins are synthesized in the ovary by growing oocytes and surrounding follicle cells, but in some cases (e.g., in fish and birds) certain VE glycoproteins are synthesized in the liver and transported to the ovary. Many other families of extracellular proteins, found in vertebrates and invertebrates, share a conserved ZP domain (~260 amino acids and 8 conserved Cys residues) with ZP and VE glycoproteins. The ZP domain enables these proteins to multimerize and perform their various functions, for example, as receptors and mechanotransducers. It is likely that the list of proteins containing a ZP domain will grow considerably in coming years and that our understanding of structural-functional relationships for such proteins will be clarified.

Contents

1. Introduction

During oogenesis, female germ cells are transformed into unfertilized eggs (Browder, 1985). In humans, this transformation begins during fetal development when germ cells enter meiosis and ends many years later with ovulation of unfertilized eggs. Oogenesis includes not only meiotic events and tremendous enlargement of the cell, but profound changes in the complement of cytoplasmic organelles and establishment of extracellular coats that surround the egg.

All eggs are surrounded by one or more extracellular coats that serve a number of vital functions during oogenesis, fertilization, and early embryogenesis (Dumont and Brummett, 1985; Yanagimachi, 1994; Wassarman, 1999). The coats have been referred to as *chorion, zona radiata (interna and externa), vitelline membrane, vitelline envelope,* or *zona pellucida,* among other names. Today, the extracellular coats surrounding mammalian and non-mammalian eggs are frequently referred to as the *zona pellucida* (ZP) and *vitelline envelope* (VE), respectively, and this terminology will be used here.

During the past ten years or so, we have learned a great deal about the glycoproteins that constitute the vertebrate egg ZP and VE. In particular, it has been noted that VE glycoproteins from fish, bird, and amphibian eggs are related to ZP glycoproteins from mammalian eggs (reviewed by Wassarman et al., 2001). This and other revelations have led to speculation about the evolution of extracellular coat glycoproteins, their assembly into a ZP and VE during oogenesis, and their functions during fertilization. For example, one mouse ZP glycoprotein, called ZP3, serves as a primary sperm receptor and acrosome reaction-inducer during mammalian fertilization (Wassarman, 1990, 1999; Arnoult et al., 1996).

Here, we review some of the molecular characteristics of ZP and VE glycoproteins and point out features common to glycoproteins from both types of extracellular coats. In addition, we review some of the features of several extracellular proteins that are related to ZP and VE glycoproteins. In the interest of brevity, not all relevant literature is presented.

2. Mammalian egg zona pellucida

The plasma membrane of all mammalian eggs is surrounded by an elastic ZP that, depending on the species, varies in thickness from ~2 to ~20 μm (Wassarman, 1988; Yanagimachi, 1994). For example, the mouse egg ZP is ~6.5 μm thick and contains ~3.5 ng of protein. The unfertilized egg ZP is very porous and can be solubilized by exposure to low pH, high temperature, or reducing agents. Transmission electron micrographs of either sectioned or solubilized ZP reveal an extensive crosslinked network of fibers/filaments punctuated by large holes (Fig. 1).

2.1. Egg zona pellucida glycoproteins

Much of what is known about ZP glycoproteins comes from research with mice, although many other eutherian mammals, from rats to humans, have been studied (see below, Section 2.5). Information about marsupial egg ZP glycoproteins also is available. Collectively, the data strongly suggests that all ZP are composed of only a few glycoproteins, often designated ZP1-3, each of which consists of a unique polypeptide that is differentially glycosylated (N- and O-linked oligosaccharides) and modified (sulfation, sialylation, etc.) (Bleil and Wassarman, 1980a; Wassarman, 1988; Wassarman et al., 1985, 2001; Liu et al., 1997; McLeskey et al., 1998; Prasad et al., 2000). ZP2 and ZP3, the most abundant ZP glycoproteins, are monomers and are present in about equimolar amounts, whereas ZP1 is a dimer composed of identical polypeptides held together by intermolecular disulfides. The glycoproteins are encoded by single-copy genes (mouse ZP1-3 genes have 12, 18, and 8 exons, respectively) located on different chromosomes (mouse ZP1-3 genes are located on chromosomes 19, 7, and 5, respectively) (Epifano et al., 1995b). Analysis of the human genome sequence suggests that genes encoding ZP1-3 are located on chromosomes 11, 16, and 7, respectively (Chamberlin and Dean, 1990; van Duin et al., 1992; Liang and Dean, 1993; Hughes and Barratt, 1999; http://www.ensembl.org/). The primary structures of polypeptides of a given ZP glycoprotein from different mammalian species are closely related to one another. For example, the polypeptide sequences of mouse and human ZP2 and ZP3, two species separated by more than 10^8 years of evolution, are 56% and 67% identical, respectively. This degree of identity is not terribly surprising since all three glycoproteins play structural roles in the egg ZP and, consequently, must interact with each other in a similar manner regardless of species.

2.2. Common features of zona pellucida genes and glycoproteins

Genes encoding ZP glycoproteins exhibit conserved organization. In mice and humans ZP1 and ZP3 consist of 12 and 8 exons, respectively, whereas ZP2 genes in mice consist of 18 exons and in humans of 19 exons. Although intron size varies extensively between species, the number and length of exons is highly conserved (Kinloch et al., 1990; Chamberlin and Dean, 1990; Liang and Dean, 1993; McLeskey et al., 1998). This reflects the presence of exon/intron boundaries that define the limits of distinct domains in ZP glycoproteins. ZP genes share TATAA boxes approximately 30 bp upstream of their transcription start-sites, as well as E-box sequences (CANNTG) at approximately −200 bp.

L. Jovine, E. S. Litscher and P. M. Wassarman

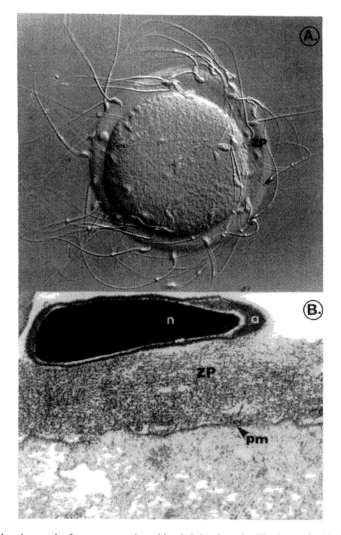

Fig. 1. (A) Light micrograph of mouse sperm bound by their heads to the ZP of an unfertilized mouse egg in vitro. The micrograph was taken using Nomarski differential interference microscopy. zp, zona pellucida. (B) Transmission electron micrograph of a thin section through an acrosome-intact mouse sperm bound by plasma membrane overlying its head to the ZP of an unfertilized mouse egg in vitro. Note the fibrillar (filamentous) nature of the egg ZP. n, sperm nucleus; a, sperm acrosome; zp, zona pellucida of an unfertilized egg; pm, plasma membrane of an unfertilized egg.

The latter apparently is involved in oocyte-specific expression of ZP genes that takes place coordinately upon binding of E12/FIGα heterodimers to the E-boxes (Epifano et al., 1995a, 1995b; Liang et al., 1997; Soyal et al., 2000).

Some biochemical characteristics of mouse ZP glycoproteins, ZP1-3, are summarized in Table 1. Several features of ZP glycoproteins have been recognized in recent years and some of these are illustrated for mouse ZP3, an ~83 kDa M_r glycoprotein consisting of an

Table 1
Biochemical features of mouse ZP glycoproteins

	ZP1	ZP2	ZP3
Number of amino acids*:	623 (528)	713 (601)	424 (331)
Number of protein subunits/molecule:	2	1	1
Predicted molecular weight (kDa)*:	69×2 (59×2)	80 (68)	46 (37)
Apparent molecular weight (kDa):	200 (120)$	120	83
Theoretical pI*:	6.1 (5.9)	6.1 (5.7)	6.1 (6.4)
Apparent pI:	4.1	5.2 (4.9–5.5)	4.7 (4.2–5.2)
Amino acid composition (%)[§]:			
Ala (A)	5.9	5.2	4.5
Arg (R)	4.5	4.5	5.4
Asn (N)	2.5	4.3	4.2
Asp (D)	4.7	5.7	4.5
Cys (C)	4.0	3.3	3.6
Gln (Q)	6.2	4.5	5.1
Glu (E)	4.5	5.8	5.1
Gly (G)	6.2	4.7	5.1
His (H)	4.0	3.0	3.9
Ile (I)	3.4	5.5	2.4
Leu (L)	9.3	9.2	9.7
Lys (K)	2.3	4.5	3.0
Met (M)	0.8	1.7	0.6
Phe (F)	5.1	5.5	4.8
Pro (P)	9.1	6.0	8.2
Ser (S)	7.4	8.7	11.2
Thr (T)	8.1	5.3	6.3
Trp (W)	1.1	1.2	1.5
Tyr (Y)	3.6	4.2	1.2
Val (V)	7.2	7.5	9.4
Number of cysteines*:	22 (21)	22 (20)	16 (12)
Potential N-glycosylation sites[‡]:	6 (6)	7 (6)	6 (5)
Available experimental data on ZP proteins glycosylation:			
N-linked sugars:	yes	yes	yes
Number of N-linked sugars:	ND**	6	3–4
Type of N-linked sugars:	complex	complex	complex
O-linked sugars:	ND	Yes	Yes
Number of O-linked sugars:	ND	ND	ND
Sialylation:	yes	yes	yes
Sulfation:	yes	yes	yes
Signal peptide[Ø]:	aa 1–20	aa 1–34	aa 1–22
ZP domain[¶]:	aa 271–542	aa 364–630	aa 45–304
Consensus furin cleavage-site:	RRRR (aa 545–548)	RSKR (aa 632–635)	RNRR (aa 350–353)
Predicted transmembrane domain[£]:	aa 595–615	aa 684–702	aa 388–409

*Values relative to the mature proteins are in parenthesis. $ Reported molecular weights are for the ZP1 dimer and monomer (in parenthesis). [§] Reported values are relative to the mature proteins only. [‡] The number of sites that conform to the general pattern N-X-[ST] is reported, together with that of the sites that conform to the more restrictive pattern N-[^P]-[ST]-[^P] (in parenthesis). **Not Determined. [Ø] Determined with SignalP (Nielsen et al., 1997). [¶] Determined with SMART (Schultz et al., 2000). [£] Determined with PHDhtm (Rost et al., 1996).

~37 kDa M_r polypeptide, 3 or 4 complex-type N-linked oligosaccharides, and an undetermined number of O-linked oligosaccharides. The N-terminal 22 amino acids of the ZP3 polypeptide constitute a so-called "signal sequence" that is missing from the secreted glycoprotein in the ZP. Just downstream of the signal sequence, beginning approximately at amino acid-45 and continuing to amino acid-304 (~260 amino acids), is a region referred to as the "ZP domain" (Bork and Sander, 1992). This domain, which was identified by pattern-based sequence analysis, is present in ZP1-3 and contains 8 conserved Cys residues. Beginning approximately at amino acid-218 and ending at amino acid-260 (~40 amino acids), within the ZP domain, lies an immunoglobulin-like "hinge region" (Wassarman and Litscher, 1995). Further downstream, encompassing amino acids 350–353, is a "consensus furin cleavage-site" (CFCS), –Arg–Asn–Arg–Arg–, that is utilized during proteolytic processing of nascent ZP3 for secretion (Litscher et al., 1999; Williams and Wassarman, 2001; Qi et al., 2002). All three ZP glycoproteins have a CFCS just downstream of their ZP domain and a potential "membrane spanning region" downstream of their CFCS, close to their C-terminus. In addition to these features, ZP1 possesses a so-called "trefoil domain" just upstream of its ZP domain (Bork, 1993).

2.3. The zona pellucida (ZP) domain

A consistent feature of ZP glycoproteins is the ZP domain introduced above (Fig. 2). Using an extensive collection of sequences assembled from the literature and the *SMART* (Schultz et al., 2000; http://smart.embl-heidelberg.de/) ZP domain database (285 sequences), a representative ZP domain sequence collection has been generated (70% identity threshold, 96 sequences). After exclusion of 14 sequences whose lengths differed $\pm 10\%$ from the average sequence length in the database (260 amino acids), the resulting database was aligned with *ClustalX* (Thompson et al., 1997; http://www-igbmc.u-strasbg.fr/BioInfo/ClustalX/Top.html). The sequence alignment was first manually optimized and then used to calculate a consensus for the ZP domain at different thresholds (*Consensus;* http://www.bork.emblheidelberg.de/Alignment/consensus.html) and to predict secondary structure (Cuff and Barton, 1999; *Jpred2;* http://jura.ebi.ac.uk:8888). Results of these analyses are presented in Fig. 3. Using an 85% threshold, the only invariant residues within the ~ 260 amino acids of the ZP domain are 8 Cys resdues, a Gly immediately upstream of the sixth Cys residue, and a Phe between the sixth and seventh Cys residues. Together with a relatively large number of positions with conserved character (~15% hydrophobic and ~6% turn-like residues followed by ~3.5% polar and 3.5% small amino acids), the conserved 8 Cys residues, present in intramolecular disulfides, are likely to be crucial for folding of the ZP domain. To date, the positions of ZP domain Cys residues involved in disulfide pairing have not been assigned.

2.4. Secondary and tertiary structure of zona pellucida glycoproteins

Predictions of the secondary structure of ZP glycoproteins have been made based on analysis of both individual sequences and aligned non-redundant sequences using *Jpred2* (Cuff et al., 1999). ZP glycoproteins are predicted to be relatively rich in β-structure (e.g., mouse ZP1 ~21%, ZP2 ~32%, and ZP3 ~28%) and have little, if any, α-helix

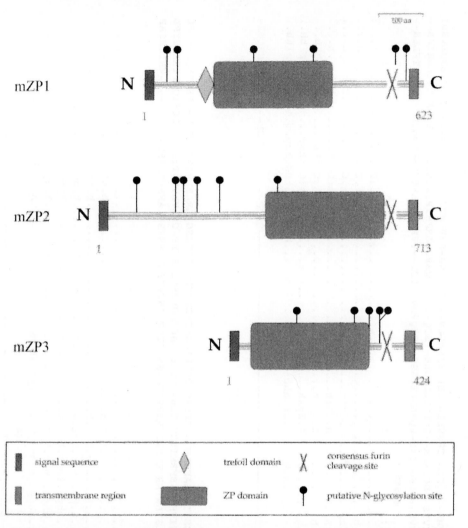

Fig. 2. Schematic representation of the overall architecture of mouse ZP glycoproteins, ZP1, ZP2, and ZP3. The polypeptide sequence of each ZP glycoprotein is depicted as a purple bar, drawn to scale, with the N- and C-termini indicated. Domains were identified with *SMART* (Schultz et al., 2000; http://smart.embl-heidelberg.de/) and signal peptides and transmembrane regions with *SignalP* (Nielsen et al., 1997; http://www.cbs.dtu.dk/services/SignalP/) and *PHDhtm* (Rost et al., 1996; http://maple.bioc.columbia.edu/pp/), respectively. Only putative N-linked glycosylation sites conforming to the strict pattern Asn-[^Pro]-Ser/Thr-[^Pro], where [^Pro] can be any amino acid other than Pro, are shown. (*For a colored version of this figure, see plate section, page 265.*)

content (<1.5% for all mouse ZP glycoproteins). These predictions are consistent with results of circular dichroism (CD) studies of Tamm-Horsfall protein (THP), a protein that has a ZP domain and, thereby, is related to ZP glycoproteins (see below). Such studies have revealed that THP adopts an overall β-structure (~33%), with a low α-helix content

```
ZP1_MUMU        QCFKSGYFTLVMS---QETALTHGVLLDNVHLAYAPNGCP--PTQKTSA-FVVFHVPLTLCGTAIQVVGEQ-LIYENQLVSDI
ZP2_MUMU        LCAQDGFMDFEVYS--HQTKPA-LNLDTLLVGNS--SCQ-PIFKVQSVGLARFHIPLNGCGTRQKFEGDK-VIYENEIHA--
ZP3_MUMU        ECLE-AELVVTVS-RDLFGTGKL-VQPGDLTLGSE--GCQPRVSVDTD--VVRFNAQLHECSSRVQMTKD-ALVYSTFLLH--

ClustalX        *  :  :  .  .  :       :   :  :.         .*          *      *  *:    * :.  :::*:.   :
Consensus 100%  .C.....................................@...............C...................h......
Consensus 95%   .C....h.h.h............................@.............h..t.C................h..h...
Consensus 90%   .C...t.h.h.h...........................@............h.h..t.Cs..............h.h..h..
Consensus 85%   .C...t.h.h.h.............tth............@............hth.httCs.h.........h.h..h..
Consensus 80%   .C..sth.l.h......h.t.h..tth.............@...........hph.httCs.hht........hhhp.t.h..
Consensus 75%   pCtt.sth.l.l..p..sh.t..l.hpsh..........tC...........hph.hstCs.shhp...tsthhpsp.hh.
Consensus 70%   pCtt.ssthtl.l..p..hshst..l.hpshhht.....tCt.........h.hpshshet.CGshhp...tsthhapsp.hh.

Conserved Cys           ①                                          ②                              ③
Jpred2          ------EEEEEE----------------------------------E----------EEEE-----EEEEEEEEE---------


ZP1_MUMU        DVQKGPQ---GSITRDSAFRLHVRCIFNAS-DFLPIQASIFSPQPP-APVTQSGPLRLELRIAT--DKT--FSSYYQGSDYPL
ZP2_MUMU        -LWENP--PSNIVFRNSEFRMTVRCYYIR--DSMLLNAHVKGHPSPEAFVKP-GPLVLVLQTYP---DQSYQRPYRKDEYPL
ZP3_MUMU        --DPRPVS-GLSILRTNRVEVPIECRYPRQGNVSSHPIQP-TWVPFRATVSSEEKLAFSLRLMEEN------WNTEKSAP

ClustalX        *          :  *  ...  .:   :.:*   :         .         * *.     * :.  *:       :.  :.  :
Consensus 100%  .....................................@.........................h.hth............
Consensus 95%   .............h.C.h.....................@.................h...h.htlh..............
Consensus 90%   ..........h....h.hpC.h.................@.........h.......h.hplht.................
Consensus 85%   ........h.p..h.h.hpC.a..th....th.......@.....h..........h.hplht.................
Consensus 80%   ......@..hhpt.thth.hpC.a.t.th..ht......@.....h.........hphpht................+..
Consensus 75%   ......@..hhpp.thth.hpC.attt.th.t.ht..h.@....sht....thphp.lhp.............p.s..
Consensus 70%   ......ps....hhppt.shtht.pC.Yttt.ph.shthps.shs....shtt.t..thphp.lhp.........ttphs.

Conserved Cys                           ④
Jpred2          ---------EEEE---EEEEEEEE--------------------------------------EEEEEE--------------
```

```
ZP1_MUMU        VRLLREPVYVEVRLL--QRTDPSLVLLLHQCWATPT--TSPFEQ-PQWPILSD-GCPFKGD-NYRTQVVAADKEAL------
ZP2_MUMU        VRYLRQPIYMEVKVLSRNDPN--IKLVLDDCWATSS--EDPAS-APQWQIVMD-GCEYELDN-YRTTFHPAG-SSA------
ZP3_MUMU        TFHLGEVAHLQAEVQTGSHLP--LQLFVDHCVATPSPLPDPNSS-PYHFIVDFHGCLVDGLSESFSAFQVP--------

ClustalX        .   * *  :::..:* *.  .   *.:..* ***.:    .*.*      :     :.      :.      *
Consensus 100%  ..........h.h.............h..............h...........h.
Consensus 95%   .h.htt.l.h.h.h..........h.h.th.h..........hh..uC.
Consensus 90%   .h.ltp.l.h.hth..........h.hhhtpChst...t.......ph.hl.t.GC.
Consensus 85%   .h.ltp.l.h.hth..........h.hhlppChsp.t...s......ph.hl.t.GC.t...s.h.
Consensus 80%   hh.ltp.lhhthph...ttt....h.hhlppChspss...s......ph.llp.GC.hp...s.h.
Consensus 75%   hh.ltp.lhhthph...ttt....hthhlppChssss...s......ph.llp.GC.hpt...sth.
Consensus 70%   hh.lsp.lhhphph...ttt....htlhlppChssss...s......ph.llpp.GC.hst...hhsphh.
Conserved Cys                                        C                       GC
Jpred2          EEE---EEEEEEEE---------EEEEEEE---------------------------EEEEE---------
                                                      (5)                                  (6)

ZP1_MUMU        PFWSHYQRFTITTFMLLDSSSQNALRGQVYFFCSASACHPLG--SDT---CSTTCDSG
ZP2_MUMU        AHSGHYQRFDVKTFAFVS--EARGLSSLIYFHCSALICNQVS--LDSP-LCSVTCPAS
ZP3_MUMU        RPRPETLQFTVDVFHFAN----SSRNTLYITCHLKVAPANQIP-DKLN--KACSFN

ClustalX        .   :*:..*.:  .   .   :*:.  *    .   .:*
Consensus 100%  .....................h.................h.
Consensus 95%   .....................a.h.........1.h.sth.hh.
Consensus 90%   .t.......h.Fth..........h.Fth.hs.1.h.Cph.hs...C.
Consensus 85%   .t.......h.Fta.t.........h.Fta.t.....lhh.Cph.hs...C.
Consensus 80%   tthhth.h.Fta.t.........tt.lhhpCph.hs...tC.
Consensus 75%   tthhthph.hFpF.s.........ts.lhhpCphths...tCs.
Consensus 70%   .tphhphphphrpfss.......ps.lahpCplphC.tt......tCs.
Conserved Cys                                  C                       C
Jpred2          -----EEEEEE----------------------EEEEEEEEEE---
                                                (7)                          (8)
```

Fig. 3. (*continued*)

```
Consensus keys
--------------

a : aromatic (F, Y, W, H)
l : aliphatic (I, V, L)
h : hydrophobic (F, Y, W, H, I, V, L, A, G, M, C, K, R, T)
p : polar (H, K, R, D, E, Q, N, S, T, C)

u : tiny (G, A, S)
s : small (G, A, S, V, T, D, N, P, C)
t : turnlike (G, A, S, H, K, R, D, E, Q, N, S, T, C)

. : any (G, A, V, I, L, M, F, Y, W, H, C, P, K, R, D, E, Q, N, S, T)

Jpred2 keys
-----------

E : sheet
- : loop
```

Fig. 3. Consensus analysis and secondary structure prediction of ZP domain sequences. The polypeptide sequences of the ZP domains of mouse ZP glycoproteins, aligned using a non-redundant 70% threshold ZP domain sequence collection (see text), are superimposed on the consensus patterns for the domain, calculated at different thresholds using the program *Consensus* (http://www.bork.embl-heidelberg.de/Alignment/consensus/html). Also shown are the *ClustalX* (Thompson et al., 1997) output for the alignment of the three ZP glycoproteins, a *Jpred2* (Cuff and Barton 1999) secondary structure prediction based on the full sequence collection, and the positions of the conserved 8 Cys residues. Identical amino acids in both the alignment and consensus patterns are indicated in red and marked with "*" in the *ClustalX* output. Conserved positions are encoded and color-coded using the *Consensus* keys reported below the alignment, and marked with either ":" or "." in the *ClustalX* output according to their degree of conservation. The *Jpred2* output is also encoded using the keys specified. (*For a colored version of this figure, see plate section, pages 266–268.*)

(<10%) (Robinson and Puett, 1973; Puett et al., 1977). Similarly, Fourier-transform infrared and circular dichroism spectroscopy results with recombinant cuticlin-1, a protein related to ZP glycoproteins, suggest that the protein contains ~50% β-structure, ~14% α-helix, and ~25% turns. Near-UV spectra of cuticlin-1 suggest that at least some of its conserved Cys residues are disulfide bonded (D'Auria et al., 1998).

Several algorithms have been employed unsuccessfully to identify known folds in ZP glycoproteins [*3D-PSSM/SAWTED* (Kelley et al., 2000; MacCallum et al., 2000; http://www.bmm.icnet.uk/~3dpssm); *DOE FOLD* (Fischer and Eisenberg, 1996; http://fold.doe-mbi.ucla.edu/); *PSIPRED/Gen THREADER* (McGuffin et al., 2000; http://bioinf.cs.ucl.ac.uk/psipred/index.html)]. The lack of success suggests that ZP glyco-proteins possess novel structural folds. On the other hand, *3D-PSSM* predicts with good reliability (E-value key, 80%) that there is a β-immunoglobulin-like fold for an N-terminal ~130 amino acid fragment of mouse ZP2.

Due to their unique chemical properties, Cys residues play a fundamental role in fold-ing of proteins. As is the case for other secreted proteins, the Cys residues of ZP2 and ZP3 are apparently all present as intramolecular disulfides; there are no free sulfhydryl groups. On the other hand, the two polypeptides of ZP1 are held together by intermolecular disulfides (Bleil and Wassarman, 1980a; Wassarman, 1988). As yet, assignments of specific Cys residues to intramolecular (ZP1-3) and intermolecular (ZP1) disulfides have not been made. Such information will be critical in assessing the tertiary structure of ZP glycoproteins.

2.5. Survey of zona pellucida glycoproteins

Investigators have characterized ZP genes and glycoproteins from a large variety of mammalian species, including rodents (e.g., mice, rats, and hamsters; Bleil and Wassarman, 1980a; Kinloch et al., 1988, 1990; Ringuette et al., 1988; Moller et al., 1990; Akatsuka et al., 1998; Scobie et al., 1999), domesticated animals (e.g., rabbits, cows, pigs, cats, and dogs; Dunbar et al., 1981; Yurewicz et al., 1987, 1993; Lee et al., 1993; Harris et al., 1994; Noguchi et al., 1994; Topper et al., 1997), marsupials (e.g., possum and dunnart; Mate and McCartney, 1998; Haines et al., 1999; McCartney and Mate, 1999; Voyle et al., 1999), and primates (e.g., marmosets, bonnet monkeys, and human beings; Chamberlin and Dean, 1990; Van Duin et al., 1992; Thillai-Koothan et al., 1993; Kolluri et al., 1995; Gupta et al., 1997; Jethanandani et al., 1998; Hughes and Barratt, 1999). While the molecular weights differ for ZP glycoproteins from different species, due in large part to differential glycosylation and modification, it is apparent that all ZP consist of only a few glycoproteins whose polypeptides are related to those of mouse ZP1-3. The primary structures of ZP2- and ZP3-related ZP glycoproteins from different species are relatively well conserved (~65–98% identity), whereas ZP1-related glycoproteins are con-served to a lesser degree (~40% identity). It also is apparent that ZP1-3 have regions of polypeptide in common, suggesting that these regions may be derived from a common ancestral gene. From a collection of all ZP1-3 sequences in the databases, representative sequence sets were generated at a 70% identity threshold (Holm and Sander, 1998). Alignment of representative homologs of each ZP glycoprotein were created with *ClustalX* (Thompson et al., 1997), manually edited, and can be found online at the URL http://www.mssm.edu/students/jovinl02/research/adbb_2002_webfig.html or can be obtained directly from jovinl02@doc.mssm.edu.

2.6. Targeted mutagenesis of zona pellucida genes

Genes encoding mouse ZP1 (Rankin et al., 1999), ZP2 (Rankin et al., 2001), and ZP3 (Liu et al., 1996; Rankin et al., 1996) have been disrupted individually by homologous recombination in embryonic stem (ES) cells. In mice that are homozygous nulls for either *ZP2* or *ZP3* unfertilized eggs lack a ZP and the mice are not fertile. On the other hand, eggs from mice that are homozygous nulls for *ZP1* have a ZP, but it is loosely organized, often permitting follicle cells to cluster between the ZP and plasma membrane. Although the mutant mice are fertile, they produce smaller litters than wild-type mice. These results are consistent with a model for ZP structure in which ZP2 and ZP3 polymerize to form long fibers/filaments that exhibit a periodicity of ~140 Å and are interconnected by ZP1 (Greve and Wassarman, 1985; Wassarman and Mortillo, 1991; Wassarman et al., 1996). Additional support for this model comes from experiments with mice that are heterozygous nulls for the ZP3 gene (Wassarman et al., 1997). Eggs from such mice have a ZP, but it contains only about one-half the normal amount of ZP2 and ZP3 and is only about one-half the width (~2.7 μm) of the ZP of eggs from wild-type mice (~6.2 μm). Despite this difference, the mutant mice are as fertile as wild-type mice.

2.7. Synthesis of zona pellucida glycoproteins

The site of synthesis of ZP glycoproteins has been investigated extensively in a variety of mammalian species. It is well established that ZP genes are expressed solely by the ovary. However, whether expression occurs solely in growing oocytes, or in both growing oocytes and surrounding follicle cells, remains a somewhat contentious issue.

Numerous studies with mice, dating back to 1980 (Bleil and Wassarman, 1980b), provide overwhelming evidence that mouse ZP glycoproteins are synthesized exclusively by growing oocytes, not by follicle cells. This evidence includes localization of ZP messenger-RNAs and localization of nascent ZP glycoproteins to growing oocytes, not to follicle cells (Greve et al., 1982; Salzmann et al., 1983; Shimizu et al., 1983; Philpott et al., 1987; Roller et al., 1989; Liang et al., 1990; Millar et al., 1993; Epifano et al., 1995a; Tong et al., 1995). Furthermore, expression of a reporter gene fused to 5′-flanking bits of a mouse ZP gene promoter is restricted solely to growing oocytes of transgenic mice (Lira et al., 1990, 1993). Similarly, experiments carried out with mammals as diverse as rats (Akatsuka et al., 1998; Scobie et al., 1999) and marsupials (Haines et al., 1999; Voyle et al., 1999) suggest that ZP genes are expressed only in growing oocytes, not in follicle cells. On the other hand, there is evidence to suggest that ZP messenger-RNAs and glycoproteins are present in oocytes and surrounding follicle (granulosa) cells of rabbit, pig, cow, and human ovaries (Lee and Dunbar, 1993; Grootenhuis et al., 1996; Kölle et al., 1996, 1998; Prasad et al., 2000; Sinowatz et al., 2001). Whether the sites of synthesis of individual ZP glycoproteins truly vary among different mammalian species remains to be examined in more detail. In any case, it is clear that in mammals all ZP glycoproteins are synthesized within the ovary.

3. Non-mammalian egg vitelline envelope

The plasma membrane of fish, bird, and amphibian eggs is surrounded by a fibrous extracellular coat of variable thickness, often consisting of multiple layers, called the VE. In some cases, a thick jelly-like layer encompasses the VE (e.g., some fish and amphibian eggs). The VE performs some of the same functions ascribed to the ZP of mammalian eggs.

3.1. Egg vitelline envelope glycoproteins

In recent years, VE glycoproteins from fish, bird, and amphibian eggs have been analyzed in some detail and found to resemble ZP glycoproteins ZP1-3. For example, *Xenopus laevis* VE glycoproteins gp37, gp41/43, and gp69/64 are homologs of ZP1 (Kubo et al., 2000), ZP3 (Kubo et al., 1997; Yang and Hedrick, 1997), and ZP2 (Tian et al., 1999), respectively; zebrafish (*Danio rerio*) VE glycoproteins are homologs of ZP2 and ZP3 (Del Giacco et al., 2000; Wang and Gong, 1999; Mold et al., 2001); rainbow trout VE glycoproteins are homologs of ZP1-3 (Hyllner et al., 2001); and two chicken VE glycoproteins are homologs of ZP1 and ZP3 (Waclawek et al., 1998; Takeuchi et al., 1999; Bausek et al., 2000). Among other fish, ZP glycoprotein homologs also have been identified in VE from carp (Chang et al., 1996, 1997), winter flounder (Lyons et al., 1993), goldfish (Chang et al., 1997), sea bream (Del Giacco et al., 1998), and medaka (Murata et al., 1995, 1997) eggs. [Note: Some fish sequences reported to be ZP2 homologs contain a trefoil domain upstream of the ZP domain and, therefore, may actually be ZP1 homologs.] In this context, it has been reported that mouse ZP glycoproteins, synthesized by *Xenopus laevis* oocytes microinjected with ZP messenger-RNAs, are incorporated with VE glycoproteins into the oocyte VE (Doren et al., 1999). Collectively, these findings suggest a clear evolutionary path over more than 400 million years from fish VE genes to mammalian ZP genes.

3.2. Synthesis of egg vitelline envelope glycoproteins

Whereas all ZP glycoproteins are synthesized in the ovary, there is considerable evidence that certain VE glycoproteins are synthesized by the liver and that their synthesis is under hormonal (estrogen) control. Two well documented examples of this are ZP1 synthesis in chickens (Bausek et al., 2000) and VE glycoprotein synthesis in rainbow trout (Hyllner et al., 2001), as well as many other fish (e.g., Oppen-Berntsen et al., 1992; Lyons et al., 1993; Del Giacco et al., 1998; Sugiyama et al., 1998). In the case of chickens, ZP1 is synthesized by the liver and is under estrogen control, but ZP3 is synthesized in the ovary by follicle (granulosa) cells (Waclawek et al., 1998); the latter is also true for ZP3 synthesis in the Japanese quail (Pan et al., 2001). All 3 trout VE glycoproteins are expressed in the livers of both male and female fish, with higher expression in females, and one VE glycoprotein is also expressed in the ovary. Expression of zebrafish ZP2 and ZP3 genes was found to be ovary-, rather than liver-specific (Wang and Gong, 1999; Mold et al., 2001). Similarly, amphibian VE glycoprotein genes are expressed in the ovary, not in the liver (Kubo et al., 1997; Yang and Hedrick, 1997). These and other findings suggest that mammalian ZP glycoproteins and amphibian VE glycoproteins are synthesized only in the ovary, by growing oocytes and follicle cells. On the other hand, fish and bird VE glycoproteins are synthesized in both the liver and ovary. Whether or not specific receptors that recognize nascent VE glycoproteins, synthesized in the liver and released into the bloodstream, are present in ovaries of fish and birds remains to be determined.

4. Related extracellular glycoproteins

In addition to their presence in ZP and VE glycoproteins, ZP domains have been recognized in a variety of extracellular proteins found in worms, flies, and mammals (Wassarman et al., 2001). A description of several of these proteins follows.

4.1. TGF-β receptor III (TGR3) and endoglin

Also called betaglycan, TGR3 is the most abundant TGF-β binding protein at the cell surface. It is a proteoglycan that exists in both membrane-bound and soluble forms that regulate TGF-β signaling in opposite fashion (Derynck and Feng, 1997; Esparza-Lopez et al., 2001). A ZP domain is also found in endoglin, a membrane glycoprotein that is structurally related to TGR3 and also binds TGF-β (Cheifetz et al., 1992).

4.2. Tamm-Horsfall protein (THP)

Also called uromodulin, THP is the most abundant protein in human urine. After processing of a GPI-linked precursor (Cavallone et al., 2001), it is secreted as a highly glycosylated species that self-assembles into filaments with molecular weights of $\sim 10^4$ kDa (Robinson and Puett, 1973). Due to their propensity to form a gel, THP filaments may ensure the water impermeability of the thick ascending limb of the loop of Henle. In addition, there is evidence to suggest that THP may play a role in kidney and systemic immunity (Kokot and Dulawa, 2000).

THP has been studied extensively by electron microscopy (Porter and Tamm, 1954; Bayer, 1964; Robinson and Puett, 1973; Puett et al., 1977; Bjugn and Flood, 1988; Wiggins, 1987). Single filaments of the protein, 10–40 Å in diameter, originate from and merge into bundles (120 Å, ave. diameter) at seemingly irregular intervals. This generates a three-dimensional matrix with 0.1–1 μm pores. Analyses of single fibrils reveal a, so-called, zig-zag course with a 100–140 Å periodicity, with the single branch of each zig-zag measuring about 60 Å in length and 20–40 Å in width. It was noted that these structural features could be interpreted as two-dimensional projections of a helical superstructure (Bayer, 1964).

4.3. GP-2

The major zymogen granule membrane glycoprotein of the exocrine pancreas, GP-2 has no known functions as yet. It is generated by proteolytic cleavage of a GPI-anchored precursor. Processing of GP-2, which exhibits extensive sequence similarity to THP, is pH-dependent and results in release of a high molecular weight, aggregated form of GP-2 into the apical ductular secretion (Fukuoka et al., 1992).

4.4. α- and β-Tectorin

Tectorins are the major non-collagenous glycoproteins of the mammalian tectorial membrane (TM), a complex fibrillogranular matrix that overlies the surface of the organ of Corti and plays a fundamental role in sound mechanotransduction (Legan et al., 1997a). Both α- and β-tectorin contain a CFCS and are released into the extracellular space from GPI-linked precursors. However, whereas β-tectorin essentially consists of a single ZP domain, α-tectorin is a larger modular molecule that is proteolytically processed into 3 polypeptides that are crosslinked by disulfides (Killick et al., 1995; Legan et al., 1997b).

The TM is constructed of three different collagens and three collagenase-insensitive glycoprotein components, α- and β-tectorin, and otogelin. The tectorins are associated with a striated matrix consisting of two types of fibrils, 7–9 nm in diameter and alternately light- and dark-staining, that are connected by staggered cross-bridges to form flat sheets. These sheets, in turn, are stacked on top of each other and wrap around bundles of collagen fibers (Hasko and Richardson, 1988; Legan et al., 2000). This organization could explain why mice that are homozygous nulls for α-tectorin have a TM that lacks a striated sheet matrix (Legan et al., 2000). Additionally, the TM of these animals lacks β-tectorin, consistent with the idea that α- and β-tectorin interact with each other and polymerize into filaments that constitute the striated matrix. This could also explain the reported link between non-syndromal hearing impairments and either single-site mutations in human α-tectorin (Verhoeven et al., 1998; Kirschhofer et al., 1998; Moreno-Pelayo et al., 2001; Steel and Kros, 2001) or deletion of the ZP domain of α-tectorin (Mustapha et al., 1999). Notably, no collagen components have been detected in the TM of birds, however, the TM contains a set of glycoproteins homologous to those of mammals (Killick et al., 1995; Legan et al., 1997a, b).

4.5. CRP-ductin gene products

Transcripts from the *CRP-ductin* gene (Cheng et al., 1996) have been shown to undergo tissue-specific alternative splicing (Takito et al., 1999), producing a group of proteins known as the mouse pancreatic acinar cell glycoprotein, pro-Muclin/CRP-ductin(-α/β) (Cheng et al., 1996; De Lisle and Ziemer 2000), the mouse vomeronasal organ protein vomeroglandin (Matsushita et al., 2000), the rat salivary gland ebnerin (Li and Snyder, 1995), the human lung proteins DMBT1 (Mollenhauer et al., 1997) and gp340 (Holmskov et al., 1999), and rabbit kidney hensin (Takito et al., 1996). All of these proteins are highly glycosylated and share a common modular organization, with different combinations of scavenger-receptor Cys-rich domains and CUB domains upstream of a ZP domain and, in certain cases, a transmembrane domain (Holmskov et al., 1999; De Lisle and Ziemer 2000).

4.6. UTCZP/ERG1

A ZP domain is present in a protein from the pregnant mouse uterus (UTCZP) whose expression is restricted to 6- to 3-days prior to birth (Kasik, 1998). The rat homolog of this protein is called ERG1 and is expressed under tight estrogen control in both the uterus and the oviduct (Chen et al., 1999). The ZP domain of UTCZP/ERG1 is downstream of two CUB domains and is followed by a CFCS and a transmembrane domain.

4.7. Cuticlin-1

A conserved component of the insoluble cuticlin residue of nematodes, cuticlin-1 is a dauer larva, stage-specific protein that forms two ribbons along the entire length of the worm (Sebastiano et al., 1991). Like β-tectorin, cuticlin-1 consists of a single ZP domain, with 12 conserved Cys residues, that is thought to be secreted after cleavage at a CFCS that is upstream of a transmembrane domain. This protein is unusually stable at high temperatures (D'Auria et al., 1998).

4.8. Dumpy

The *dumpy* gene of *Drosophila* encodes a gigantic 2.5 MDa extracellular protein. The protein contains 308 EGF domains, 185 DPY modules, and a single ZP domain upstream of a membrane anchor sequence. *dumpy* is expressed at sites of cuticle-epidermal cell attachment. Its gene product is thought to constitute long membrane-bound fibers that insert into the cuticle to allow local maintenance of mechanical tension (Wilkin et al., 2000).

4.9. NompA

The product of the *no-mechanoreceptor-potential A* gene, NompA, is a *Drosophila* peripheral nervous system protein. The protein contains a large extracellular fragment composed of several plasminogen N-terminal (PAN) modules, followed by a single ZP domain, a CFCS, a transmembrane domain, and the longest cytoplasmic tail of all known ZP domain proteins (Chung et al., 2001). After cleavage at the CFCS, the protein is incorporated into a specialized extracellular matrix, called the dendritic cap. This structure separates the mechanosensory bristles, a cuticular structure of the fly, from neuronal sensory endings where transduction of the external signal takes place. As *Nomp* mutations are known to disrupt the transduction of tactile, proprioceptive, and auditory stimuli, it is likely that NompA functions as part of a system linking mechanical stimuli to the transduction apparatus.

4.10. Summary

Overall, it should be noted that these ZP domain-containing proteins have several features in common. (1) They are all found in multicellular eukaryotes, contain a signal sequence that allows them to be secreted, and most are extensively glycosylated. (2) Most are modular proteins, with various combinations of extracellular modules upstream of a single C-terminal ZP domain. (3) The ZP domain is often found just upstream of either a transmembrane domain or a GPI-anchor sequence. (4) In most cases a CFCS or dibasic sequence is present between the ZP domain and the transmembrane domain/GPI-anchor. (5) The proteins can be membrane bound, but generally are proteolytically processed C-terminal to their ZP domain to generate species that diffuse into the extracellular space to assemble into filaments or matrices. (6) Most have been shown to act as receptors and/or have mechanical functions.

5. Final comments

A very similar constellation of glycoproteins is found in the ZP of eggs from all mammals, including marsupials, and in the VE of eggs from fish, birds, and amphibians. Furthermore, these egg glycoproteins have many counterparts elsewhere; for example, in ears (tectorins), noses (vomeroglandin), and urine (THP) of humans, to cuticles (cuticulin-1) of dauer-stage worms (*C. elegans*), and to the epidermis (Dumpy) and nervous system (NompA) of flies (*D. melanogaster*). All of these extracellular or membrane-bound glycoproteins possess the ZP domain first identified in ZP glycoproteins, TGF-β type III receptor, and related proteins by Bork and Sander (1992).

The modular structure of proteins containing a ZP domain can be used to identify new members of this class of proteins. For example, in fish, each of the VE components can be easily identified as a homolog of one of the mouse ZP glycoproteins. ZP1 homologs are characterized by a single trefoil domain immediately preceding the ZP domain. ZP2 homologs share a long N-terminal segment that is not similar to any other known domain. ZP3 homologs typically have the shortest polypeptides with either four, or occassionally two, additional Cys residues following their ZP domain.

Much of the evidence summarized here suggests that the ZP domain of proteins is responsible for their common ability to assemble into filaments and three-dimensional matrices (Killick et al., 1995; Legan et al., 1997b). In this context, the apparent discrepancy between systems with multiple components (e.g., ZP, VE, and TM) and those with a single component (e.g., THP) can be resolved by proposing that the former systems evolved by duplication of a single gene encoding an ancestral protein. Accordingly, THP would assemble into homodimers in much the same way that ZP2 and ZP3 or α- and β-tectorin assemble into heterodimers. It is interesting to note that, to date, no ZP2 homolog has been identified in the bird VE, suggesting that a ZP3 homodimer may substitute for a ZP2–ZP3 heterodimer in birds.

It is clear that results of X-ray crystallographic and NMR studies of ZP domain-containing proteins would have a significant impact on our understanding of the biological activities of this functionally diverse group of proteins. Hopefully, the next decade of research will extend many of the ideas presented in this review and impact in a constructive manner on aspects of human health.

Acknowledgments

We thank Huayu Qi and Zev Williams for daily discussions and helpful advice. We also thank Franco Cotelli and Rosaria de Santis for discussion and for communicating results before publication. L.J. is especially grateful to Jong Park for many useful discussions about protein sequence analysis. L.J. is a postdoctoral fellow supported by the Human Frontier Science Program Organization. Not all relevant references are quoted here and we apologize to those affected. Some of the research from the authors' laboratory was supported in part by the NIH, most recently by HD-35105.

Note added in proof

Recently, a ZP domain has been indentified in VERL, the abalone (*Haliotis rufescens*) egg VE sperm receptor (Galindo et al., 2002. Gene 288, 111–117), the HrVC120 sperm receptor from an ascidian (*Halocynthia roretzi*) (Sawada et al., 2002. Proc. Natl. Acad. Sci. USA 99, 1223–1228), and a major egg coat protein from another ascidian (*Ciona intestinalis*) (Rosaria de Santis, personal communication). In VERL and HrVC120, a consensus furin cleavage-site and a hydrophobic domain are found C-terminal to the ZP domain. Also, Jovine et al. have reported that the ZP domain is a module that allows different extracellular proteins to polymerize into filaments of similar three-dimensional structure (Jovine et al., 2002. Nat. Cell Biol. 6, 457–461). It was also reported that the C-terminal transmembrane and short cytoplasmic tails of ZP2 and ZP3 are not required for their secretion, but are essential for their incorporation into the ZP. This provides a rationale for the conservation of C-terminal hydrophobic sequences in ZP domain proteins. Finally, analysis of epitope-tagged ZP2 mutants carrying human α-tectorin ZP domain mutations suggests that an experimental system is now available for evaluating mutations that affect conserved ZP domain amino acids.

References

Akatsuka, K., Yoshida-Komiya, H., Tulsiani, D.R., Orgebin-Crist, M.C., Hiroi, M., Araki, Y. 1998. Rat zona pellucida glycoproteins: Molecular cloning and characterization of the three major components. Mol. Reprod. Dev. 51, 454–467.

Arnoult, C., Zeng, Y., Florman, H.M. 1996. ZP3-dependent activation of sperm cation channels regulates acrosomal secretion during mammalian fertilization. J. Cell. Biol. 134, 637–645.

Bausek, N., Waclawek, M., Schneider, W.J., Wohlrab, F. 2000. The major chicken egg envelope protein ZP1 is different from ZPB and is synthesized in the liver. J. Biol. Chem. 275, 28866–28872.

Bayer, M.E. 1964. An electron microscope examination of urinary mucoprotein and its interaction with influenza virus. J. Cell. Biol. 21, 265–274.

Bjugn, R., Flood, P.R. 1988. Scanning electron microscopy of human urine and purified Tamm-Horsfall's glycoprotein. Scand. J. Urol. Nephrol. 22, 313–315.

Bleil, J.D., Wassarman, P.M. 1980a. Structure and function of the zona pellucida: Identification and characterization of the proteins of the mouse oocyte's zona pellucida. Dev. Biol. 76, 185–202.

Bleil, J.D., Wassarman, P.M. 1980b. Synthesis of zona pellucida proteins by denuded and follicle-enclosed mouse oocytes during culture in vitro. Proc. Natl. Acad. Sci. USA 77, 1029–1033.

Bork, P. 1993. A trefoil domain in the major rabbit zona pellucida protein. Protein Sci. 2, 669–670.

Bork, P., Sander, C.A. 1992. A large domain common to sperm receptors (ZP2 and ZP3) and TGF-β type III receptor. FEBS Lett. 300, 237–240.

Browder, L. (Ed.) 1985. *Developmental Biology: A Comprehensive Synthesis*, Vol. 1, *Oogenesis*, New York: Plenum Press.

Cavallone, D., Malagolini, N., Serafini-Cessi, F. 2001. Mechanism of release of urinary Tamm-Horsfall glycoprotein from the kidney GPI-anchored counterpart. Biochem. Biophys. Res. Commun. 280, 110–114.

Chamberlin, M.E., Dean, J. 1990. Human homolog of the mouse sperm receptor. Proc. Natl. Acad. Sci. USA 87, 6014–6018.

Chang, Y.S., Hsu, C.C., Wang, S.C., Tsao, C.C., Huang, F.L. 1997. Molecular cloning, structural analysis, and expression of carp ZP2 gene. Mol. Reprod. Dev. 46, 258–267.

Chang, Y.S., Wang, S.C., Tsao, C.C., Huang, F.L. 1996. Molecular cloning, structural analysis, and expression of carp ZP3 gene. Mol. Reprod. Dev. 44, 295–304.

Cheifetz, S., Bellon, T., Cales, C., Vera, S., Bernabeu, C., Massague, J., Letarte, M. 1992. Endoglin is a component of the transforming growth factor-β receptor system in human endothelial cells. J. Biol. Chem. 267, 19027–19030.

Chen, D., Xu, X., Zhu, L.J., Angervo, M., Li, Q., Bagchi, M.K., Bagchi, I.C. 1999. Cloning and uterus/oviduct-specific expression of a novel estrogen-regulated gene (ERG1). J. Biol. Chem. 274, 32215–32224.

Cheng, H., Bjerknes, M., Chen, H. 1996. CRP-ductin: a gene expressed in intestinal crypts and in pancreatic and hepatic ducts. Anat. Rec. 244, 327–343.

Chung, Y.D., Zhu, J., Han, Y., Kernan, M.J. 2001. *nompA* encodes a PNS-specific, ZP domain protein required to connect mechanosensory dendrites to sensory structures. Neuron 29, 415–428.

Cuff, J.A. Barton, G.J. 1999. Evaluation and improvement of multiple sequence methods for protein secondary structure prediction. Proteins 34, 508–519.

D'Auria, S., Rossi, M., Tanfani, F., Bertoli, F., Parise, G., Bazzicalupo, P. 1998. Structural analysis of ASCUT-1, a protein component of the cuticle of the parasitic nematode *Ascaris lubricoides*. Eur. J. Biochem. 255, 588–594.

De Lisle, R.C., Ziemer, D. 2000. Processing of pro-Muclin and divergent trafficking of its products to zymogen granules and the apical plasma membrane of pancreatic acinar cells. Eur. J. Cell. Biol. 79, 892–904.

Del Giacco, L., Diani, S., Cotelli, F. 2000. Identification and spatial distribution of the mRNA encoding an egg envelope component of the Cyprinid zebrafish, *Danio rerio,* homologous to the mammalian ZP3(ZPC). Dev. Genes Evol. 210, 41–46.

Del Giacco, L., Vanoni, C., Bonsignorio, D., Duga, S., Mosconi, G., Santucci, A., Cotelli, F. 1998. Identification and spatial distribution of the mRNA encoding the gp49 component of the gilthead sea bream, *Sparus aurata,* egg envelope. Mol. Reprod. Dev. 49, 58–69.

Derynck, R., Feng, X.-H. 1997. TGF-β receptor signaling. Biochim. Biophys. Acta 1333, F105–F150.

Doren, S., Landsberger, N., Dwyer, N., Gold, L., Blanchette-Mackie, J., Dean, J. 1999. Incorporation of mouse zona pellucida proteins into the envelope of *Xenopus laevis* oocytes. Dev. Genes Evol. 209, 330–339.

Dumont, J.N., Brummett, A.R. 1985. Egg envelopes in vertebrates. In: *Developmental Biology: A Comprehensive Synthesis,* Vol. 1, Oogenesis (L. Browder, Ed.), New York: Plenum Press, pp. 235–288.

Dunbar, B.S., Liu, C., Simmons, D.W. 1981. Identification of the three major proteins of porcine and rabbit zonae pellucidae by two-dimensional gel electrophoresis: Comparison with follicular fluid, sera, and ovarian cell proteins. Biol. Reprod. 24, 1111–1124.

Epifano, O., Liang, L.F., Dean, J. 1995b. Mouse Zp1 encodes a zona pellucida protein homologous to egg envelope proteins in mammals and fish. J. Biol. Chem. 270, 27254–27258.

Epifano, O., Liang, L.F., Familiari, M., Moos, M.C., Dean, J. 1995a. Coordinate expression of the three zona pellucida genes during mouse oogenesis. Development 121, 1947–1956.

Esparza-Lopez, J., Montiel, J.L., Vilchis-Landeros, M.M., Okadome, T., Miyazono, K., Lopez-Casillas, F. 2001. Ligand binding and functional properties of betaglycan, a co-receptor of the transforming growth factor-β superfamily. Specialized binding regions for transforming growth factor-beta and inhibin A. J. Biol. Chem. 276, 14588–14596.

Fischer, D., Eisenberg, D. 1996. Protein fold recognition using sequence-derived predictions. Protein Sci. 5, 947–955.

Fukuoka, S., Freedman, S.D., Yu, H., Sukhatme, V.P., Scheele, G.A. 1992. GP-2/THP gene family encodes self-binding glycosylphosphatidylinositol-anchored proteins in apical secretory compartments of pancreas and kidney. Proc. Natl. Acad. Sci. USA 89, 1189–1193.

Greve, J.M., Wassarman, P.M. 1985. Mouse egg extracellular coat is a matrix of interconnected filaments possessing a structural repeat. J. Mol. Biol. 181, 253–264.

Greve, J.M., Salzmann, G.S., Roller, R.J., Wassarman, P.M. 1982. Biosynthesis of the major zona pellucida glycoprotein secreted by oocytes during mammalian oogenesis. Cell 31, 749–759.

Grootenhuis, A.J., Philipsen, H.L.A., de Breet-Grijsbach, J.T.M., van Duin, M. 1996. Immuno-cytochemical localization of ZP3 in primordial follicles of rabbit, marmoset, rhesus monkey, and human ovaries using antibodies against human ZP3. J. Reprod. Fertil. 50, 43–54.

Gupta, S.K., Sharma, M., Behera, A.K., Bisht, R., Kaul, R. 1997. Sequence of complementary deoxyribonucleic acid encoding Bonnet monkey zona pellucida glycoprotein-ZP1 and its high level expression in *Escherichia coli*. Biol. Reprod. 57, 532–538.

Haines, B.P., Rathjen, P.D., Hope, R.M., Whyatt, L.M., Holland, M.K., Breed, W.G. 1999. Isolation and characterization of a cDNA encoding a zona pellucida protein from the marsupial *Trichosurus vulpecula* (brushtail possum). Mol. Reprod. Dev. 52, 174–182.

Harris, J.D., Hibler, D.W., Fontenot, G.K., Hsu, K.T., Yurewicz, E.C., Sacco, A.G. 1994. Cloning and characterization of zona pellucida genes and cDNA from a variety of mammalian species: The ZPA, ZPB and ZPC gene families. DNA Sequence 4, 361–393.

Hasko, J.A., Richardson, G.P. 1988. The ultrastructural organization and properties of the mouse tectorial membrane matrix. Hear. Res. 35, 21–38.

Holm, L., Sander, C. 1998. Removing near-neighbour redundancy from large protein sequence collections. Bioinformatics 14, 423–429.

Holmskov, U., Mollenhauer, J., Madsen, J., Vitved, L., Gronlund, J., Tornoe, I., Kliem, A., Reid, K.B., Poustka, A., Skjodt, K. 1999. Cloning of gp-340, a putative opsonin receptor for lung surfactant protein D. Proc. Natl. Acad. Sci. USA 96, 10794–10799.

Hughes, D.C., Barratt, C.L.R. 1999. Identification of the true human orthologue of the mouse ZP1 gene: Evidence for greater complexity in the mammalian zona pellucida? Biochim. Biophys. Acta 1447, 303–306.

Hyllner, S.J., Westerlund, L., Olsson, P.-E., Schopen, A. 2001. Cloning of rainbow trout egg envelope proteins: Members of a unique group of structural proteins. Biol. Reprod. 64, 805–811.

Jethanandani, P., Santhanam, R., Gupta, S.K. 1998. Molecular cloning and expression in *Escherichia coli* of cDNA encoding Bonnet monkey (*Macaca radiata*) zona pellucida glycoprotein-ZP2. Mol. Reprod. Dev. 50, 229–239.

Kasik, J.W. 1998. A cDNA cloned from pregnant mouse uterus exhibits temporo-spatial expression and predicts a novel protein. Biochem. J. 330, 947–950

Kelley, L.A., MacCallum, R.M., Sternberg, M.J. 2000. Enhanced genome annotation using structural profiles in the program 3D-PSSM. J. Mol. Biol. 299, 499–520.

Killick, R., Legan, P.K., Malenczak, C., Richardson, G.P. 1995. Molecular cloning of chick β-tectorin, an extracellular matrix molecule of the inner ear. J. Cell. Biol. 129, 535–547.

Kinloch, R.A., Roller, R.J., Fimiani, C.M., Wassarman, D.A., Wassarman, P.M. 1988. Primary structure of the mouse sperm receptor's polypeptide chain determined by genomic cloning. Proc. Natl. Acad. Sci. USA 85, 6409–6413.

Kinloch, R.A., Ruiz-Seiler, B., Wassarman, P.M. 1990. Genomic organization and polypeptide primary structure of zona pellucida glycoprotein hZP3, the hamster sperm receptor. Dev. Biol. 142, 414–421.

Kirschhofer, K., Kenyon, J.B., Hoover, D.M., Franz, P., Weipoltshammer, K., Wachtler, F., Kimberling, W.J. 1998. Autosomal-dominant, prelingual, nonprogressive sensorineural hearing loss: Localization of the gene (*DFNA8*) to chromosome 11q by linkage in an Austrian family. Cytogenet. Cell. Genet. 82, 126–130.

Kokot, F., Dulawa, J. 2000. Tamm-Horsfall protein updated. Nephron 85, 97–102.

Kölle, S., Sinowatz, F., Boie, G., Palma, G. 1998. Differential expression of ZPC in the ovary, oocyte and embryo. Mol. Reprod. Dev. 49, 435–443.

Kölle, S., Sinowatz, F., Boie, G., Totzauer, I., Amselgruber, W., Plendl, J. 1996. Localization of messenger RNA encoding zona protein ZP3-α in the porcine ovary, oocyte and embryo by nonradioactive in situ hybridization. Histochem. J. 28, 441–447.

Kolluri, S., Kaul, R., Banerjee, K., Gupta, S.K. 1995. Nucleotide sequence of cDNA encoding Bonnet monkey (*Macaca radiata*) zona pellucida glycoprotein-ZP3. Reprod. Fertil. Dev. 7, 1209–1212.

Kubo, H., Kawano, T., Tsubuki, S., Kawashima, S., Katagiri, C., Suzuki, A. 1997. A major glycoprotein of *Xenopus* egg vitelline envelope, gp41, is a frog homolog of mammalian ZP3. Dev. Growth Diff. 39, 405–417.

Kubo, H., Kawano, T., Tsubuki, S., Kotani, M., Kawasaki, H., Kawashima, S. 2000. Egg envelope glycoprotein gp37 as a *Xenopus* homolog of mammalian ZP1, based on cDNA cloning. Dev. Growth Diff. 42, 419–427.

Lee, V.H., Dunbar, B.S. 1993. Developmental expression of the rabbit 55-kDa zona pellucida protein and messenger RNA in ovarian follicles. *Dev. Biol.* 155, 371–382.

Lee, V.H., Schwoebel, E., Prasad, S., Cheung, P., Timmons, T.M., Cook, R., Dunbar, B.S. 1993. Identification and structural characterization of the 75-kDa rabbit zona pellucida protein. J. Biol. Chem. 268, 12412–12417.

Legan, P.K., Richardson, G.P. 1997a. Extracellular matrix and cell adhesion molecules in the developing inner ear. Sem. Cell. Dev. Biol. 8, 217–224.

Legan, P.K., Rau, A., Keen, J.N., Richardson, G.P. 1997b. The mouse tectorins. Modular matrix proteins of the inner ear homologous to components of the sperm-egg adhesion system. J. Biol. Chem. 272, 8791–8801.

Legan, P.K., Lukashkina, V.A., Goodyear, R.J., Kossi, M., Russell, I.J., Richardson, G.P. 2000. A targeted deletion in alpha-tectorin reveals that the tectorial membrane is required for the gain and timing of cochlear feedback. Neuron 28, 273–285.

Li, X.J., Snyder, S.H. 1995. Molecular cloning of Ebnerin, a von Ebner's gland protein associated with taste buds. J. Biol. Chem. 270, 17674–17679.

Liang, L.F., Dean, J. 1993. Conservation of mammalian secondary sperm receptor genes enables the promoter of the human gene to function in mouse oocytes. Dev. Biol. 156, 399–408.

Liang, L., Soyal, S.M., Dean, J. 1997. FIGα, a germ cell specific transcription factor involved in the coordinate expression of the zona pellucida genes. Development 124, 4939–4947.

Liang, L.-F., Chamow, S.M., Dean, J. 1990. Ooctye-specific expression of mouse ZP2: Developmental regulation of zona pellucida genes. Mol. Cell. Biol. 10, 1507–1515.

Lira, S.A., Kinloch, R.A., Mortillo, S., Wassarman, P.M. 1990. An upstream region of the mouse *ZP3* gene directs expression of firefly luciferase specifically to growing oocytes in transgenic mice. Proc. Natl. Acad. Sci. USA 87, 7215–7219.

Lira, S.A., Schickler, M., Wassarman, P.M. 1993. Cis-acting DNA elements involved in oocyte-specific expression of mouse sperm receptor gene *mZP3* are located close to the gene's transcription start-site. Mol. Reprod. Dev. 36, 494–499.

Litscher, E.S., Qi, H., Wassarman, P.M. 1999. Mouse zona pellucida glycoproteins mZP2 and mZP3 undergo carboxy-terminal proteolytic processing in growing oocytes. Biochemistry 38, 12280–12287.

Liu, C., Litscher, E.S., Wassarman, P.M. 1997. Zona pellucida glycoprotein mZP3 bioactivity is not dependent on the extent of glycosylation of its polypeptide or on sulfation and sialylation of its oligosaccharides. J. Cell. Sci. 110, 745–752.

Liu, C., Litscher, E.S., Mortillo, S., Sakai, Y., Kinloch, R.A., Stewart, C.L., Wassarman, P.M. 1996. Targeted disruption of the *mZP3* gene results in production of eggs lacking a zona pellucida and infertility in female mice. Proc. Natl. Acad. Sci. USA 93, 5431–5436.

Lyons, C.E., Payette, K.L., Price, J.L., Huang, R.C.C. 1993. Expression and structural analyis of a teleost homolog of a mammalian zona pellucida gene. J. Biol. Chem. 268, 21351–21358.

MacCallum, R.M., Kelley, L.A., Sternberg, M.J. 2000. SAWTED: structure assignment with text description-enhanced detection of remote homologues with automated SWISS-PROT annotation comparisons. Bioinformatics 16, 125–129.

McCartney, C.A., Mate, K.E. 1999. Cloning and characterization of a zona pellucida 3 cDNA from a marsupial, the brushtail possum *Trichosurus vulpecula*. Zygote 7, 1–9.

McGuffin, L.J., Bryson, K., Jones, D.T. 2000. The PSIPRED protein structure prediction server. Bioinformatics 16, 404–405.

McLeskey, S.B., Dowds, C., Carballada, R., White, R.R., Saling, P.M. 1998. Molecules involved in mammalian sperm-egg interaction. Int. Rev. Cytol. 177, 57–113.

Mate, K.E., McCartney, C.A. 1998. Sequence and analysis of the zona pellucida 2 (ZP2) cDNA from a marsupial, the brushtail possum, *Trichosurus vulpecula*. Mol. Reprod. Dev. 51, 322–329.

Matsushita, F., Miyawaki, A., Mikoshiba, K. 2000. Vomeroglandin/CRP-Ductin is strongly expressed in the glands associated with the mouse vomeronasal organ: identification and characterization of mouse vomeroglandin. Biochem. Biophys. Res. Commun. 268, 275–281.

Millar, S.E., Lader, E.S., Dean, J. 1993. ZAP-1 DNA binding activity is first detected at the onset of zona pellucida gene expression in embryonic mouse oocytes. Dev. Biol. 158, 410–413.

Mold, D.E., Kim, I.F., Tsai, C-M., Lee, D., Chang, C-Y., Huang, R.C.C. 2001. Cluster of genes encoding the major egg envelope protein of zebrafish. Mol. Reprod. Dev. 58, 4–14.

Mollenhauer, J., Wiemann, S., Scheurlen, W., Korn, B., Hayashi, Y., Wilgenbus, K.K., von Deimling, A., Poustka, A. 1997. DMBT1, a new member of the SRCR superfamily, on chromosome 10q25.3–26.1 is deleted in malignant brain tumours. Nat. Genet. 17, 32–39.

Moller, C.C., Bleil, J.D., Kinloch, R.A., Wassarman, P.M. 1990. Structural and functional relationships between mouse and hamster zona pellucida glycoproteins. Dev. Biol. 137, 276–286.

Moreno-Pelayo, M.A., del Castillo, I., Villamar, M., Romero, L., Hernandez-Calvin, F.J., Herraiz, C., Barbera, R., Navas, C., Moreno, F. 2001. A cysteine substitution in the zona pellucida domain of α-tectorin results in autosomal dominant, postlingual, progressive, mid frequency hearing loss in a Spanish family. J. Med. Genet. 38, E13.

Murata, K., Sakaki, T., Yasamasu, S., Iuchi, I., Enami, J., Yasumasu, I., Yanagami, K. 1995. Cloning of cDNAs for the precursor protein of a low molecular-weight subunit of the inner layer of the egg envelope (chorion) of the fish *Oryzias latipes*. Dev. Biol. 167, 9–17.

Murata, K., Sugiyama, H., Yasumasu, S., Iuchi, I., Yasumasu, I., Yamagami, K. 1997. Cloning of cDNA and estrogen-induced hepatic gene expression for choriogenin H, a precursor protein of the fish egg envelope (chorion). Proc. Natl. Acad. Sci. USA 94, 2050–2055.

Mustapha, M., Weil, D., Chardenoux, S., Elias, S., El-Zir, E., Beckmann, J.S., Loiselet, J., Petit, C. 1999. An α-tectorin gene defect causes a newly identified autosomal recessive form of sensorineural pre-lingual non-syndromic deafness, DNFB21. Hum. Mol. Genet. 8, 409–412.

Nielsen, H., Engelbrecht, J., Brunak, S., von Heijne, G. 1997. Identification of prokaryotic and eukaryotic signal peptides and prediction of their cleavage sites. Protein Eng. 10, 1–6.

Noguchi, S., Yonezawa, N., Katsumata, T., Hashimuze, K., Kuwayama, M., Hamano, S., Watanabe, S., Nakano, M. 1994. Characterization of the zona pellucida glycoproteins from bovine ovarian and fertilized eggs. Biochim. Biophys. Acta 1201, 7–14.

Oppen-Berntsen, D.O., Gram-Jensen, E., Walther, B.T. 1992. Zona radiata proteins are synthesized by rainbow trout (*Oncorhynchus mykiss*) hepatocytes in response to oestradiol-17β. J. Endocrinol. 135, 293–302.

Pan, J., Sesanami, T., Kono, Y., Matsuda, T., Mori, M. 2001. Effects of testosterone on production of perivitelline membrane glycoprotein ZPC by granulosa cells of Japanese quail (*Coturnix japonica*). Biol. Reprod. 64, 310–316.

Philpott, C.C., Ringuette, M.J., Dean, J. 1987. Oocyte-specific expression and developmental regulation of ZP3, the sperm receptor of the mouse zona pellucida. Dev. Biol. 121, 568–575.

Porter, K.R., Tamm, I. 1954. Direct visualisation of a mucoprotein component of urine. J. Biol. Chem. 212, 135–139.

Prasad, S.V., Skinner, S.M., Carino, C., Wang, N., Cartwright, J., Dunbar, B.S. 2000. Structure and function of the proteins of the mammalian zona pellucida. Cells, Tissues, Organs 166, 148–164.

Puett, D., Holladay, L.A., Robinson, J.P. 1977. Circular dichroism of human urinary Tamm-Horsfall glycoprotein. Mol. Cell. Biochem. 15, 109–116.

Qi, H., Williams, Z., Wassarman, P.M. 2002. Secretion and assembly of zona pellucida glycoproteins by growing mouse oocytes microinjected with epitope-tagged cDNAs for mZP2 and mZP3. Mol. Biol. Cell 13, 540–541.

Rankin, T., Familiari, M., Lee, E., Ginsberg, A.M., Dwyer, N., Blanchette-Mackie, J., Drago, J., Westphal, H., Dean, J. 1996. Mice homozygous of an insertional mutation in the *Zp3* gene lack a zona pellucida and are infertile. Development 122, 2903–2910.

Rankin, T., Talbot, P., Lee, E., Dean, J. 1999. Abnormal zonae pellucidae in mice lacking ZP1 result in early embryonic loss. Development 126, 3847–3855.

Rankin, T., O'Brien, M., Lee, E., Wigglesworth, K., Eppig, J., Dean, J. 2001. Defective zonae pellucidae in *ZP2*-null mice disrupt folliculogenesis, fertility and development. Development 128, 1119–1126.

Ringuette, M.J., Chamberlin, M.E., Baur, A.W., Sobieski, D.A., Dean, J. 1988. Molecular analysis of cDNA coding for ZP3, a sperm binding protein of the mouse zona pellucida. Dev. Biol. 127, 287–295.

Robinson, J.P., Puett, D. 1973. Morphological and conformational studies of Tamm-Horsfall urinary glycoprotein. Arch. Biochem. Biophys. 159, 615–621.

Roller, R.J., Kinloch, R.A., Hiraoka, B.Y., Li, S.S.-L., Wassarman, P.M. 1989. Gene expression during mammalian oogenesis and early embryogenesis: Quantification of three messenger-RNAs abundant in fully-grown mouse oocytes. Development 106, 251–261.

Rost, B., Fariselli, P., Casadio, R. 1996. Topology prediction for helical transmembrane proteins at 86% accuracy. Protein Sci. 5, 1704–1718.

Salzmann, G.S., Greve, J.M., Roller, R.J., Wassarman, P.M. 1983. Biosynthesis of the sperm receptor during oogenesis in the mouse. EMBO J. 2, 1451–1457.

Schultz, J., Copley, R.R., Doerks, T., Ponting, C.P., Bork, P. 2000. SMART: a web-based tool for the study of genetically mobile domains. Nucleic Acids Res. 28, 231–234.

Scobie, G.A., Kerr, L.E., MacDuff, P., Aitken, R.J. 1999. Cloning, sequencing and site of origin of the rat sperm receptor protein, ZP3. Zygote 7, 27–35.

Sebastiano, M., Lassandro, F., Bazzicalupo, P. 1991. *cut-1*, a *Caenorhabditis elegans* gene coding for a dauer-specific noncollagenous component of the cuticle. Dev. Biol. 146, 519–530.

Shimizu, S., Tsuji, M., Dean, J. 1983. *In vitro* biosynthesis of three sulphated glycoproteins of murine zonae pellucidae by oocytes grown in culture. J. Biol. Chem. 258, 5858–5863.

Sinowatz, F., Kölle, S., Töpfer-Petersen, E. 2001. Biosynthesis and expression of zona pellucida glycoproteins in mammals. Cells, Tissues, Organs 168, 24–35.

Soyal, S.M., Amleh, A., Dean, J. 2000. FIGα, a germ cell-specific transcription factor required for ovarian follicle formation. Development 127, 4645–4654.

Steel, K.P., Kros, C.J. 2001. A genetic approach to understanding auditory function. Nat. Genet. 27, 143–149.

Sugiyama, H., Yasumasu, S., Murata, K., Iuchi, I., Yamagami, K. 1998. The third egg envelope subunit in fish: cDNA cloning and analysis, and gene expression. Dev. Growth Differ. 40, 35–45.

Takeuchi, Y., Nishimura, K., Aoki, N., Adachi, T., Sato, C., Kitajima, K., Matsuda, T. 1999. A 42-kDa glycoprotein from chicken egg-envelope, an avian homolog of the ZPC family glycoproteins in mammalian zona pellucida. Its first identification, cDNA cloning and granulosa cell-specific expression. Eur. J. Biochem. 260, 736–742.

Takito, J., Hikita, C., Al-Awqati, Q. 1996. Hensin, a new collecting duct protein involved in the *in vitro* plasticity of intercalated cell polarity. J. Clin. Invest. 98, 2324–2331.

Takito, J., Yan, L., Ma, J., Hikita, C., Vijayakumar, S., Warburton, D., Al-Awqati, Q. 1999. Hensin, the polarity reversal protein, is encoded by *DMBT1*, a gene frequently deleted in malignant gliomas. Am. J. Physiol. 277, F277–F289.

Thillai-Koothan, P., van Duin, M., Aitken, R.J. 1993. Cloning, sequencing and oocyte-specific expression of the marmoset sperm receptor protein, ZP3. Zygote 1, 93–101.

Thompson, J.D., Gibson, T.J., Plewniak, F., Jeanmougin, F., Higgins, D.G. 1997. The CLUSTAL_X windows interface: flexible strategies for multiple sequence alignment aided by quality analysis tools. Nucleic Acids Res. 25, 4876–4882.

Tian, J., Gong, H., Lennarz, W.J. 1999. *Xenopus laevis* sperm receptor gp69/64 glycoprotein is a homolog of the mammalian sperm receptor ZP2. Proc. Natl. Acad. Sci. USA 96, 829–834.

Tong, Z., Nelson, L.M., Dean, J. 1995. Inhibition of zona pellucida gene expression by antisense oligonucleotides injected into mouse oocytes. J. Biol. Chem. 270, 849–853.

Topper, E.K., Kruijt, L., Calvette, J., Mann, K., Topfer-Petersen, E., Woelders, H. 1997. Identification of bovine zona pellucida glycoproteins. Mol. Reprod. Dev. 46, 344–350.

Van Duin, M., Polman, J., Verkoelen, C., Bunschoten, H., Meyerink, J., Olijve, W., Aitken, R.J. 1992. Cloning a characterization of the human sperm receptor ligand ZP3: Evidence for a second polymorphic allele with a different frequency in the Caucasian and Japanese populations. Genomics 14, 1064–1070.

Verhoeven, K., Van Laer, L., Kirschhofer, K., Legan, P.K., Hughes, D.C., Schatteman, I., Verstreken, M., Van Hauwe, P., Coucke, P., Chen, A., Smith, R.J., Somers, T., Offeciers, F.E., Van de Heyning, P., Richardson, G.P., Wachtler, F., Kimberling, W.J., Willems, P.J., Govaerts, P.J., Van Camp, G. 1998. Mutations in the human α-tectorin gene cause autosomal dominant non-syndromic hearing impairment. Nat. Genet. 19, 60–62.

Voyle, R.B., Haines, B.P., Loffler, K.A., Hope, R.M., Rathjen, P.D., Breed, W.G. 1999. Isolation and characterization of zona pellucida A (ZPA) cDNAs from two species of marsupial: Regulated oocyte-specific expression of ZPA transcripts. Zygote 7, 239–248.

Waclawek, M., Foisner, R., Nimpf, J., Schneider, W.J. 1998. The chicken homologue of zona pellucida glycoprotein-3 is synthesized by granulosa cells. Biol. Reprod. 59, 1230–1239.

Wang, H, Gong, Z. 1999. Characterization of two zebrafish cDNA clones encoding egg envelope proteins ZP2 and ZP3. Biochim. Biophys. Acta 1446, 156–160.

Wassarman, P.M. 1999. Mammalian fertilization: Molecular aspects of gamete adhesion, exocytosis, and fusion. Cell 96, 175–183.

Wassarman, P.M. 1990. Profile of a mammalian sperm receptor. Development 108, 1–17.

Wassarman, P.M. 1988. Zona pellucida glycoproteins. Ann. Rev. Biochem. 57, 415–442.

Wassarman, P.M., Litscher, E.S. 1995. Sperm-egg recognition mechanisms in mammals. Curr. Topics Dev. Biol. 30, 1–19.

Wassarman, P.M., Mortillo, S. 1991. Structure of the mouse egg extracellular coat, the zona pellucida. Intl. Rev. Cytol. 130, 85–109.

Wassarman, P.M., Bleil, J.D., Florman, H.M., Greve, J.M., Roller, R.J., Salzmann, G.S., Samuels, F.G. 1985. The mouse egg's sperm receptor: What is it and how does it work? Cold Spring Harbor Symp. Quant. Biol. 50, 11–18.

Wassarman, P.M., Jovine, L., Litscher, E.S. 2001. A profile of fertilization in mammals. Nat. Cell. Biol. 3, E59–E64.

Wassarman, P.M., Liu, C., Litscher, E.S. 1996. Constructing the mammalian egg zona pellucida: Some new pieces of an old puzzle. J. Cell. Sci. 109, 2001–2004.

Wassarman, P.M., Qi, H., Litscher, E.S. 1997. Mutant female mice carrying a single *mZP3* allele produce eggs with a thin zona pellucida, but reproduce normally. Proc. Roy. Soc., Lond. B 26, 323–328.

Wiggins, R.C. 1987. Uromucoid (Tamm-Horsfall glycoprotein) forms different polymeric arrangements on a filter surface under different physicochemical conditions. Clin. Chim. Acta 162, 329–340.

Wilkin, M.B., Becker, M.N., Mulvey, D., Phan, I., Chao, A., Cooper, K., Chung, H.J., Campbell, I.D., Baron, M., MacIntyre, R. 2000. *Drosophila* Dumpy is a gigantic extracellular protein required to maintain tension at epidermal-cuticle attachment sites. Curr. Biol. 10, 559–567.

Williams, Z., Wassarman, P.M. 2001. Secretion of mZP3, the sperm receptor, requires cleavage of its polypeptide at a consensus furin cleavage-site. Biochemistry 40, 929–937.

Yanagimachi, R. 1994. Mammalian fertilization. In: *The Physiology of Reproduction*, Vol. 1, (E. Knobil, J.D. Neill, Eds.), New York: Raven Press, pp. 189–317.

Yang, J, Hedrick, J.L. 1997. cDNA cloning and sequence analysis of the *Xenopus laevis* egg envelope glycoprotein gp43. Dev. Growth Diff. 39, 457–467.

Yurewicz, E.C., Hibler, D., Fontenot, G.K., Sacco, A.G., Harris, J. 1993. Nucleotide sequence of cDNA encoding ZP3 alpha, a sperm-binding glycoprotein from zona pellucida of pig oocytes. Biochim. Biophys. Acta 1174, 211–214.

Yurewicz, E.C., Sacco, A.G., Subramanian, M.G. 1987. Structural characterization of Mr=55,000 antigen (ZP3) of porcine oocyte zona pellucida. J. Biol. Chem. 262, 564–571.

Advances in Developmental Biology and Biochemistry, Vol. 12
M. DePamphilis (Editor)

Activation of zygotic gene expression in mammals

Melvin L. DePamphilis, Kotaro J. Kaneko and Alex Vassilev

*National Institute of Child Health and Human Development, National Institutes of Health,
Bethesda, MD 20892-2753, USA*

Summary

Preimplantation development in the mouse involves expression of about 11,000 genes, only a few hundred of which appear during the transition from maternal to zygotic gene expression. Transcription begins in most, if not all, mammals during the late 1-cell stage (phase I), but expression of most zygotic genes is delayed until the 2-cell to the 16-cell stage, depending on the mammal. In mice, a small group of genes are expressed immediately after the first mitosis, while a larger group of genes are expressed during the subsequent G2 phase. Zygotic gene activation (ZGA) is delayed by a time dependent mechanism ("zygotic clock") that regulates both transcription and translation. It appears to involve post-translational modification of RNA polymerases, translational control of maternal gene expression, developmental acquisition of chromatin mediated repression as well as the ability to alleviate this repression with sequence-specific enhancers, and changes in DNA methylation. This delay allows remodeling of chromatin into a form that globally represses gene activity so that selected genes can then be activated in a temporally and spatially specific program. Some transcription factors such as Sp1 and TBP function from oocyte to embryo, while others are selectively expressed during ZGA. OCT-4(OCT-3) regulates the production and subsequent differentiation of embryonic stem cells, a prerequisite for embryo development. TEAD-2(TEF-4), whose activity is regulated by the transcriptional coactivator YAP65, appears to regulate contact inhibition during cell proliferation, a prerequisite for tissue formation. Thus, identification and characterization of genes whose expression is activated by fertilization opens the door to understanding how mammalian development begins.

Contents

1. Introduction

Mouse development begins when an egg is fertilized by a sperm. Fertilization activates the first of six to seven cell cleavage cycles over a four day period that culminate in the formation of a blastocyst containing 64–128 cells comprising two cell types (Fig. 1). The outer layer of cells comprises the trophectoderm that will form the placenta, while the inner cell mass consists of about 25 totipotent embryonic stem cells that will produce the embryo. The first round of DNA replication occurs in both the maternal and paternal pronucleus of the fertilized egg (or 1-cell embryo). About 4 h later the first mitosis occurs to produce a 2-cell embryo in which each "zygotic nucleus" contains a complete set of parental chromosomes. This cleavage event is rapidly followed by the second S-phase and expression of a small number of zygotic genes. Changes in totipotency of individual cells are first detected at the 8-cell stage, although clearly differentiated cells first appear at the blastocyst stage. Here the blastomeres of the embryo exhibit different cellular features that presumably lead to the formation of embryonic (inner cell mass) and extraembryonic (trophectoderm) lineages (Johnson and Maro, 1986).

All of the genes in sperm and eggs are transcriptionally silent. Therefore, the first one or two cell cleavage events and the activation of zygotic gene expression must rely on the reservoir of mRNAs and proteins stored in the egg. This transition from maternal to zygotic gene dependence and its associated changes in chromatin organization, gene expression, DNA replication, and cell division is a phenomenon characteristic of most, if not all, of the metazoa. The problems of limited availability (~30 eggs/female) and small size (mouse eggs are 100–1000 times smaller than those from frogs or flies) of mouse zygotes have been overcome by characterizing endogenous gene expression in the presence and absence of metabolic inhibitors, by quantifying transient expression of reporter genes injected into the nuclei of oocytes and early embryos, by analyzing expression of integrated reporter genes in the oocytes and early embryos of transgenic mice, and by transplantation of nuclei from one developmental stage to another [reviewed in (Majumder and DePamphilis, 1995; Nothias et al., 1995; Renard, 1998; Latham, 1999; Schultz et al., 1999)]. The results from these varied approaches are remarkably consistent and reflect the physiological state of the cell.

Replication and expression of genes encoded in extrachromosomal DNA respond to the same signals that regulate these functions in cellular DNA: They require specific *cis*-acting regulatory sequences and the trans-acting proteins that activate them, and occur only when

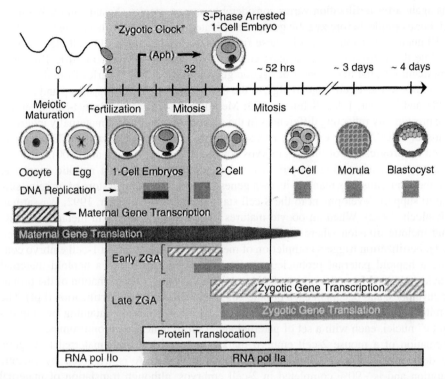

Fig. 1. Maternal to zygotic gene transition in the mouse. Maternal events are indicated in *red*, paternal events in *blue*, and zygotic events in *green*. Open bars apply to both. Embryonic stem cells ("inner cell mass") are indicated in *yellow*; trophectodermal cells in *orange*. Periods of transcription are indicated by hatched bars; translation by solid bars. (*For a colored version of this figure, see plate section, page 269.*)

the host cell executes the same function with its own genome. For example, oocytes can express plasmid-encoded genes but cannot replicate plasmid DNA, whereas embryos can do both. However, embryos express plasmid-encoded genes or the chromosomal genes in a transplanted nucleus only when the host cell is transcriptionally competent, and only if the correct promoter/enhancer sequences are present together with the correct transcription factors. The results of these studies have revealed a series of specific morphological, biochemical and molecular events that occur post-fertilization and that allow selected genes to be expressed [reviewed recently in (Latham, 1999; Schultz et al., 1999; Latham and Schultz, 2001).

2. Maternal to zygotic gene transition

2.1. Timing

While transcription of the haploid genome in oocytes from all species stops when they mature into eggs, the extent of development and the time that elapses before transcription begins again after fertilization varies considerably among species (Yasuda and Schubiger, 1992). For example, before zygotic genes are activated in flies, the fertilized egg has undergone 10 nuclear divisions in 1.5 h, those from frogs have undergone 11 cell cleavages in 6 h, while those from mice have undergone only one cleavage event in 24 h. But even among mammals, the major onset of zygotic gene activation (ZGA) varies from the 2-cell stage in mice to the 4-cell stage in pigs, to the 8 to 16-cell stage in cows, sheep and rabbits (Schultz and Heyner, 1992; Schultz, 1993; Memili and First, 1999; Brunet-Simon et al., 2001), presumably reflecting differences in the amount of one or more maternally inherited proteins or mRNAs. For example, mouse oocytes express a protein (MATER) that is required for progression of mouse embryos beyond two cells (Tong et al., 2000).

In the mouse, a growing oocyte, arrested at diplotene of its first meiotic prophase, transcribes and translates many of its own genes, thereby producing a store of proteins sufficient to support development to the 8-cell stage (Schultz and Heyner, 1992; Wassarman and Kinloch, 1992). When an oocyte matures into an egg, it arrests in metaphase of its second meiotic division where transcription stops and translation of mRNA is reduced (Fig. 1). Fertilization triggers completion of meiosis and formation of a 1-cell embryo containing a haploid paternal pronucleus derived from the sperm and a haploid maternal pronucleus derived from the oocyte. DNA replication begins after formation of the pronuclear envelopes, and occurs in each pronucleus from 10 to 16 h post-fertilization (hpf). The first mitosis occurs from 17 to 20 hpf to produce a 2-cell embryo containing two diploid "zygotic" nuclei, each with a set of paternal and a set of maternal chromosomes.

Formation of a mouse 2-cell embryo marks the transition from maternal to zygotic gene dependence (Fig. 1). Degradation of maternal mRNA is triggered by meiotic maturation and is ~90% completed in 2-cell embryos, although translation of maternal mRNA continues into the 8-cell stage (Schultz, 1993). Injection of plasmid DNA (Nothias et al., 1996) and transplantation of nuclei from 2-cell stage embryos into 1-cell embryos (Latham et al., 1992) has shown that late 1-cell embryos are transcriptionally competent. Transcription of endogenous genes also has been detected in late 1-cell mouse embryos [(Matsumoto et al., 1994; Temeles et al., 1994; Bouniol et al., 1995; Christians et al., 1995) and references therein] where it begins at the end of S-phase (Aoki et al., 1997; Bouniol-Baly et al., 1997; Beaujean et al., 2000). However, transcription-dependent protein synthesis does not begin until 2–4 h after completion of the first mitosis and the beginning of S-phase in 2-cell embryos. Eight to 10 h later during G2-phase of 2-cell embryos, expression of zygotic genes increases in both amount and complexity [(Latham et al., 1991; Wiekowski et al., 1991; Christians et al., 1995; Nothias et al., 1996) and references therein].

Thus, ZGA has at least three recognizable phases: transcription without translation begins in late 1-cell embryos (phase I), but transcription coupled to translation does not begin until the early 2-cell stage in development (phase II), and robust transcription coupled to translation does not begin until the late 2-cell stage (phase III). The existence of the first and third phases can be demonstrated clearly by injection of a plasmid encoded reporter gene. The reporter gene is transcribed as soon as the 1-cell embryo becomes transcriptionally competent, but it is not translated until the third phase of ZGA begins in late 2-cell embryos (Nothias et al., 1996). The second phase of ZGA is specific for a subset of zygotic genes. An early phase of ZGA consisting of a small group of genes appears to occur shortly after fertilization in all vertebrates; it is the late, robust phase of ZGA that is delayed to different extents in different species (Memili and First, 1999).

2.2. Genes

The available data suggest that from 38 to 282 genes are specifically expressed during ZGA in mice, but only a few of these genes have been identified. Unfertilized mouse eggs contain 258 expressed sequence tags not found in adult mouse cDNA libraries (Rajkovic et al., 2001). Analysis of cDNA expression libraries have identified from 9,718 to 11,483 genes expressed from the oocyte to the blastocyst stage, 798–1585 of which were not found elsewhere in the mouse (Ko et al., 2000; Stanton and Green, 2001). About 2.5% of newly synthesized proteins show a transient increase at the 2-cell stage (Latham et al., 1991), and about 2.9% of unique expressed sequence tags in preimplantation embryos sharply increased in amount at the 2-cell stage (Ko et al., 2000). Early ZGA genes (Fig. 1; Table I) include heat shock genes, and the translation initiation factor eIF-1A, a gene of unknown function, and three as yet unidentified proteins known as the "transcription-requiring complex" (Schultz, 1993; Wiekowski et al., 1991). Another candidate, Fas, is a gene involved in apoptosis whose mRNA appears to be expressed specifically at the 2-cell stage in rats and the 4-cell stage in humans (Kawamura et al., 2001). Examples of late ZGA genes include linker histone H1, transcription factors OCT-4(OCT-3) and TEAD-2(TEF-4), and splicing factor SRp20 (Table 1). In the rabbit, ribosomal proteins S20 and L5, splicing factor p55, α-tubulin, a membrane transport protein VDAC2, and chaperonin are transcribed at the 8 to 16-cell stage (late ZGA, (Brunet-Simon et al., 2001)). Expression of all of these genes is α-amanitin sensitive, and therefore presumed to be RNA polymerase II promoters.

Although less well characterized, RNA polymerase I and III dependent gene expression follows the same time course as RNA polymerase II dependent gene expression. Pol I, II and III promoters are all recognized in late 1-cell mouse embryos (Nothias et al., 1996). Small nuclear RNAs U1 to U5 whose synthesis is dependent on pol II, as well as U6

Table 1
Genes expressed at the beginning of mouse development

Gene	Expression begins	Imprinting	Reference
Hsp70.1	early 2-cell		(Christians et al., 1995)
Hsp68	early 2-cell		(Bevilacqua et al., 1995)
U2afbp	early 2-cell	paternal	(Hatada et al., 1995; Latham et al., 1995)
EIF-1A	early 2-cell		(Davis et al., 1996)
Histone H1	late 2-cell		(Wiekowski et al., 1997)
Oct4(Oct3)	late 2-cell		(Palmieri et al., 1994)
Tead2(Tef4)	late 2-cell		(Kaneko et al., 1997)
Cyclin-A	late 2-cell		(Fuchimoto et al., 2001)
Xist	late 2-cell	paternal	(Zuccotti et al., 2002)
Sry	late 2-cell		(Zwingman et al., 1993)
Zfy	late 2-cell		(Zwingman et al., 1993)
3 Zn^{++} finger proteins	2-cell on; 8-cell off		(Choo et al., 2001)
SRp20	2-cell to 4-cell		(Jumaa et al., 1999)
Snrpn	4-cell	paternal	(Szabo and Mann, 1995)

whose synthesis depends on pol III, begin to accumulate at the 2-cell stage in mice (Dean et al., 1989), concurrent with late ZGA. Similarly, U2 begins to accumulate at the 8 to 16-cell stage in cows (Watson et al., 1992). Changes in nucleolar morphology and nucleolar protein markers that provide a sensitive indicator of pol I dependent ribosomal gene expression are localized to the middle of the 2-cell stage during mouse development, and the earliest time at which rRNA synthesis has been detected is the mid- to late 2-cell stage [reviewed in (Flechon and Kopecny, 1998; Baran et al., 2001)]. These results are consistent with the hypothesis that the high rate of ribosome assembly that begins in 2-cell embryos requires the coordinate expression of pol I dependent rRNA synthesis and pol II dependent expression of genes for ribosomal proteins (Taylor and Piko, 1992). In pigs and cows, rRNA genes are also activated at the 4-cell and 8-cell stages, respectively, concurrent with late ZGA (Hyttel et al., 2000). Thus, it is clear that the transition from maternal to zygotic gene expression is a highly orchestrated sequence of events that span the first 40 h in the life of a fertilized mouse egg, and up to several days in other animals.

3. Zygotic clock

In organisms that undergo rapid nuclear division after fertilization (e.g. amphibians, fish, echinoderms, flies), ZGA is determined by the ratio of nuclei to cytoplasm (Masui and Wang, 1998). Thus, ZGA occurs after a preset number of nuclear divisions have occurred, regardless of whether or not cellularization occurs (e.g. 13 cell cleavages in *Xenopus*; 10 nuclear divisions in the syncytium of *Drosophila*). However, in mice, ZGA is a time-dependent event that is delayed for about 24 h post-fertilization, and therefore begins after formation of a 2-cell embryo (Fig. 1). The fact that ZGA is independent of DNA synthesis, cytokinesis, or the ratio of cytoplasm to nuclei [reviewed in (Wiekowski et al., 1991)] allows ZGA to begin in S-phase arrested 1-cell embryos after they have advanced temporally to the "2-cell stage" (Fig. 1). The mechanism that determines when ZGA begins (referred to as the "zygotic clock") is not simply the time required to convert sperm and egg chromatin into a transcribable form, because it also regulates expression of injected plasmid encoded genes utilizing RNA polymerase I, II or III dependent promoters (Nothias et al., 1996). Therefore, the zygotic clock must involve delayed expression or activation of one or more transacting factors required for transcription and/or translation. Nevertheless, early ZGA genes are unique in that they begin translation immediately following the first mitosis, whereas plasmid encoded genes are not translated until late ZGA, even though they are transcribed when injected into late 1-cell embryos (Nothias et al., 1996). Therefore, the zygotic clock regulates both transcription and translation of mRNA. Furthermore, the fact that plasmid encoded genes are expressed during late ZGA reveals that early ZGA genes are not simply more accessible to transcription factors than late ZGA genes; early ZGA genes are in some way predestined for immediate expression, perhaps by association with the transcription machinery.

3.1. Nuclear translocation

Following fertilization, transcription factor Sp1, TATA-box binding protein, and RNA polymerase II all undergo a time dependent concentration in the pronuclei (Worrad et al., 1994;

Bellier et al., 1997), and both Sp1 mRNA and transcription factor activity increases markedly from the 1-cell to the 4-cell stage (Majumder et al., 1993; Worrad and Schultz, 1997). Both Sp1 and HSF1 are present in the pronuclei of 1-cell embryos and are required for Hsp70.1 gene expression during early ZGA (Bevilacqua et al., 1997; Christians et al., 1997), but translocation of transcription factors to the nucleus cannot be the only zygotic clock component, because in contrast to ZGA (Wang and Latham, 1997), translocation of these proteins does not require concomitant protein synthesis.

3.2. Protein phosphorylation

In mice, rabbits and frogs, the carboxyl-terminal domain (CTD) of the RNA pol-II catalytic subunit is hyperphosphorylated and transcriptionally inactive in mature eggs (RNA pol IIo) (Bellier et al., 1997; Palancade et al., 2001). Fertilization triggers dephosphorylation of the CTD into a hypophosphorylated form IIa that is required for initiation of transcription and a reduced phosphorylated form IIe characteristic of somatic cells. Form IIa appears in late 1-cell embryos prior to early ZGA in both mice and rabbits, but form IIe is not observed until late ZGA. Late ZGA occurs in late 2-cell mouse embryos, at the 8 to 16-cell stage in rabbits, and the mid-blastula transition in frogs. Like ZGA, these changes in phosphorylation require concomitant protein synthesis, and they may also account for the observation that ZGA in the mouse requires protein kinase activity (Schultz, 1993). Post-translational modifications of other proteins required for transcription such as Sp1 and TBP may also contribute to the zygotic clock mechanism (Latham, 1999), but this remains more speculative.

3.3. Translational control

Maternal mRNA degradation is triggered by meiotic maturation, and is about 90% completed in 2-cell embryos, although maternal protein synthesis continues into the 8-cell stage (Schultz, 1993). Nevertheless, some of the newly synthesized proteins are made from maternally inherited transcripts that are polyadenylated after fertilization and then translated (Oh et al., 2000). For example, newly synthesized histone H1 (Wiekowski et al., 1997), TEAD-2 (Wang and Latham, 2000), Sp1, TATA-binding protein (Worrad and Schultz, 1997), and cyclin-A (Fuchimoto et al., 2001) are first translated from maternal mRNA in late 1-cell or early 2-cell embryos and then from zygotic mRNA during the late 2-cell to 4-cell stage. Such proteins presumably are required for chromatin remodeling and cell proliferation. As much as one third of the mRNAs in 2-cell embryos appear to contain a cytoplasmic polyadenylation element (Oh et al., 2000).

Despite the fact that transcription is first detected in G2-phase 1-cell embryos, fertilized mouse eggs can delay expression of zygotic genes by uncoupling translation from transcription (Nothias et al., 1996). Injection of plasmid encoded genes revealed that the time courses for nascent mRNA accumulation from pol I, II or III dependent promoters are biphasic, with the second phase of transcription occurring just prior to late ZGA. An RNA polymerase II dependent gene can be transcribed prior to the first mitosis, but it is not translated until zygotic gene expression begins (up to 15 h later). At this time, translation of RNA pol II transcripts becomes tightly linked to transcription (the second phase).

Thus, the small amount of RNA synthesis that occurs in late 1-cell embryos may simply represent premature transcription resulting from the appearance of transcription machinery before the parental genomes are completely masked by chromatin structure. In fact, one or more factors capable of repressing DNA transcription and replication are absent from 1-cell embryos; they are produced just prior to ZGA (Henery et al., 1995; Wiekowski et al., 1997). Thus, by delaying transcription and by preventing any transcripts that are made from being translated, the zygotic clock provides a window of opportunity for remodeling parental chromosomes in the absence of gene expression.

4. Developmental acquisition of transcriptional regulation

4.1. Chromatin-mediated repression

Transient expression from plasmid-encoded reporter genes and nuclear transplantation experiments have revealed the presence of transacting factors that can repress the activities of promoters and replication origins from 20 to >500-fold as fertilized mouse eggs develop into 2-cell and 4-cell embryos [reviewed in (Wiekowski et al., 1993, 1997; Henery et al., 1995; Majumder and DePamphilis, 1995; Nothias et al., 1995, 1996; Majumder et al., 1997). The same appears to be true in rabbits (Christians et al., 1994), despite the fact that ZGA in rabbits does not occur until the 8 to 16-cell stage. While maternal nuclei in both oocytes and 1-cell embryos exhibit this repression, the paternal pronucleus exhibits repression only when 1-cell embryos develop beyond S-phase (Fig. 2). Sperm chromatin, which contains protamines but not core histones H2B and H3, becomes dispersed about 1 h after fertilization, and then within the following 7 h protamines are replaced by core histones and the DNA condensed, a process that is completed prior to DNA replication (McLay and Clarke, 1997). Paternal pronuclei generally replicate more quickly than maternal pronuclei (Ferreira and Carmo-Fonseca, 1997; Aoki and Schultz, 1999), which inherit a full set of histones from the oocyte (McLay and Clarke, 1997; Wiekowski et al., 1997) and exhibit a different pattern of histone acetylation (Adenot et al., 1997; Aoki and Schultz, 1999). Thus, paternal pronuclei may exhibit an "unrepressed" chromatin state while maternal pronuclei exhibit a "repressed" chromatin state.

This hypothesis is supported by observations that expression of microinjected genes in paternal pronuclei is ~5-fold greater than in maternal pronuclei, and inhibitors of histone deacetylase stimulate transcription in maternal pronuclei while reducing transcription in paternal pronuclei (Wiekowski et al., 1993, 1997; Nothias et al., 1996; Aoki et al., 1997). The factor(s) responsible for this repression is absent from the cytoplasm of early 1-cell embryos; it is not simply excluded from the paternal pronucleus. It is produced in the cytoplasm sometime between S-phase in a 1-cell embryo and formation of a 2-cell embryo where it can operate within any nucleus, regardless of its parental origin or ploidy (Henery et al., 1995). Repression increases as development proceeds from the 2-cell to the 4-cell stage. Such changes in chromatin structure could account for the fact that Hsp70.1 expression, which is constitutive in 2-cell embryos, is repressed to a basal level during the 4 to 8-cell stage (Thompson et al., 1995).

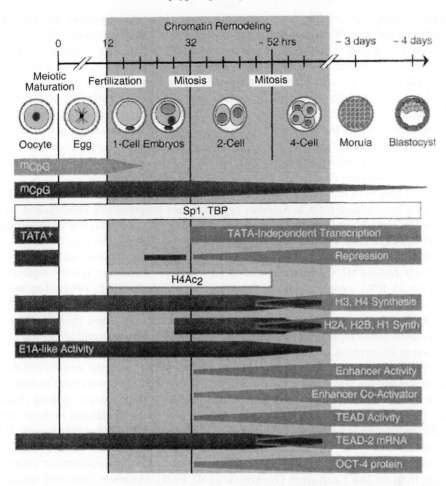

Fig. 2. Developmental changes affecting regulation of gene expression at the beginning of mouse development. These include changes in DNA methylation of the paternal and maternal genomes, changes in histone synthesis and modification that lead to chromatin-mediated repression, acquisition of chromatin-mediated repression, and acquisition of enhancer function. Some transcription factors such as Sp1 and TBP are ubiquitous. TATA-box function appears to be restricted to differentiated cells. Color coding and bar coding are the same as in Fig. 1. *(For a colored version of this figure, see plate section, page 270.)*

These observations correlate well with changes in the synthesis and modification of chromatin-bound histones at the beginning of mouse development [Fig. 2, (Wiekowski et al., 1997)]. Oocytes, which can repress microinjected promoter activity, synthesize a full complement of histones; in addition, histone synthesis, which originates from mRNA inherited from the oocyte, can be observed up to the early 2-cell stage. However, while histones H3 and H4 continue to be synthesized in early 1-cell embryos, synthesis of histones H2A, H2B and H1 (proteins required for chromatin condensation) is delayed until the late 1-cell stage, reaching their maximum rate in early 2-cell embryos. Moreover, histone H4 in both 1-cell and 2-cell embryos is predominantly diacetylated (a modification that facilitates DNA transcription and chromatin assembly). Deacetylation towards the unacetylated and monoacetylated H4 population observed in differentiated cells, such as the fibroblasts, begins at the late 2-cell to 4-cell stage. These changes correlate with the establishment of chromatin-mediated repression during formation of a 2-cell embryo as well as the increase in repression from the 2-cell to 4-cell stage, where accumulation of linker histone H1 and deacetylation of core histones are observed. Arresting development at the beginning of S-phase in 1-cell embryos prevents both the appearance of chromatin-mediated repression of transcription in paternal pronuclei and synthesis of new histones. The pattern of hyperacetylated histones throughout the nuclei also changes as embryos progress from the 1-cell to the 4-cell stage, but the significance of these changes remains speculative (Thompson et al., 1995; Worrad et al., 1995; Stein et al., 1997).

Similar changes in histone composition occur during frog development. Histone H4 is stored in a diacetylated form in *Xenopus* eggs and then progressively deacetylated after ZGA begins at the blastula stage (Dimitrov et al., 1993). Histone deacetylase inhibitors can then induce expression of specific genes (Almouzni et al., 1994). The type of linker histone changes from the maternal histone H1 variant (B4) at the mid-blastula transition to the somatic histone H1 variant at the end of gastrulation, resulting in specific repression of oocyte 5S RNA gene expression (Bouvet et al., 1994; Kandolf, 1994). The initial activation of H1 synthesis occurs entirely by activation of maternal transcripts that then disappear by the early gastrula stage (Woodland et al., 1979). Thus, the induction of chromatin-mediated repression may be a common feature of the maternal to zygotic gene transition in all metazoa.

Oocytes and early embryos of many non-mammalian species lack the somatic cell form of the linker histone H1; somatic H1 was originally believed to be absent from mouse and bovine oocytes and then first expressed at the 4 to 8-cell stage in development (Smith et al., 1995; Clarke et al., 1997). mRNA for H1 variants could be detected in these cells, and $H1^0$ protein could be detected in oocytes, suggesting that $H1^0$, rather than somatic H1, was present during ZGA. However, incorporation of radio-labeled amino acids revealed that somatic H1 is indeed synthesized in mouse oocytes and in preimplantation embryos starting from the late 1-cell stage (Wiekowski et al., 1997). Moreover, improved immunological detection confirmed that somatic H1 is present throughout this period, although it is not present on maternal metaphase II chromatin (Adenot et al., 2000). Upon formation of pronuclear envelopes, somatic H1 is rapidly incorporated onto maternal and paternal chromatin, and the amount of somatic H1 on embryonic chromatin steadily increases to the 8-cell stage. These data suggest that the major changes in histone H1 that accompany ZGA involve changes in the amount and location of somatic H1, rather than in selective expression of H1 variants. Nevertheless, a maternally derived H1 variant (H1oo) that bears significant homology with the oocyte-specific *Xenopus* B4 histone and sea urchin cs-H1 histone is specifically expressed in mouse oocytes, 1-cell and 2-cell embryos, but disappears by the 4 to 8-cell stages (Tanaka et al., 2001). Thus, while somatic H1 appears responsible for global changes in chromatin structure, selected genes may be regulated by oocyte specific H1 variants.

Manipulating the ratio of chromatin bound proteins to DNA can alter ZGA. For example, injection of somatic cell histone H1 into mouse 1-cell embryos retards the onset of RNA polymerase II transcription in late 1-cell embryos (phase I of ZGA), presumably by increasing chromatin condensation, while injection of HMG-1 (a protein that binds AT-rich sequences and that can modulate gene expression) advances it (Beaujean et al., 2000). Similarly, depleting HMG-14 and 17, two chromosomal proteins that generally stimulate transcription, by injecting 1-cell embryos with antisense oligonucleotides delays, but does not prevent, preimplantation development (Mohamed et al., 2001). The fact that these experimental insults did not prevent development of 1-cell embryos into blastocysts (Beaujean et al., 2000; Lin and Clarke, 1996; Mohamed et al., 2001) suggests that cells recover by degrading injected proteins and antisense oligonucleotides.

Global inhibition of ZGA by H1 does not affect all genes, because injection of histone H1 into 1-cell embryos does not inhibit expression of the "transcription-requiring complex" of proteins [phase II of ZGA, (Stein and Schultz, 2000)]. This may explain why injection of neither H1 nor HMG-1 had a significant effect on the development of 1-cell embryos. Nevertheless, co-injection of histones H1, H2A, H2B, H3 and H4 together with a plasmid-encoded reporter gene into the paternal pronuclei of 1-cell embryos can recreate the behavior observed when plasmid encoded genes are injected into the nuclei of two-cell embryos: the injected gene is repressed and this repression can be relieved by including either a functional enhancer or histone deacetylase inhibitors (Rastelli et al., 2001). In these experiments, the effect of H1 was to reduce the amount of core histones needed to repress gene expression. Thus, chromatin-mediated repression that can be relieved by embryo responsive enhancers (phase III of ZGA) can be made to occur in 1-cell paternal pronuclei simply by assembling the DNA into somatic cell chromatin.

4.2. Enhancer function

Direct evidence that repression of promoter activity at the beginning of mouse development is mediated through chromatin structure comes from the effects of inhibitors of histone deacetylases and enhancers on gene expression. Repression can be relieved either by treating cells with inhibitors of histone deacetylase (Majumder et al., 1993; Wiekowski et al., 1993; Henery et al., 1995; Thompson et al., 1995; Worrad et al., 1995; Aoki et al., 1997) or, in cleavage stage embryos (2-cell embryos to morula), by linking the promoter or replication origin to an embryo-responsive enhancer (Majumder and DePamphilis, 1995; Majumder et al., 1997). Inhibitors of histone deacetylase, such as trichostatin A and sodium butyrate, increase the fraction of hyperacetylated core histones and thereby stimulate transcription from specific genes, consistent with the fact that transcriptionally active eukaryotic genes are generally associated with acetylated core histones (Wade et al., 1997). In mouse 1-cell embryos, butyrate can stimulate promoter activity on genes injected into the maternal pronucleus ("repressed"), but not into the paternal pronucleus ("unrepressed") (Wiekowski et al., 1993). Thus, stimulation results from changes in chromatin structure on the injected plasmid rather than from increased synthesis of transcription factors, which would imply that butyrate would affect both pronuclei.

Enhancers can substitute for butyrate in stimulation of promoter activity, but only after formation of a 2-cell embryo (Fig. 2). Enhancers are inactive in oocytes and 1-cell embryos, even when the appropriate sequence-specific activation protein is present that is required for enhancer activity (e.g. GAL4:VP16, TEAD, Sp1), because an enhancer-specific coactivator activity is not produced until ZGA (Majumder et al., 1997; Lawinger et al., 1999).

This coactivator activity works on most, if not all, enhancers and likely represents one or more of the proteins identified in both yeast and mammalian cells that appears to mediate the interaction of upstream sequence specific transcription factors with the RNA pol II complex (Malik and Roeder, 2000; Vogel and Kristie, 2000).

In vitro, enhancers do not stimulate transcription unless the DNA substrate is organized into chromatin [reviewed in (Majumder et al., 1993; Paranjape et al., 1994)]. In vivo, enhancers have little, if any, effect on promoters injected into 2-cell embryos cultured in butyrate (Majumder et al., 1993; Wiekowski et al., 1993). Therefore, the primary role of enhancers is not simply to provide additional transcription factors to facilitate formation of an active initiation complex, but to relieve repression of weak promoters by chromatin (Fig. 3). Direct proof for this hypothesis comes from coinjection of purified histones and a plasmid-encoded reporter gene into the paternal pronuclei of 1-cell embryos (Rastelli et al., 2001). Acquisition of chromatin-mediated repression of the injected promoter and subsequent relief of this repression either by functional enhancers or by histone deacetylase inhibitors occurred at a specific ratio of histones to DNA. The extent of the enhancer-mediated stimulation was inversely related to the acetylation status of the histones and on the developmental acquisition of the enhancer-specific coactivator activity.

The results described above reveal that the "zygotic clock" delays ZGA in mice by preventing the bulk of transcription and translation until 2-cell embryos have entered G2-phase (36–40 h post-fertilization). By that time, chromatin-mediated repression has begun and the ability to use enhancers has been acquired. The net result is a global repression of mammalian gene expression, thus permitting specific genes to be activated at specific times and in specific tissues through the expression of specific promoter and enhancer activation proteins that activate their promoter/enhancer as well as by specific proteins that relieve chromatin-mediated repression (e.g. histone acetyltransferase, GAGA, SWI/SNF).

4.3. DNA replication

As discussed below, the ability of chromatin structure to repress DNA transcription or replication, and the ability of enhancers to relieve this repression first appear in mouse development with formation of 2-cell and 4-cell embryos. However, enhancers alone cannot always relieve chromatin-mediated repression. Once a repressed state is formed, it may be necessary for DNA to replicate in order to reprogram itself into a transcriptionally active state (Fig. 3). When DNA is injected into either pronucleus of 1-cell embryos, and the injected embryo then undergoes mitosis to form a 2-cell embryo, the injected promoter becomes "irreversibly" repressed, in that neither enhancers nor butyrate can restore its activity (Wiekowski et al., 1993; Henery et al., 1995). This is not due to loss of plasmid DNA from the injected pronucleus during mitosis, because repression is reversible when the injected pronucleus is transplanted to a 2-cell embryo that then undergoes mitosis (Henery et al., 1995). Therefore, something happens to DNA between completion of S-phase in a 1-cell embryo and formation of a 2-cell embryo that prevents activation of injected genes, while allowing embryonic genes to undergo ZGA. One explanation is that plasmid DNA does not replicate when injected into mouse embryos unless it contains a

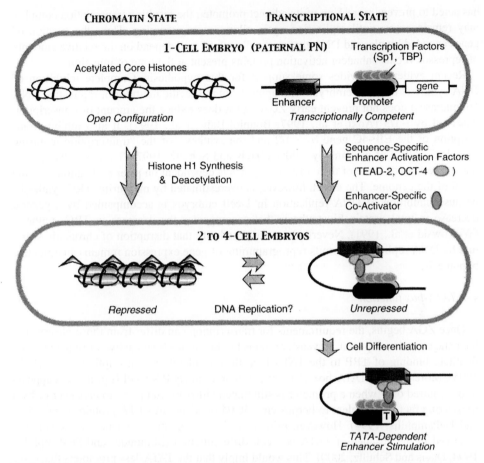

Fig. 3. Relationships between chromatin structure, promoter/enhancer activity and DNA replication during the maternal to zygotic gene transition in the mouse. Chromatin in the paternal pronucleus is in an open configuration that does not suppress promoter activity. Formation of a 2-cell embryo is accompanied by changes in chromatin structure that repress promoter activity. This repression can be relieved by enhancer activity which requires sequence specific-enhancer binding proteins such as TEAD-2 or OCT-4, and an as yet unidentified enhancer specific coactivator, all of which first become available during ZGA. DNA replication may facilitate activation of some genes. Cell differentiation is accompanied by the need for a TATA-box element in the promoter in order for enhancers to function. (*For a colored version of this figure, see plate section, page 271.*)

viral replication origin (DePamphilis et al., 1988), whereas the genome of a 1-cell embryo undergoes one round of replication prior to early ZGA and two rounds prior to late ZGA. DNA replication may be required to restore the newly remodeled zygotic genome to a transcriptionally competent state. Chromatin assembly in 1-cell embryos occurs in the absence of at least one factor required for enhancer function that does not appear until the 2-cell stage. Therefore, if chromatin-mediated repression begins in late 1-cell embryos, before enhancers are functional, DNA replication may be required to disrupt the repressed state so that appropriate transcription factors can bind (Wolffe, 1996). Conversely, once an enhancer has acted to prevent repression of its adjunct promoter, the resulting transcription complex may remain active until replication again allows reprogramming. Thus, the fraction of genes encoded by plasmid DNA that are "on" or "off" will depend on the relative amounts of repressor versus enhancer activation proteins present at the time of injection.

Recent evidence provides some support for this hypothesis. Complete inhibition of DNA synthesis in 1-cell embryos with aphidicolin (a specific inhibitor of replicative DNA polymerases) does not prevent early ZGA, but it does reduce the amount of transcription observed by ~1/3 (Aoki et al., 1997; Bouniol-Baly et al., 1997). It also inhibits transcription of the EIF-4C gene by ~1/2 and the synthesis of the "transcription-requiring complex" (TRC) of proteins by ~90% (Davis and Schultz, 1997). These results suggest that expression of some of the early ZGA genes is dependent on prior replication of some or all of the genome. The effect, however, is overestimated by measuring TRC synthesis, because inhibition of DNA replication in 1-cell embryos is accompanied by a general decrease in total protein biosynthesis that accompanies the decrease in TRC synthesis (Wiekowski et al., 1991). Nevertheless, the possibility that disruption of chromatin structure by DNA replication permits reprogramming of gene expression remains an attractive hypothesis.

4.4. TATA-box function

Once ZGA begins, the requirements for transcription can differ from those observed at later stages in development. In yeast, proteins binding to upstream activator sequences can facilitate binding of TBP to the TATA-box, thus accelerating transcription (Xiao et al., 1995). Moreover, the TATA-box element common to many RNA pol II promoters appears to be required only when a promoter is stimulated either by an upstream enhancer or by a transacting factor equivalent to herpes virus ICP4 or adenovirus E1A proteins (Majumder and DePamphilis, 1994). However, prior to cell differentiation, enhancer stimulation is observed in the absence of TATA-box cell differentiation (Majumder and DePamphilis, 1994; Davis and Schultz, 2000). This would imply that the TATA-less promoters that drive transcription of typical housekeeping genes can be stimulated selectively in cleavage stage embryos, but not in differentiated cells (Figs. 2, 3). In fact, oocytes are unique in that they express an E1A-like activity (Dooley et al., 1989) that may serve to stimulate genes in place of enhancer activity (Fig. 2).

4.5. DNA methylation

Methylation of specific DNA sites has been extensively related to genomic imprinting (monoallelic expression of specific genes) and to repression of selected genes during animal development (see chapter on "Genomic Imprinting"). In fact, three of the genes identified thus far that are expressed during ZGA are expressed only from the paternal allele (Table I). The bulk of DNA methylation in eukaryotes consists of 5-methyl cytosines within CpG dinucleotides. In mammals, three families of DNA methyltransferases (Dnmt 1, 2 and 3) have been identified (Bestor, 2000). Dnmt-1 is the enzyme responsible for maintaining DNA methylation patterns during DNA replication by converting hemimethylated CpG dinucleotides into fully methylated CpG dinucleotides. Dnmt-3 is involved in *de novo* methylation. The role of Dnmt-2 is not yet known. In addition to these enzymes, five proteins have been identified that bind to ^{m}CpGs (MeCP1 & 2, and MBD proteins 1 to 4; MBD-2 is the ^{m}CpG binding protein in the multiprotein complex, MeCP1). These proteins are associated with ^{m}CpGs and are believed to be involved in gene silencing (Hendrich et al., 2001).

DNA methylation is clearly required for proper embryonic development. For example, loss of Dnmt-1 results in embryonic stem (ES) cells that proliferate normally with their DNA highly demethylated but that die upon differentiation, with the consequence that embryos lacking this enzyme are delayed in development, and do not survive past mid-gestation (Li et al., 1992). Similarly, mice lacking MBD-3 die during early embryogenesis (Hendrich et al., 2001). The precise reasons for these effects, however, and the identity of the factors that control methylation pattern dynamics during gametogenesis and early development are largely unknown.

Global methylation patterns are erased in primordial germ cells and then reestablished in sex-specific patterns in mature male and female germ cells. There is an additional wave of demethylation during preimplantation development, although some sequences, notably certain imprinted genes, retain gametic methylation patterns at all stages (Brandeis et al., 1993; Reik et al., 2001). While primordial germ cells are highly methylated when compared to somatic cells, both male and female primordial germ cells undergo global demethylation that is completed by embryonic day 13 or 14. Both imprinted genes and single-copy genes are demethylated at this stage. Whether or not this demethylation step involves active or passive demethylation is unknown. However, when embryonic germ cells are fused with somatic cells, somatic nuclei undergo demethylation suggesting the existence of factor(s) in germs cells that confer demethylation in "trans" (Tada et al., 1997). In the male germ cells, remethylation takes place at the prospermatogonia stage (E15/16 onward), which precedes the mitosis/meiosis step. In contrast, remethylation of female primordial germ cells occurs during the growth of oocytes in young animals. This demethylation/remethylation steps are thought to be required in order to reset the imprinting cues and reprogram epigenetic modifications in germ cells.

Demethylation and remethylation also take place during embryogenesis (Fig. 2). The paternal genome is actively demethylated within the first 8 h after fertilization, prior to DNA replication (Mayer et al., 2000; Oswald et al., 2000). Although the nature of the demethylating mechanism is not yet clear (Wolffe et al., 1999), it would appear that chromatin remodeling of sperm DNA exposes it to enzymes that can demethylate it, while the chromatin structure of oocyte DNA protects it from demethylation. In contrast to the paternal genome, the maternal genome is passively demethylated by DNA replication in the absence of Dnmt-1 activity as cells undergo cleavage from the 2-cell to the 32-cell (morula) to blastula stages (Rougier et al., 1998). CpG methylation activity returns after implantation, and confers methylation patterns similar to what is observed in somatic cells.

DNA methylation patterns associated with imprinted genes appear to be shielded from demethylation during preimplantation development (Tremblay et al., 1997; Warnecke et al., 1998). Part of the mechanism that prevents loss of these methylation imprints involves a novel form of Dnmt1. Mouse oocytes and preimplantation embryos lack Dmnt1, but they express a variant called Dnmt1o. After fertilization, this extremely abundant form of Dnmt1 is relegated to the cytoplasm until the 8-cell stage when, for a single cell cycle, it enters the nucleus. Dnmt1o is then excluded from the nucleus in subsequent cell cycles until post-implantation day 8 (Carlson et al., 1992).

Although genomic methylation patterns are established normally in Dnmt1o-deficient oocytes, embryos derived from such oocytes show a loss of allele-specific expression and methylation at certain imprinted loci (Howell et al., 2001). Experiments with mice lacking functional Dnmt-1o suggest that the maintenance of imprinted methylation is critically dependent on the role of this enzyme during the 8-cell stage. Here the absence of Dnmt1o results in failure of the newly replicated paternal strand to become methylated. Subsequently, maintenance methylation occurs normally, but there is a 50% reduction of methylation on newly replicated paternal alleles. This heritable change results in some cells having either differentially methylated parental alleles or ones that are unmethylated. Thus, Dnmt1o is not required for the establishment of maternal genomic imprints, but instead appears to be required specifically for maintenance methylation of imprinted loci.

While CpG methylation has been directly linked to a number of biological phenomena including X-chromosome inactivation, imprinting, and silencing of repeat sequences and retroviruses, its direct role in regulating gene expression has yet to be demonstrated (Walsh and Bestor, 1999; Smith, 2000; Reik et al., 2001). Methylation of CpG dinucleotides is commonly correlated with a loss of gene expression both in vivo and in vitro (Eden and Cedar, 1994; Pikaart et al., 1998; Siegfried et al., 1999). Nevertheless, the absence of a change in the DNA methylation pattern of several tissue specific genes during early mouse development has led to the hypothesis that CpG methylation is a consequence of the absence of transcription rather than the cause, and thereby serves to insure that repressed genes remain silent. Methylation appears to be directly involved in preventing gene expression only in specialized cases such as imprinting, X-chromosome inactivation, and silencing transposable elements that might interfere with genome stability (Walsh and Bestor, 1999). In support of this hypothesis, recent studies on somatic cell cloning suggest that subtle epigenetic changes can be tolerated (Reik et al., 2001; Rideout et al., 2001). Not only were epigenetic markers unstable in embryonic stem cells, which still can give rise to "normal" animals, variations in imprinted gene expression were seen in mice generated from the same subclone of ES cells (Humpherys et al., 2001). These observations suggest that such epigenetic instability and resulting changes in gene expression do not have catastrophic effects on development.

Apart from imprinted genes, genes that are destined to be expressed during ZGA may be inherited from both parents in an unmethylated state. For example, the mouse TEAD-2 promoter region and gene is unmethylated in both sperm and oocytes and remains unmethylated throughout preimplantation development (K. Kaneko, unpublished data). Compare this with the *Sgy* gene located only 3.8 kb upstream (Kaneko and DePamphilis, 2000). *Sgy* is transcribed at basal level in some cells, but at full activity only in develop-ing spermatocytes and lymphocytes. The *Sgy* promoter region and gene is unmethylated in sperm but methylated in oocytes. Demethylation of the oocyte copy occurs by the 2-cell stage, and then both copies remain unmethylated except for a specific site within the second intron that is methylated in all cells (including gametes) except those that express *Sgy* at full activity. This site appears to be a cell-specific, methylation-sensitive, enhancer of *Sgy* gene activity (K. Kaneko, unpublished data).

What is the purpose of these demethylation/remethylation events in preimplantation embryos? If DNA methylation's primary function is to suppress expression of parasitic repeat sequences (Walsh and Bestor, 1999), then it is curious to note that oocytes and cleavage stage embryos contain an abundance of mRNA transcripts for B1 and B2 repeat elements (Alu repeats in humans) (Taylor and Piko, 1987), consistent with the notion that global demethylation would result in an increase in transcription of repeated sequences. Thus, oocytes and early embryos can tolerate expression of these potentially harmful sequences during the first 3–4 days of development. One general consequence of global demethylation is to remove one of the major obstacles to binding of proteins to DNA. With the exception of the five proteins known to bind specifically to ^{m}CpG dinucleotides, the affinity for DNA of all other DNA binding proteins appears to be reduced by methylation of their DNA binding sites. Thus, demethylation may facilitate remodeling of sperm chromatin into somatic cell chromatin in 1-cell embryos, and it may allow more subtle remodeling to take place in both paternal and maternal genomes during the preimplantation period.

5. Transcription factors selectively expressed at ZGA

Most promoters are active to some extent when injected into growing mouse oocytes, or fertilized mouse eggs or 2-cell embryos, and transcription factors such as Sp1 and TBP have been shown to be present physically and functionally throughout early development (Fig. 2). However, one would anticipate that ZGA will produce specific transcription factors that will, in turn, activate genes that are not essential to initiate fertilization, DNA replication, chromatin assembly, and transcription, but that are essential for cell proliferation and cell differentiation. Two such transcription factors have been identified so far.

5.1. TEAD-2/TEF-4

Mice and humans (as well as other mammals) contain a transcription factor family consisting of four highly conserved genes referred to originally as TEF (Transcription Enhancer Factor) but which have been redesignated by the mouse genome project (www.informatics.jax.org) as TEAD (TEA Domain) [(Jacquemin et al., 1996, 1999; Kaneko and DePamphilis, 1998) and references therein]. These genes express proteins of 426 to 445 amino acids whose sequences are 67% to 76% identical, and contain a 72 amino acid DNA binding domain (the "TEA" DNA binding domain) that is 94% to 100% identical. The TEA domain is found in proteins from mammals, birds, flies, fungi and yeast and recognizes a canonical M-CAT motif (5'-CATTCCT-3') found in promoters specific for transcription in muscle, as well as similar motifs found in the SV40 enhancer [GT-IIc (CATTCCA), Sph-I (CATGCTT), Sph-II (CATACTT)] or the PyV F9 enhancer (ACATTCCAG) (Jiang et al., 2000). All four TEAD proteins specifically bind the PyV F9 enhancer site with a K_d of 16–38 nM (Kaneko and DePamphilis, 1998). In addition, all four human or mouse TEAD proteins bind to the Sph-II site less tightly than to the PyV F9 site (Xiao et al., 1991; Kaneko and DePamphilis, 1998). Thus, all four TEAD proteins exhibit similar, if not identical, affinities for the TEAD DNA binding site.

Polyomavirus (PyV) host range mutants have been isolated that can replicate in undifferentiated mouse embryonal carcinoma (EC) cells or embryonic stem cells, and the enhancer in these PyV mutants contains at least one TEAD-1(TEF-1) binding site (Xiao et al., 1991). These PyV mutant enhancers, as well as one or more tandem copies of the SV40 GT-IIc site, can enhance promoter activity in mouse cleavage stage embryos, EC and ES cells up to 600-fold (Martinez-Salas et al., 1989; Melin et al., 1993). However, TEAD-1 (TEF-1) is not responsible for the observed enhancer promoter activity at this stage.

At least one TEAD protein is expressed in most adult tissues, and all four are abundantly expressed in some tissues, such as lung (Yockey et al., 1996; Kaneko et al., 1997). Nevertheless, the primary expression pattern for each member of this family differs significantly. TEAD-1(TEF-1) is required for mouse cardiac development by day 10 of embryo development (Chen et al., 1994) and for gene expression in cardiac muscle cells (Butler and Ordahl, 1999; Chen et al., 1994; Gupta et al., 1997; Ueyama et al., 2000). TEAD-4(TEF-3) appears to play a specific role in activating skeletal muscle genes (Jacquemin et al., 1996; Yockey et al., 1996). TEAD-3(TEF-5) is expressed primarily in the placenta (Jacquemin et al., 1998; Jiang et al., 1999) and in cardiac muscle (Azakie et al., 1996). In adult mice, TEAD-2(TEF-4) is expressed strongly in heart and lung tissues, the granulosa cells of the ovary, and weakly in several other tissues (Kaneko et al., 1997; Kaneko and DePamphilis, 2000). However, TEAD-2(TEF-4) is the only TEAD gene expressed in mouse embryos during the first seven days of development (Kaneko et al., 1997; Wang and Latham, 2000). Thus, TEAD-2(TEF-4) appears to play a unique role at the beginning of mammalian development, and is presumed to be responsible for TEAD(TEF)-dependent transcription in preimplantation embryos.

TEAD-specific transcription factor activity is not detected in mouse oocytes but is detected as soon as zygotic gene expression begins at the 2-cell stage in mouse development and then increases as embryos progress to the 4-cell stage [Fig. 2 (Kaneko et al., 1997)]. These changes in TEAD transcription factor activity are accompanied by changes in TEAD-2 RNA, but not by changes in the expression of other TEAD genes. While TEAD-1 and TEAD-3 RNAs are also detected in oocytes, they are 2–5 times less abundant than TEAD-2, and they are completely degraded after fertilization. In fact, embryos homozygous for a disruption in TEAD-1 survive until day 11 (Chen et al., 1994), consistent with the initial appearance of significant levels of TEAD-1 RNA in embryos at day eight (Kaneko et al., 1997). In contrast, about 5,000 copies of TEAD-2 mRNA are present in each oocyte, and this mRNA is recruited to polysomes only when development proceeds to the 2-cell stage (Wang and Latham, 2000), coincident with the initial appearance of TEAD transcription factor activity (Kaneko et al., 1997). TEAD-2 RNA is degraded in 2-cell and 4-cell embryos, but then increases 50-fold as preimplantation embryos progress from 4-cells to blastocysts (Kaneko et al., 1997), consistent with degradation of maternally inherited mRNA in 2-cell embryos and subsequent transcription of the zygotic TEAD-2 gene during preimplantation development. Thus, TEAD-2 is one of the first genes expressed at the beginning of mouse development where it presumably plays a role in activating transcription of other genes during preimplantation development. Protein kinase A, which facilitates ZGA, also stimulates transcription by TEAD-1 (Gupta et al., 2000), and therefore, by analogy, PKA may stimulate TEAD-2 activity during preimplantation development.

TEAD-1 appears to require one or more transcriptional coactivator proteins, because ectopic expression in non-TEAD-1 expressing cells does not elicit enhancer function from appropriate DNA sequences, and over-expression of TEAD-1 in TEAD-1 expressing cells results in down-regulation of enhancer function, consistent with titration of a coactivator activity ("squelching") (Hwang et al., 1993). Several candidates have been reported. TATA-box binding protein (TBP) alleviates squelching by TEAD-1 in some cells (Jiang and Eberhardt, 1996). TEAD-1 stimulation of cardiac muscle troponin T expression depends on flanking sequences binding poly(ADP-ribose) polymerase (Butler and Ordahl, 1999), a general coactivator of transcription (Meisterernst et al., 1997; Cervellera and Sala, 2000). MAX, a nuclear phosphoprotein that forms a heterodimer with MYC, binds to TEAD-1 to stimulate expression of the cardiac muscle α–myosin heavy-chain gene (Gupta et al., 1997). SRC1, a nuclear receptor coactivator protein, can interact with TEAD-2 and stimulate TEAD-dependent transcription (Belandia and Parker, 2000), but stimulation is modest. TONDU has been suggested to function as a general coactivator for TEAD, because it binds to all four of the human TEAD proteins, and it can substitute for *Vestigial* (Vg) in *Drosophila* (Vaudin et al., 1999). Vg is a transcriptional coactivator of *Scalloped* (Sd), a *Drosophila* transcription factor that contains the TEA DNA-binding domain. Therefore, TONDU may be a transcriptional coactivator of TEAD proteins in tissues such as lung, kidney, placenta and heart where TONDU is expressed, although no direct evidence for this has been published.

Perhaps the strongest candidate for a TEAD transcriptional coactivator is YAP65, a strong transcriptional coactivator that is expressed in 2-cell mouse embryos as well as in most proliferating, non-lymphocytic cells of the adult animal (Vassilev et al., 2001). The C-terminus of YAP65 contains an acidic transcriptional activation domain that can substitute for the herpes virus VP16 transcription activation domain both in yeast and in mammalian cells (Yagi et al., 1999). The N-terminus of YAP65 contains a novel 108 amino acid domain that interacts specifically with the C-terminal half of all four TEAD proteins, and both this interaction and sequence-specific DNA binding by TEAD proteins are required to elicit TEAD-dependent transcription in mouse cells (Vassilev et al., 2001). In fact, a chimeric protein containing the TEAD DNA binding domain fused to the YAP activation domain was sufficient to elicit TEAD-dependent transcription (A. Vassilev, unpublished data). Therefore, the C-terminal acidic activation domain in YAP is the transcriptional activation domain for all four TEAD transcription factors.

TEAD-dependent transcription in mouse fibroblasts is regulated by the availability of YAP65 (Fig. 4). TEAD proteins are localized in the nucleus, but YAP is localized in the cytoplasm. When YAP is overexpressed in mouse fibroblasts, most of it forms a complex with 14-3-3 proteins (Vassilev et al., 2001; and unpublished results). The 14-3-3 family of proteins binds a multitude of functionally diverse signaling proteins, and are involved in shuttling these proteins out of the nucleus and thereby localizing them in the cytoplasm (Fu et al., 2000). The 14-3-3 binding domain in YAP requires phosphorylation of a single serine residue (Kanai et al., 2000) that lies within the TEAD binding domain, suggesting that 14-3-3 and TEAD proteins compete for the same binding site on YAP protein. A smaller fraction of YAP is associated with the multi-PDZ domain protein, MUPP1 (Vassilev et al., 2001). PDZ domain proteins are involved in the assembly of protein signaling complexes on the membranes of synaptic junctions (Garner et al., 2000).

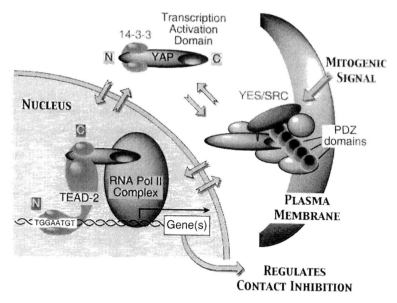

Fig. 4. Regulation of TEAD-dependent transcription in mammals depends on association of TEAD protein (localized in the nucleus) with YAP65, a transcriptional coactivator localized in the cytoplasm. See text for details. (*For a colored version of this figure, see plate section, page 272.*)

For example, YAP is concentrated at the apical membrane in human epithelial cells as a consequence of its interaction with PDZ domain proteins (Mohler et al., 1999). Thus, MUPP1 may sequester some YAP at the plasma membrane, while the remaining YAP is retained in the cytoplasm by association with 14-3-3 proteins.

Given the fact that YAP is the rate limiting factor for TEAD-dependent transcription (Vassilev et al., 2001), and the fact that YAP binds to the SRC/YES protein tyrosine kinase family via its SH3 binding domain (Sudol, 1994), we suggest that TEAD-dependent transcription is regulated by the availability of YAP65, and the availability of YAP65 is regulated by the SRC/YES signal transduction pathway (Fig. 4) (Vassilev et al., 2001). One way this might occur is if mitogenic signals are relayed via the SRC/YES family of proteins to release YAP from the plasma membrane and allow it to translocate into the nucleus where it activates TEAD-dependent transcription. Conversely, TEAD-dependent transcription can be terminated by releasing YAP to translocate back to the cytoplasm where it is sequestered by 14-3-3 proteins. The putative YAP/MUPP1 plasma membrane site could be reloaded from the YAP/14-3-3 pool. Obviously, movement of YAP in and out of the nucleus could be regulated in other ways as well.

Recent unpublished data (A. Vassilev and M. DePamphilis) suggests that the function of TEAD-2:YAP65 dependent transcription in mouse cells is to regulate contact inhibition of cell proliferation. Manipulations that increased TEAD-2:YAP65-dependent transcription did not affect either the rate of cell proliferation or the distribution of cells throughout their cell cycle, but did allow cells to form foci when confluent and to proliferate in soft agar. Conversely, decreasing TEAD-2:YAP65-dependent transcription reduced the density of cells at confluence and did not allow growth in soft agar. Moreover, these changes correlated with changes in the expression of several proteins known to be involved in cell adhesion. Genetic analyses suggested that TEAD-dependent transcription of specific cell adhesion proteins is part of the pathway for transformation of fibroblasts into cancer cells.

5.2. OCT-4

The blastocyst stage consists of three tissues. A hollow epithelial sphere (the tro-phoblast) that will eventually form the placenta. The trophoblast encloses the inner cell mass (ICM) that will form the embryo proper, and the layer of cells lying on the ICM is the primitive endoderm that will contribute to the fetal membranes and regulate ICM dif-ferentiation. OCT-4 (also called OCT-3) is pivotal in regulating formation of all three tis-sues. OCT-4 is a member of the POU-domain transcription-factor family which binds to an 8 bp DNA sequence found in the promoters and enhancer regions of many genes. In the mouse, OCT-4 is first detected in the oocyte and, like other maternally inherited genes, its expression declines during the first two cleavage divisions, but reappears at the 4 to 8-cell stage where it is expressed in all nuclei (Fig. 2). By the blastocyst stage, OCT-4 becomes restricted to the inner cell mass. However, in pigs and cows, OCT-4 is detected in both the ICM and the trophectoderm (Kirchhof et al., 2000).

Relative levels of OCT-4 influence ES cell fate, consistent with its expression in the preimplantation embryo. OCT-4 is first detected at the 4-cell stage. In late blastocysts, OCT-4 is not expressed in the trophectoderm, but is found in the inner cell mass and, at higher levels, in the primitive ectoderm. In post-implantation embryos, OCT-4 disappears as cells undergo differentiation, with expression persisting in the germ cells (Palmieri et al., 1994).

OCT-4 expression is restricted to pluripotent and totipotent cells. Embryos in which the OCT-4 gene has been inactivated die following implantation, because they produce trophoblast cells but not the ICM (Nichols et al., 1998). OCT-4 levels regulate the differ-entiation of ES cells (totipotent cells derived from the ICM) into three different cell types that correspond to those found in embryos (Niwa et al., 2000). Starting with an ES clone heterozygous for functional OCT-4 gene, the authors introduced an inducible OCT-4 trans-gene. Upon induction of the additional copy, OCT-4 levels increased by twofold compared with those of ordinary ES cells. This triggered their differentiation into cells with the mor-phology and gene-expression patterns characteristic of extra-embryonic endoderm and mesoderm. Using ES cells with an inducible OCT-4 transgene, clones were derived that totally lacked OCT-4 expression. Stem-cell characteristics were retained due to the con-tinued expression of the OCT-4 transgene. Withdrawal of the inducing agent inhibited transgene expression and, now completely lacking OCT-4, the ES cells differentiated into cells both resembling and expressing genes typical of trophectoderm. Therefore, OCT-4 is not simply a repressor of trophectoderm formation in the embryo. Normal levels of OCT-4 are compatible with the formation of ES cells (ICM), while lack of OCT-4 expres-sion drives the formation of trophectoderm, and elevated levels result in extra-embryonic endoderm and mesoderm (Fig. 5). Taken together, the results described above reveal that OCT-4 regulates tissue formation in preimplantation embryos.

OCT-4 does not function alone. Activation of transcription by OCT-4 from remote binding sites requires a cofactor that is restricted to embryonal stem cells. The adenovirus E1A protein can mimic the activity of this stem cell-specific factor and stimulates OCT-4

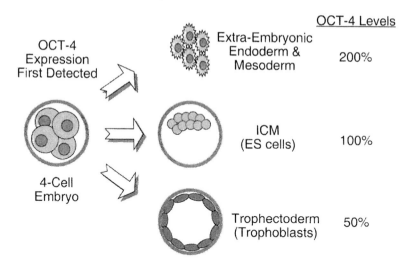

Fig. 5. Regulation of mouse ES cell development depends on the level of OCT-4 expression. Normal pre-implan-tation levels stabilize totipotent ES cells. Higher levels cause differentiation into endoderm and mesoderm, while lower levels induce differentiation into trophectoderm. See text for details. (*For a colored version of this figure, see plate section, page 272.*)

activity in differentiated cells (Brehm et al., 1999). Fibroblast growth factor (FGF)-4 gene expression in the inner cell mass of the blastocyst, which helps sustain cell proliferation, requires the combined activity of both OCT-4 and Sox-2. Sox-2 is a transcription factor that contains an HMG box and is coexpressed with OCT-4 in the early mouse embryo. Cooperative binding of these proteins to the enhancer DNA, mediated by their binding domains, stably tethers each factor to DNA and increases the activity of intrinsic activa-tion domains within each protein (Ambrosetti et al., 2000). Sox-2 also represses OCT-4 mediated activation of osteopontin by way of a canonical Sox element that is located close to the OCT-4 DNA binding site. Osteopontin is a protein secreted by cells of the pre-implantation embryo and that can bind to specific integrin subtypes and modulate cell adhesion/migration. Repression depends on a carboxyl-terminal region of Sox-2 that is outside of the HMG box (Botquin et al., 1998). It appears that the ratios between OCT-4 protein and the other transcription factors ultimately determine cell fate.

5.3. Regulation of TEAD-2 and OCT-4 expression

So far, regulatory regions for only two of the four TEAD genes have been analyzed. Interestingly, the promoters for both the human TEAD-1 and mouse TEAD-2 lack a conventional TATA-box, and exhibit multiple transcription start sites (Boam et al., 1995; Suzuki et al., 1996). The hTEAD-1 promoter (−365 to −30) is sufficient to direct tissue-specific expression in vivo, and contains multiple DNA binding sites for transcription factors Sp1 and ATF-1. Mutations in either the Inr element or the proximal Sp1 site abolish transcription from the principle start site.

The 5'-flanking region of mTEAD-2 contains many potential binding sites for transcription factors, such as Sp1, GATA binding factors, steroid hormone receptors, octamer binding factors, and other transcription factors involved in myogenesis and hematopoeisis (such as those that bind the E-box); but so far none of these factors have been shown to be directly involved in regulating this gene. Interestingly, CpG's beyond ~600 bp upstream of potential start sites are methylated, suggesting that potential regulatory sequences exist within this short promoter region or in regions downstream of the start site (K. Kaneko, T. Rein, M.L. DePamphilis, unpublished results). Several B1/B2 short interspersed element (SINE) as well as other direct and inverted repeat sequences exist further upstream, suggesting that methylation of these sequences might be required in order to prevent interference with TEAD-2 expression.

Similarly, the promoter region of OCT-4 gene also involves a minimal TATA-less promoter and contains many similar binding sites, such as Sp1, hormone response elements and E-box elements (Nordhoff et al., 2001; Yeom et al., 1996). In addition, two enhancer regions have been identified that specifically activate OCT-4 expression in morula, ICM, PGCs and epiblasts (Yeom et al., 1996). In contrast to mTEAD-2, however, OCT-4 expression is strictly limited to embryonic and germ cell types and is down-regulated in somatic cells. Therefore, while TEAD-2 and OCT-4 may share common regulatory region that allow their expression in preimplantation embryos, much of the enhancer sequences will be expected to be different.

It is worth noting that Sp1-dependent transcription, as well as Sp1 mRNA and protein, occurs in mouse oocytes and preimplantation embryos where it increases several fold during the first three cell divisions (Chalifour et al., 1987; Majumder et al., 1993; Worrad and Schultz, 1997; Lawinger et al., 1999;). Moreover, transcriptional activation of the Hsp70.1 gene involves Sp1 and a novel murine GAGA box-binding factor. In addition, mouse unfertilized eggs, 1-cell and 2-cell embryos display a GAGA box-binding activity of maternal origin that disappears at the 4-cell stage (Bevilacqua et al., 2000). Thus, either Sp1 or another member of the Sp1 family (Marin et al., 1997) together with a nucleosome disruption factor such as GAGA may activate transcription of specific chromatin-repressed genes during ZGA.

Further investigation of the regulatory region of the mTEAD-2 gene in our lab has led to the surprising discovery of another gene only 3.8 kb upstream of mTEAD-2 (Kaneko and DePamphilis, 2000). This new gene is a single copy, spermatocyte/ T-lymphocyte specific gene called *Soggy* (mSgy) that is transcribed in the direction opposite to mTEAD-2, thus placing the regulatory elements of these two genes in close proximity. Minimal promoter for Sgy is contained within 400 bp upstream of the start sites and like mTEAD-2 and Oct-4, consists of a TATA-less promoter.. mSgy mRNA is not maternally inherited but small amount of expression (~15% the level of mTEAD-2) is detected in late preimplantation embryos (K. Kaneko, K. Latham and M. DePamphilis, unpublished results). However, unlike mTEAD-2 , extensive Sgy expression is detected only after day 15, and in adult tissues only in the developing spermatocytes of seminiferous tubules and T-cell lymphocytes. Remarkably, little, if any, expression of mTEAD-2 is detected in spermatocytes or in lymphocytes. Moreover, in all cell lines tested to date, expression of these two genes appears to be mutually exclusive. Since mTEAD-2 and mSgy do not appear to be expressed in the same cells, the mSgy/mTEAD-2 locus provides a unique paradigm for differential regulation of gene expression during mammalian development.

References

Adenot, P.G., Campion, E., Legouy, E., Allis, C.D., Dimitrov, S., Renard, J., Thompson, E.M. 2000. Somatic linker histone H1 is present throughout mouse embryogenesis and is not replaced by variant H1 degrees. J. Cell. Sci. 113, 2897–2907.

Adenot, P.G., Mercier, Y., Renard, J.P., Thompson, E.M. 1997. Differential H4 acetylation of paternal and maternal chromatin precedes DNA replication and differential transcriptional activity in pronuclei of one-cell mouse embryos. Development 124, 4615–4625.

Almouzni, G., Khochbin, S., Dimitrov, S., Wolffe, A.P. 1994. Histone acetylation influences both gene expression and development of Xenopus laevis. Dev. Biol. 165, 654–669.

Ambrosetti, D.C., Scholer, H.R., Dailey, L., Basilico, C. 2000. Modulation of the activity of multiple transcriptional activation domains by the DNA binding domains mediates the synergistic action of Sox2 and Oct-3 on the fibroblast growth factor-4 enhancer. J. Biol. Chem. 275, 23387–23397.

Aoki, E., Schultz, R.M. 1999. DNA replication in the one-cell mouse embryo: stimulatory effect of histone acetylation. Zygote 7, 165–172.

Aoki, F., Worrad, D.M., Schultz, R.M. 1997. Regulation of transcriptional activity during the first and second cell cycles in the preimplantation mouse embryo. Dev. Biol. 181, 296–307.

Azakie, A., Larkin, S.B., Farrance, I.K., Grenningloh, G., Ordahl, C.P. 1996. DTEF-1, a novel member of the transcription enhancer factor-1 (TEF-1) multigene family. J. Biol. Chem. 271, 8260–8265.

Baran, V., Brochard, V., Renard, J.P., Flechon, J.E. 2001. Nopp 140 involvement in nucleologenesis of mouse preimplantation embryos. Mol. Reprod. Dev. 59, 277–284.

Beaujean, N., Bouniol-Baly, C., Monod, C., Kissa, K., Jullien, D., Aulner, N., Amirand, C., Debey, P., Kas, E. 2000. Induction of early transcription in one-cell mouse embryos by microinjection of the nonhistone chromosomal protein HMG-I. Dev. Biol. 221, 337–354.

Belandia, B., Parker, M.G. 2000. Functional interaction between the p160 coactivator proteins and the transcriptional enhancer factor family of transcription factors. J. Biol. Chem. 275, 30801–30805.

Bellier, S., Chastant, S., Adenot, P., Vincent, M., Renard, J.P., Bensaude, O. 1997. Nuclear translocation and carboxyl-terminal domain phosphorylation of RNA polymerase II delineate the two phases of zygotic gene activation in mammalian embryos. Embo. J. 16, 6250–6262.

Bestor, T.H. 2000. The DNA methyltransferases of mammals. Hum. Mol. Genet. 9, 2395–2402.

Bevilacqua, A., Fiorenza, M.T., Mangia, F. 1997. Developmental activation of an episomic hsp70 gene promoter in two-cell mouse embryos by transcription factor Sp1. Nucleic Acids Res. 25, 1333–1338.

Bevilacqua, A., Fiorenza, M.T., Mangia, F. 2000. A developmentally regulated GAGA box-binding factor and Sp1 are required for transcription of the hsp70.1 gene at the onset of mouse zygotic genome activation. Development 127, 1541–1551.

Bevilacqua, A., Kinnunen, L.H., Bevilacqua, S., Mangia, F. 1995. Stage-specific regulation of murine Hsp68 gene promoter in preimplantation mouse embryos. Dev. Biol. 170, 467–478.

Boam, D.S., Davidson, I., Chambon, P. 1995. A TATA-less promoter containing binding sites for ubiquitous transcription factors mediates cell type-specific regulation of the gene for transcription enhancer factor-1 (TEF-1), J. Biol. Chem. 270, 19487–19494.

Botquin, V., Hess, H., Fuhrmann, G., Anastassiadis, C., Gross, M.K., Vriend, G., Scholer, H.R. 1998. New POU dimer configuration mediates antagonistic control of an osteopontin preimplantation enhancer by Oct-4 and Sox-2. Genes Dev. 12, 2073–2090.

Bouniol, C., Nguyen, E., Debey, P. 1995. Endogenous transcription occurs at the one-cell stage in the mouse embryo. Exp. Cell. Res. 218, 57–62.

Bouniol-Baly, C., Nguyen, E., Besombes, D., Debey, P. 1997. Dynamic organization of DNA replication in one-cell mouse embryos: relationship to transcriptional activation. Exp. Cell. Res. 236, 201–211.

Bouvet, P., Dimitrov, S., Wolffe, A.P. 1994. Specific regulation of Xenopus chromosomal 5S rRNA gene transcription in vivo by histone H1. Genes Dev. 8, 1147–1159.

Brandeis, M., Ariel, M., Cedar, H. 1993. Dynamics of DNA methylation during development. BioEssays 15, 709–713.

Brehm, A., Ohbo, K., Zwerschke, W., Botquin, V., Jansen-Durr, P., Scholer, H.R. 1999. Synergism with germ line transcription factor Oct-4: viral oncoproteins share the ability to mimic a stem cell-specific activity. Mol. Cell. Biol. 19, 2635–2643.

Brunet-Simon, A., Henrion, G., Renard, J.P., Duranthon, V. 2001. Onset of zygotic transcription and maternal transcript legacy in the rabbit embryo. Mol. Reprod. Dev. 58, 127–136.

Butler, A.J., Ordahl, C.P. 1999. Poly(ADP-ribose) polymerase binds with transcription enhancer factor 1 to MCAT1 elements to regulate muscle-specific transcription. Mol. Cell. Biol. 19, 296–306.

Carlson, L.L., Page, A.W., Bestor, T.H. 1992. Properties and localization of DNA methyltransferase in preimplantation mouse embryos: implications for genomic imprinting. Genes Dev. 6, 2536–2541.

Cervellera, M.N., Sala, A. 2000. Poly(ADP-ribose) polymerase is a B-MYB coactivator. J. Biol. Chem. 275, 10692–10696.

Chalifour, L.E., Wirak, D.O., Hansen, U., Wassarman, P.M., DePamphilis, M.L. 1987. cis- and trans-acting sequences required for expression of simian virus 40 genes in mouse oocytes. Genes Dev. 1, 1096–1106.

Chen, Z., Friedrich, G.A., Soriano, P. 1994. Transcriptional enhancer factor 1 disruption by a retroviral gene trap leads to heart defects and embryonic lethality in mice. Genes Dev. 8, 2293–2301.

Choo, K.B., Chen, H.H., Cheng, W.T., Chang, H.S., Wang, M. 2001. In silico mining of EST databases for novel pre-implantation embryo-specific zinc finger protein genes. Mol. Reprod. Dev. 59, 249–255.

Christians, E., Campion, E., Thompson, E.M., Renard, J.P. 1995. Expression of the HSP 70.1 gene, a landmark of early zygotic activity in the mouse embryo, is restricted to the first burst of transcription. Development 121, 113–122.

Christians, E., Michel, E., Adenot, P., Mezger, V., Rallu, M., Morange, M., Renard, J.P. 1997. Evidence for the involvement of mouse heat shock factor 1 in the atypical expression of the HSP70.1 heat shock gene during mouse zygotic genome activation. Mol. Cell. Biol. 17, 778–788.

Christians, E., Rao, V.H., Renard, J.P. 1994. Sequential acquisition of transcriptional control during early embryonic development in the rabbit. Dev. Biol. 164, 160–172.

Clarke, H.J., Bustin, M., Oblin, C. 1997. Chromatin modifications during oogenesis in the mouse: removal of somatic subtypes of histone H1 from oocyte chromatin occurs post-natally through a post-transcriptional mechanism. J. Cell. Sci. 110, 477–487.

Davis, W., Jr., De Sousa, P.A., Schultz, R.M. 1996. Transient expression of translation initiation factor EIF-4C during the 2-cell stage of the preimplantation mouse embryo: identification by mRNA differential display and the role of DNA replication in zygotic gene activation. Dev. Biol. 174, 190–201.

Davis, W., Jr., Schultz, R.M. 1997. Role of the first round of DNA replication in reprogramming gene expression in the preimplantation mouse embryo, Mol. Reprod. Dev. 47, 430–434.

Davis, W., Jr., Schultz, R.M. 2000. Developmental change in TATA-Box utilization during preimplantation mouse development. Dev. Biol. 218, 275–283.

Dean, W.L., Seufert, A.C., Schultz, G.A., Prather, R.S., Simerly, C., Schatten, G., Pilch, D.R., Marzluff, W.F. 1989. The small nuclear RNAs for pre-mRNA splicing are coordinately regulated during oocyte maturation and early embryogenesis in the mouse. Development 106, 325–334.

DePamphilis, M.L., Martinez-Salas, E., Cupo, D., Hendrickson, E.A., Fritze, C.E., Folk, W.R., Heine, U. 1988. Initiation of Polyomavirus and SV40 DNA Replication, and the Requirements for DNA Replication During Mammalian Development. In: *Eukaryotic DNA Replication*, (B. Stillman, T. Kelly, Eds.), Cold Spring Harbor, NY, Cold Spring Harbor Laboratory Press, pp. 165–175.

Dimitrov, S., Almouzni, G., Dasso, M., Wolffe, A.P. 1993. Chromatin transitions during early Xenopus embryogenesis: changes in histone H4 acetylation and in linker histone type. Dev. Biol. 160, 214–227.

Dooley, T.P., Miranda, M., Jones, N.C., DePamphilis, M.L. 1989. Transactivation of the adenovirus EIIa promoter in the absence of adenovirus E1A protein is restricted to mouse oocytes and preimplantation embryos. Development 107, 945–956.

Eden, S., Cedar, H. 1994. Role of DNA methylation in the regulation of transcription. Curr. Opin. Genet. Dev. 4, 255–259.

Ferreira, J., Carmo-Fonseca, M. 1997. Genome replication in early mouse embryos follows a defined temporal and spatial order. J. Cell. Sci. 110, 889–897.

Flechon, J.E., Kopecny, V. 1998. The nature of the 'nucleolus precursor body' in early preimplantation embryos: a review of fine-structure cytochemical, immunocytochemical and autoradiographic data related to nucleolar function. Zygote 6, 183–191.

Fu, H., Subramanian, R.R., Masters, S.C. 2000. 14-3-3 proteins: structure, function, and regulation. Annu. Rev. Pharmacol. Toxicol. 40, 617–647.

Fuchimoto, D., Mizukoshi, A., Schultz, R.M., Sakai, S., Aoki, F. 2001. Post-transcriptional regulation of cyclin A1 and cyclin A2 during mouse oocyte meiotic maturation and preimplantation development. Biol. Reprod. 65, 986–993.

Garner, C.C., Nash, J., Huganir, R.L. 2000. PDZ domains in synapse assembly and signalling. Trends Cell Biol. 10, 274–280.

Gupta, M.P., Amin, C.S., Gupta, M., Hay, N., Zak, R. 1997. Transcription enhancer factor 1 interacts with a basic helix-loop-helix zipper protein, max, for positive regulation of cardiac alpha-myosin heavy-chain gene expression. Mol. Cell. Biol. 17, 3924–3936.

Hatada, I., Kitagawa, K., Yamaoka, T., Wang, X., Arai, Y., Hashido, K., Ohishi, S., Masuda, J., Ogata, J., Mukai, T. 1995. Allele-specific methylation and expression of an imprinted U2af1-rs1 (SP2) gene. Nucleic Acids Res. 23, 36–41.

Hendrich, B., Guy, J., Ramsahoye, B., Wilson, V.A., Bird, A. 2001. Closely related proteins MBD2 and MBD3 play distinctive but interacting roles in mouse development. Genes Dev. 15, 710–723.

Henery, C.C., Miranda, M., Wiekowski, M., Wilmut, I., DePamphilis, M.L. 1995. Repression of gene expression at the beginning of mouse development. Dev. Biol. 169, 448–460.

Howell, C.Y., Bestor, T.H., Ding, F., Latham, K.E., Mertineit, C., Trasler, J.M., Chaillet, J.R. 2001. Genomic imprinting disrupted by a maternal effect mutation in the Dnmt1 gene. Cell 104, 829–838.

Humpherys, D., Eggan, K., Akutsu, H., Hochedlinger, K., Rideout, W.M., III, Biniszkiewicz, D., Yanagimachi, R., Jaenisch, R. 2001. Epigenetic instability in ES cells and cloned mice. Science 293, 95–97.

Hwang, J.J., Chambon, P., Davidson, I. 1993. Characterization of the transcription activation function and the DNA binding domain of transcriptional enhancer factor-1. Embo. J. 12, 2337–2348.

Hyttel, P., Laurincik, J., Viuff, D., Fair, T., Zakhartchenko, V., Rosenkranz, C., Avery, B., Rath, D., Niemann, H., Thomsen, P.D., et al. 2000. Activation of ribosomal RNA genes in preimplantation cattle and swine embryos. Anim. Reprod. Sci. 60–61, 49–60.

Jacquemin, P., Depetris, D., Mattei, M.G., Martial, J.A., Davidson, I. 1999. Localization of human transcription factor TEF-4 and TEF-5 (TEAD2, TEAD3) genes to chromosomes 19q13.3 and 6p21.2 using fluorescence in situ hybridization and radiation hybrid analysis. Genomics 55, 127–129.

Jacquemin, P., Hwang, J.J., Martial, J.A., Dolle, P., Davidson, I. 1996. A novel family of developmentally regulated mammalian transcription factors containing the TEA/ATTS DNA binding domain. J. Biol. Chem. 271, 21775–21785.

Jacquemin, P., Sapin, V., Alsat, E., Evain-Brion, D., Dolle, P., Davidson, I. 1998. Differential expression of the TEF family of transcription factors in the murine placenta and during differentiation of primary human trophoblasts in vitro. Dev. Dyn. 212, 423–436.

Jiang, S.W., Desai, D., Khan, S., Eberhardt, N.L. 2000. Cooperative binding of TEF-1 to repeated GGAATG-related consensus elements with restricted spatial separation and orientation, DNA Cell Biol. 19, 507–14.

Jiang, S.W., Eberhardt, N.L. 1996. TEF-1 transrepression in BeWo cells is mediated through interactions with the TATA-binding protein, TBP. J. Biol. Chem. 271, 9510–9518.

Jiang, S.W., Wu, K., Eberhardt, N.L. 1999. Human placental TEF-5 transactivates the human chorionic somatomammotropin gene enhancer. Mol. Endocrinol. 13, 879–889.

Johnson, M.H., Maro, B. 1986. Time and space in the mouse early embryo: a cell biological approach to cell diversification. In: Experimental Approaches to Mammalian Embryonic Development. (R.J. a. P.R., Eds.), Cambridge University Press, pp. 35–65.

Jumaa, H., Wei, G., Nielsen, P.J. 1999. Blastocyst formation is blocked in mouse embryos lacking the splicing factor SRp20. Curr. Biol. 9, 899–902.

Kanai, F., Marignani, P.A., Sarbassova, D., Yagi, R., Hall, R.A., Donowitz, M., Hisaminato, A., Fujiwara, T., Ito, Y., Cantley, L.C., Yaffe, M.B. 2000. TAZ: a novel transcriptional coactivator regulated by interactions with 14-3-3 and PDZ domain proteins. Embo. J. 19, 6778–6791.

Kandolf, H. 1994. The H1A histone variant is an in vivo repressor of oocyte-type 5S gene transcription in Xenopus laevis embryos. Proc. Natl. Acad. Sci. USA 91, 7257–7261.

Kaneko, K.J., Cullinan, E.B., Latham, K.E., DePamphilis, M.L. 1997. Transcription factor mTEAD-2 is selectively expressed at the beginning of zygotic gene expression in the mouse. Development 124, 1963–1973.

Kaneko, K.J., DePamphilis, M.L. 1998. Regulation of gene expression at the beginning of mammalian development and the TEAD family of transcription factors. Dev. Genet. 22, 43–55.

Kaneko, K.J., DePamphilis, M.L. 2000. Soggy, a spermatocyte-specific gene, lies 3.8 kb upstream of and anti-podal to TEAD-2, a transcription factor expressed at the beginning of mouse development. Nucleic Acids Res. 28, 3982–3990.

Kawamura, K., Fukuda, J., Kodama, H., Kumagai, J., Kumagai, A., Tanaka, T. 2001. Expression of Fas and Fas ligand mRNA in rat and human preimplantation embryos. Mol. Hum. Reprod. 7, 431–436.

Kirchhof, N., Carnwath, J.W., Lemme, E., Anastassiadis, K., Scholer, H., Niemann, H. 2000. Expression pattern of Oct-4 in preimplantation embryos of different species. Biol. Reprod. 63, 1698–1705.

Ko, M.S., Kitchen, J.R., Wang, X., Threat, T.A., Hasegawa, A., Sun, T., Grahovac, M.J., Kargul, G.J., Lim, M.K., Cui, Y., et al. 2000. Large-scale cDNA analysis reveals phased gene expression patterns during preimplanta-tion mouse development. Development 127, 1737–1749.

Latham, K.E. 1999. Mechanisms and control of embryonic genome activation in mammalian embryos. Int. Rev. Cytol. 193, 71–124.

Latham, K.E., Garrels, J.I., Chang, C., Solter, D. 1991. Quantitative analysis of protein synthesis in mouse embryos. I. Extensive reprogramming at the one- and two-cell stages. Development 112, 921–932.

Latham, K.E., Rambhatla, L., Hayashizaki, Y., Chapman, V.M. 1995. Stage-specific induction and regulation by genomic imprinting of the mouse U2afbp-rs gene during preimplantation development. Dev. Biol. 168, 670–676.

Latham, K.E., Schultz, R.M. 2001. Embryonic genome activation. Front Biosci. 6, D748–759.

Latham, K.E., Solter, D., Schultz, R.M. 1992. Acquisition of a transcriptionally permissive state during the one-cell stage of mouse embryogenesis. Dev. Biol. 149, 457–462.

Lawinger, P., Rastelli, L., Zhao, Z., Majumder, S. 1999. Lack of enhancer function in mammals is unique to oocytes and fertilized eggs. J. Biol. Chem. 274, 8002–8011.

Li, E., Bestor, T.H., Jaenisch, R. 1992. Targeted mutation of the DNA methyltransferase gene results in embry-onic lethality. Cell 69, 915–926.

Lin, P., Clarke, H.J. 1996. Somatic histone H1 microinjected into fertilized mouse eggs is transported into the pronuclei but does not disrupt subsequent preimplantation development. Mol. Reprod. Dev. 44, 185–192.

Majumder, S., DePamphilis, M.L. 1994. TATA-dependent enhancer stimulation of promoter activity in mice is developmentally acquired. Mol. Cell. Biol. 14, 4258–4268.

Majumder, S., DePamphilis, M.L. 1995. A unique role for enhancers is revealed during early mouse development. BioEssays 17, 879–889.

Majumder, S., Miranda, M., DePamphilis, M.L. 1993. Analysis of gene expression in mouse preimplantation embryos demonstrates that the primary role of enhancers is to relieve repression of promoters [published erra-tum appears in EMBO J. 1993 Oct;12(10):4042]. Embo. J. 12, 1131–1140.

Majumder, S., Zhao, Z., Kaneko, K., DePamphilis, M.L. 1997. Developmental acquisition of enhancer function requires a unique coactivator activity. Embo. J. 16, 1721–1731.

Malik, S., Roeder, R.G. 2000. Transcriptional regulation through Mediator-like coactivators in yeast and metazoan cells. Trends Biochem. Sci. 25, 277–283.

Marin, M., Karis, A., Visser, P., Grosveld, F., Philipsen, S. 1997. Transcription factor Sp1 is essential for early embryonic development but dispensable for cell growth and differentiation. Cell 89, 619–628.

Martinez-Salas, E., Linney, E., Hassell, J., DePamphilis, M.L. 1989. The need for enhancers in gene expression first appears during mouse development with formation of a zygotic nucleus. Genes & Development 3, 1493–1506.

Masui, Y., Wang, P. 1998. Cell cycle transition in early embryonic development of Xenopus laevis. Biol. Cell. 90, 537–548.

Matsumoto, K., Anzai, M., Nakagata, N., Takahashi, A., Takahashi, Y., Miyata, K. 1994. Onset of paternal gene activation in early mouse embryos fertilized with transgenic mouse sperm. Mol. Reprod. Dev. 39, 136–140.

Mayer, W., Niveleau, A., Walter, J., Fundele, R., Haaf, T. 2000. Demethylation of the zygotic paternal genome. Nature 403, 501–502.

McLay, D.W., Clarke, H.J. 1997. The ability to organize sperm DNA into functional chromatin is acquired during meiotic maturation in murine oocytes. Dev. Biol. 186, 73–84.

Meisterernst, M., Stelzer, G., Roeder, R.G. 1997. Poly(ADP-ribose) polymerase enhances activator-dependent transcription in vitro. Proc. Natl. Acad. Sci. USA 94, 2261–2265.

Melin, F., Miranda, M., Montreau, N., DePamphilis, M.L., Blangy, D. 1993. Transcription enhancer factor-1 (TEF-1) DNA binding sites can specifically enhance gene expression at the beginning of mouse development. Embo. J. 12, 4657–4666.

Memili, E., First, N.L. 1999. Control of gene expression at the onset of bovine embryonic development. Biol. Reprod. 61, 1198–1207.

Mohamed, O.A., Bustin, M., Clarke, H.J. 2001. High-mobility group proteins 14 and 17 maintain the timing of early embryonic development in the mouse. Dev. Biol. 229, 237–249.

Mohler, P.J., Kreda, S.M., Boucher, R.C., Sudol, M., Stutts, M.J., Milgram, S.L. 1999. Yes-associated protein 65 localizes p62(c-Yes) to the apical compartment of airway epithelia by association with EBP50. J. Cell. Biol. 147, 879–890.

Nichols, J., Zevnik, B., Anastassiadis, K., Niwa, H., Klewe-Nebenius, D., Chambers, I., Scholer, H., Smith, A. 1998. Formation of pluripotent stem cells in the mammalian embryo depends on the POU transcription factor Oct4. Cell 95, 379–391.

Niwa, H., Miyazaki, J., Smith, A.G. 2000. Quantitative expression of Oct-3/4 defines differentiation, dedifferentiation or self-renewal of ES cells [see comments]. Nat. Genet. 24, 372–376.

Nordhoff, V., Hubner, K., Bauer, A., Orlova, I., Malapetsa, A., Scholer, H.R. 2001. Comparative analysis of human, bovine, and murine Oct-4 upstream promoter sequences. Mamm. Genome. 12, 309–317.

Nothias, J.Y., Majumder, S., Kaneko, K.J., DePamphilis, M.L. 1995. Regulation of gene expression at the beginning of mammalian development. J. Biol. Chem. 270, 22077–22080.

Nothias, J.Y., Miranda, M., DePamphilis, M.L. 1996. Uncoupling of transcription and translation during zygotic gene activation in the mouse. Embo. J. 15, 5715–5725.

Oh, B., Hwang, S., McLaughlin, J., Solter, D., Knowles, B.B. 2000. Timely translation during the mouse oocyte-to-embryo transition. Development 127, 3795–3803.

Oswald, J., Engemann, S., Lane, N., Mayer, W., Olek, A., Fundele, R., Dean, W., Reik, W., Walter, J. 2000. Active demethylation of the paternal genome in the mouse zygote. Curr. Biol. 10, 475–478.

Palancade, B., Bellier, S., Almouzni, G., Bensaude, O. 2001. Incomplete RNA polymerase II phosphorylation in Xenopus laevis early embryos. J. Cell. Sci. 114, 2483–2489.

Palmieri, S.L., Peter, W., Hess, H., Scholer, H.R. 1994. Oct-4 transcription factor is differentially expressed in the mouse embryo during establishment of the first two extraembryonic cell lineages involved in implantation. Dev. Biol. 166, 259–267.

Paranjape, S.M., Kamakaka, R.T., Kadonaga, J.T. 1994. Role of chromatin structure in the regulation of transcription by RNA polymerase II. Annu. Rev. Biochem. 63, 265–297.

Pikaart, M.J., Recillas-Targa, F., Felsenfeld, G. 1998. Loss of transcriptional activity of a transgene is accompanied by DNA methylation and histone deacetylation and is prevented by insulators. Genes Dev. 12, 2852–2862.

Rajkovic, A., Yan, M.S.C., Klysik, M., Matzuk, M. 2001. Discovery of germ cell-specific transcripts by expressed sequence tag database analysis. Fertil. Steril. 76, 550–554.

Rastelli, L., Robinson, K., Xu, Y., Majumder, S. 2001. Reconstitution of enhancer function in paternal pronuclei of one-cell mouse embryos. Mol. Cell. Biol. 21, 5531–5540.

Reik, W., Dean, W., Walter, J. 2001. Epigenetic reprogramming in mammalian development. Science 293, 1089–1093.

Renard, J.P. 1998. Chromatin remodelling and nuclear reprogramming at the onset of embryonic development in mammals. Reprod. Fertil. Dev. 10, 573–580.

Rideout, W.M., III, Eggan, K., Jaenisch, R. 2001. Nuclear cloning and epigenetic reprogramming of the genome. Science 293, 1093–1098.

Rougier, N., Bourc'his, D., Gomes, D.M., Niveleau, A., Plachot, M., Paldi, A., Viegas-Pequignot, E. 1998. Chromosome methylation patterns during mammalian preimplantation development. Genes Dev. 12, 2108–2113.

Schultz, G.A., Heyner, S. 1992. Gene expression in preimplantation mammalian embryos. Mutat. Res. 296, 17–31.

Schultz, R.M. 1993. Regulation of zygotic gene activation in the mouse. BioEssays 15, 531–538.

Schultz, R.M., Davis, W., Jr., Stein, P., Svoboda, P. 1999. Reprogramming of gene expression during preimplantation development. J. Exp. Zool. 285, 276–282.

Siegfried, Z., Eden, S., Mendelsohn, M., Feng, X., Tsuberi, B.Z., Cedar, H. 1999. DNA methylation represses transcription in vivo. Nat. Genet. 22, 203–206.

Smith, L.C., Meirelles, F.V., Bustin, M., Clarke, H.J. 1995. Assembly of somatic histone H1 onto chromatin during bovine early embryogenesis. J. Exp. Zool. 273, 317–326.

Smith, S.S. 2000. Gilbert's conjecture: the search for DNA (cytosine-5) demethylases and the emergence of new functions for eukaryotic DNA (cytosine-5) methyltransferases. J. Mol. Biol. 302, 1–7.

Stanton, J.L., Green, D.P. 2001. Meta-analysis of gene expression in mouse preimplantation embryo development. Mol. Hum. Reprod. 7, 545–552.

Stein, P., Schultz, R.M. 2000. Initiation of a chromatin-based transcriptionally repressive state in the preimplantation mouse embryo: lack of a primary role for expression of somatic histone H1. Mol. Reprod. Dev. 55, 241–248.

Stein, P., Worrad, D.M., Belyaev, N.D., Turner, B.M., Schultz, R.M. 1997. Stage-dependent redistributions of acetylated histones in nuclei of the early preimplantation mouse embryo. Mol. Reprod. Dev. 47, 421–429.

Sudol, M. 1994. Yes-associated protein (YAP65) is a proline-rich phosphoprotein that binds to the SH3 domain of the Yes proto-oncogene product. Oncogene 9, 2145–2152.

Suzuki, K., Yasunami, M., Matsuda, Y., Maeda, T., Kobayashi, H., Terasaki, H., Ohkubo, H. 1996. Structural organization and chromosomal assignment of the mouse embryonic TEA domain-containing factor (ETF) gene. Genomics 36, 263–270.

Szabo, P.E., Mann, J.R. 1995. Allele-specific expression and total expression levels of imprinted genes during early mouse development: implications for imprinting mechanisms. Genes Dev. 9, 3097–3108.

Tada, M., Tada, T., Lefebvre, L., Barton, S.C., Surani, M.A. 1997. Embryonic germ cells induce epigenetic reprogramming of somatic nucleus in hybrid cells. Embo. J. 16, 6510–6520.

Tanaka, M., Hennebold, J.D., Macfarlane, J., Adashi, E.Y. 2001. A mammalian oocyte-specific linker histone gene H1oo: homology with the genes for the oocyte-specific cleavage stage histone (cs-H1) of sea urchin and the B4/H1M histone of the frog. Development 128, 655–664.

Taylor, K.D., Piko, L. 1987. Patterns of mRNA prevalence and expression of B1 and B2 transcripts in early mouse embryos. Development 101, 877–892.

Taylor, K.D., Piko, L. 1992. Expression of ribosomal protein genes in mouse oocytes and early embryos. Mol. Reprod. Dev. 31, 182–188.

Temeles, G.L., Ram, P.T., Rothstein, J.L., Schultz, R.M. 1994. Expression patterns of novel genes during mouse preimplantation embryogenesis. Mol. Reprod. Dev. 37, 121–129.

Thompson, E.M., Legouy, E., Christians, E., Renard, J.P. 1995. Progressive maturation of chromatin structure regulates HSP70.1 gene expression in the preimplantation mouse embryo. Development 121, 3425–3437.

Tong, Z.B., Gold, L., Pfeifer, K.E., Dorward, H., Lee, E., Bondy, C.A., Dean, J., Nelson, L.M. 2000. Mater, a maternal effect gene required for early embryonic development in mice. Nat. Genet. 26, 267–268.

Tremblay, K.D., Duran, K.L., Bartolomei, M.S. 1997. A 5' 2-kilobase-pair region of the imprinted mouse H19 gene exhibits exclusive paternal methylation throughout development. Mol. Cell. Biol. 17, 4322–4329.

Ueyama, T., Zhu, C., Valenzuela, Y.M., Suzow, J.G., Stewart, A.F. 2000. Identification of the functional domain in the transcription factor RTEF-1 that mediates alpha 1-adrenergic signaling in hypertrophied cardiac myocytes. J. Biol. Chem. 275, 17476–17480.

Vassilev, A., Kaneko, K.J., Shu, H., Zhao, Y., DePamphilis, M.L. 2001. TEAD/TEF transcription factors utilize the activation domain of YAP65, a Src/Yes-associated protein localized in the cytoplasm. Genes Dev. 15, 1229–1241.

Vaudin, P., Delanoue, R., Davidson, I., Silber, J., Zider, A. 1999. TONDU (TDU), a novel human protein related to the product of vestigial (vg) gene of Drosophila melanogaster interacts with vertebrate TEF factors and substitutes for Vg function in wing formation. Development 126, 4807–4816.

Vogel, J.L., Kristie, T.M. 2000. The novel coactivator C1 (HCF) coordinates multiprotein enhancer formation and mediates transcription activation by GABP. Embo. J. 19, 683–690.

Wade, P.A., Pruss, D., Wolffe, A.P. 1997. Histone acetylation: chromatin in action. Trends Biochem. Sci. 22, 128–132.

Walsh, C.P., Bestor, T.H. 1999. Cytosine methylation and mammalian development. Genes Dev. 13, 26–34.

Wang, Q., Latham, K.E. 1997. Requirement for protein synthesis during embryonic genome activation in mice. Mol. Reprod. Dev. 47, 265–270.

Wang, Q., Latham, K.E. 2000. Translation of maternal messenger ribonucleic acids encoding transcription factors during genome activation in early mouse embryos. Biol. Reprod. 62, 969–978.

Warnecke, P.M., Mann, J.R., Frommer, M., Clark, S.J. 1998. Bisulfite sequencing in preimplantation embryos: DNA methylation profile of the upstream region of the mouse imprinted H19 gene. Genomics 51, 182–190.

Wassarman, P.M., Kinloch, R.A. 1992. Gene expression during oogenesis in mice. Mutat. Res. 296, 3–15.

Watson, A.J., Wiemer, K.E., Arcellana-Panlilio, M., Schultz, G.A. 1992. U2 small nuclear RNA localization and expression during bovine preimplantation development. Mol. Reprod. Dev. 31, 231–240.

Wiekowski, M., Miranda, M., DePamphilis, M.L. 1991. Regulation of gene expression in preimplantation mouse embryos: effects of the zygotic clock and the first mitosis on promoter and enhancer activities. Dev. Biol. 147, 403–414.

Wiekowski, M., Miranda, M., DePamphilis, M.L. 1993. Requirements for promoter activity in mouse oocytes and embryos distinguish paternal pronuclei from maternal and zygotic nuclei. Dev. Biol. 159, 366–378.

Wiekowski, M., Miranda, M., Nothias, J.Y., DePamphilis, M.L. 1997. Changes in histone synthesis and modification at the beginning of mouse development correlate with the establishment of chromatin mediated repression of transcription. J. Cell. Sci. 110, 1147–1158.

Wolffe, A.P. 1996. Chromatin structure and DNA replication: implications for transcriptional activity. In: _DNA Replication In Eukaryotic Cells_, (M.L. DePamphilis, Ed.), Cold Spring Harbor, NY, Cold Spring Harbor Laboratory Press, pp. 271–293.

Wolffe, A.P., Jones, P.L., Wade, P.A. 1999. DNA demethylation. Proc. Natl. Acad. Sci. USA 96, 5894–5896.

Woodland, H.R., Flynn, J.M., Wyllie, A.J. 1979. Utilization of stored mRNA in Xenopus embryos and its replacement by newly synthesized transcripts: histone H1 synthesis using interspecies hybrids. Cell 18, 165–171.

Worrad, D.M., Ram, P.T., Schultz, R.M. 1994. Regulation of gene expression in the mouse oocyte and early preimplantation embryo: developmental changes in Sp1 and TATA box-binding protein, TBP. Development 120, 2347–2357.

Worrad, D.M., Schultz, R.M. 1997. Regulation of gene expression in the preimplantation mouse embryo: temporal and spatial patterns of expression of the transcription factor Sp1. Mol. Reprod. Dev. 46, 268–277.

Worrad, D.M., Turner, B.M., Schultz, R.M. 1995. Temporally restricted spatial localization of acetylated isoforms of histone H4 and RNA polymerase II in the 2-cell mouse embryo. Development 121, 2949–2959.

Xiao, H., Friesen, J.D., Lis, J.T. 1995. Recruiting TATA-binding protein to a promoter: transcriptional activation without an upstream activator. Mol. Cell. Biol. 15, 5757–5761.

Xiao, J.H., Davidson, I., Matthes, H., Garnier, J.M., Chambon, P. 1991. Cloning, expression, and transcriptional properties of the human enhancer factor TEF-1. Cell 65, 551–568.

Yagi, R., Chen, L.F., Shigesada, K., Murakami, Y., Ito, Y. 1999. A WW domain-containing yes-associated protein (YAP) is a novel transcriptional coactivator. Embo. J. 18, 2551–2562.

Yasuda, G.K., Schubiger, G. 1992. Temporal regulation in the early embryo: is MBT too good to be true? Trends Genet. 8, 124–127.

Yeom, Y.I., Fuhrmann, G., Ovitt, C.E., Brehm, A., Ohbo, K., Gross, M., Hubner, K., Scholer, H.R. 1996. Germline regulatory element of Oct-4 specific for the totipotent cycle of embryonal cells. Development 122, 881–894.

Yockey, C.E., Smith, G., Izumo, S., Shimizu, N. 1996. cDNA cloning and characterization of murine transcriptional enhancer factor-1-related protein 1, a transcription factor that binds to the M- CAT motif. J. Biol. Chem. 271, 3727–3736.

Zuccotti, M., Boiani, M., Ponce, R., Guizzardi, S., Scandroglio, R., Garagna, S., Redi, C.A. 2002. Mouse Xist expression begins at zygotic genome activation and is timed by a zygotic clock. Mol. Reprod. Devel. 61, 14–20.

Zwingman, T., Erickson, R.P., Boyer, T., Ao, A. 1993. Transcription of the sex-determining region genes Sry and Zfy in the mouse preimplantation embryo. Proc. Natl. Acad. Sci. USA 90, 814–817.

Advances in Developmental Biology and Biochemistry, Vol. 12
M. DePamphilis (Editor)

Early embryonic gene transcription in *Xenopus*

Gert Jan C. Veenstra

Department of Molecular Biology, Nijmegen Center for Molecular Life Sciences, University of Nijmegen,
Geert Grooteplein 26–28, 6525 GA Nijmegen, The Netherlands

Summary

Early embryonic development in *Xenopus laevis* is characterized by global transcriptional repression until the mid-blastula stage. This repression is mediated by chromatin, but is enhanced by a variety of other mechanisms, such as limitation of the general transcription factor TBP, cytoplasmic retention of other transcription factors, and additional constraints on transcriptional activator function. These multiple layers of transcriptional regulation are controlled in dynamic and temporally distinct ways between the early blastula and gastrula stages of development.

Contents

1. Introduction

Upon fertilization, the egg of metazoan organisms starts a dynamic and intricate program of cleavage, induction, morphogenetic movements, patterning, proliferation and differentiation, all of which are tightly regulated in a spatio-temporal fashion. In the amphibian *Xenopus laevis*, the externally developing embryo is provided with considerable maternal stores of RNA and protein, allowing the synthesis of structural and regulatory proteins in the absence of *de novo* RNA synthesis during the initial stages of development. A major increase in embryonic transcription occurs in *Xenopus* at the midblastula stage (Nieuwkoop-Faber stage 8.5). This transition from a transcriptionally repressed to a transcriptionally active state occurs following the embryo's first 12 cell cycles. At this stage, a progressive loss of cell cycle synchrony and the acquisition of cell motility is observed along with the major increase in embryonic transcription. These changes are collectively referred to as the mid-blastula transition (Newport and Kirschner, 1982a, 1982b), even though similar changes do not necessarily coincide in other species (Yasuda and Schubiger, 1992; Andéol, 1994). Using sensitive methods, low levels of transcription can be observed prior to the MBT in *Xenopus* (Kimelman et al., 1987; Nakakura et al., 1987). However, the rate of transcription per cell increases approximately 200-fold at the mid-blastula stage (Kimelman et al., 1987), indicative of a major gene-regulatory transition taking place at this stage.

Shortly after the MBT, other regulatory changes are observed as well, including changes referred to as the early gastrula transition and the gastrula-neurula transition (resp. EGT and GNT) (Andrews et al., 1991; Howe et al., 1995). At the onset of gastrulation (Nieuwkoop-Faber stage 10), embryonic gene products for the first time regulate the cell cycle (Newport and Dasso, 1989; Howe et al., 1995; Howe and Newport, 1996), and control programmed cell death (Hensey and Gautier, 1997; Stack and Newport, 1997). This stage is also marked by the earliest detectable endogenous apoptosis (Hensey and Gautier, 1998; Veenstra et al., 1998). Some of the molecular events underlying the EGT are independent of the onset of transcription at the MBT, gastrulation itself however requires embryonic transcription (Newport and Kirschner, 1982a; Sible et al., 1997), identifying both the mid-blastula and early gastrula transitions as important gene-regulatory milestones. Later during development, many other gene-regulatory changes take place, including the coordinate inactivation of a number of RNA polymerase III-dependent genes, a process referred to as the gastrula-neurula transition (Andrews et al., 1991).

This chapter will first explore the general characteristics of these gene-regulatory milestones, and then focus on the mechanisms that might contribute to them. It has become increasingly evident that no single molecule or mechanism controls the onset of transcription, and several of the known mechanisms will be discussed.

2. Transcription dynamics during early development

At the mid-blastula transition, rates of transcription increase several orders of magnitude (Newport and Kirschner, 1982a, 1982b), not only when whole embryos are considered, but also when rates of transcription are normalized for cell number (Kimelman et al., 1987). At this stage significant levels of the hyperphosphorylated form of the large subunit of RNA polymerase II (IIo) appear (Veenstra et al., 1999a; Palancade et al., 2001), consistent with the unphosphorylated and phosphorylated forms of RNA Pol II being involved in preinitiation and transcript elongation respectively (reviewed in Dahmus, 1995; Shilatifard, 1998).

Initially, some genes are expressed irrespective of their regular spatial or temporal regulation. For example, the myogenic transcription factor MyoD is ubiquitously expressed after the MBT, only to be down-regulated shortly thereafter in all cells except prospective muscle, in which the gene is further induced (Rupp and Weintraub, 1991). Also, the oocyte-specific 5S rRNA genes are temporarily expressed after the mid-blastula transition, and are subsequently repressed during gastrulation to be expressed only in developing oocytes in mature females (Wormington and Brown, 1983; Bouvet et al., 1994). Like the oocyte 5S genes, satellite I and an oocyte-type tRNA gene are expressed in the oocyte and for a short period of time after the MBT, whereas the somatic 5S genes are transcribed from the MBT onwards (Andrews et al., 1991). Therefore, some genes are transcribed at low levels at the MBT with little if any regulatory specificity, in contrast to the regulation of these genes at other stages of development. At least in some instances, the modest levels of transcription in mid-to-late blastula embryos reflect basal transcription driven by the core promoter, whereas the nature and extent of targeted activation and repression mechanisms change subsequently, between the MBT and the EGT (Fig. 1). For example, histone H2B transcription is exclusively driven by the core promoter in the late blastula embryo, whereas upstream proximal promoter elements start to contribute to transcription efficiency between the MBT and EGT in a progressive fashion (Veenstra et al., 1999a). The extent to which histone deacetylation plays a role in regulating the hsp70 promoter changes around the EGT as well; the histone deacetylase inhibitor TSA , while capable of derepressing the hsp70 promoter in late gastrula embryos, has no effect on hsp70 transcription in early gastrula embryos (Strouboulis et al., 1999). Concordantly, sodium butyrate, another histone deacetylase inhibitor, did not induce hyperacetylation of chromatin-associated histone H4 until gastrulation (Dimitrov et al., 1993), suggesting the absence or constrained function of the relevant histone deacetylases (HDACs) and histone acetyl transferases (HATs) before the onset of gastrulation.

Therefore, both the mid-blastula and early gastrula transitions mark quantitative and qualitative changes in transcription regulation. The dynamics of transcription regulation in this developmental window could be characterized by repression of transcription before the MBT, relatively low levels of transcription—possibly mediated predominantly by the general transcription machinery—at the MBT, and gene-selective activation and repression of transcription playing more dominant roles later, starting between late blastula and early gastrula stages (Fig. 1). It should be noted however, that this model, especially as it relates to targeted activation and repression during development, is based on a limited number of observations, and additional work is needed to enhance our understanding of the nature

G. J. C. Veenstra

A. Early blastula

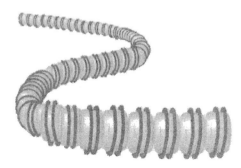

B. Late blastula

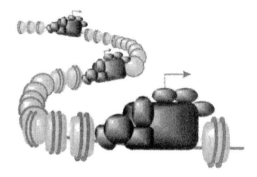

C. Gastrula

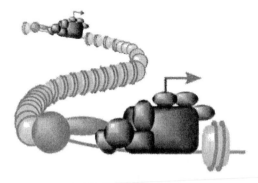

Fig. 1. Model of transcriptional regulation during early embryonic development in *Xenopus*. A. The embryonic genome is transcriptionally quiescent before the MBT, due to a repressive chromatin structure (symbolized by a high density of nucleosomes in this panel), and constraints on the transcription machinery. Constraints on the transcription machinery include rate-limitation of transcription initiation due to low levels of TATA binding protein (TBP), cytoplasmic retention of transcriptional regulators, and other constraints on the transcriptional activation.

and dynamics of early embryonic gene regulation in *Xenopus*. We will now consider what is known about the mechanisms that drive the gene-regulatory changes during early embryogenesis.

3. The cytoplasm-to-nucleus ratio regulates the MBT, ... or not?

Work by Newport and Kirschner (1982a, 1982b) has indicated that the changes at the MBT—including the onset of embryonic transcription—are not regulated by the number of cleavages, by a count of rounds of DNA synthesis, or by elapsed time after fertilization. Instead, the ratio of the number of nuclei to the fixed amount of cytoplasm in the early embryo was found to be of regulatory importance, and a model was postulated in which a maternally derived repressor of transcription is titrated by the exponentially increasing amount of genomic DNA (Newport and Kirschner, 1982a, 1982b). This model has been influential, and in the subsequent paragraphs both evidence supportive of the model, and evidence suggestive of more complicated mechanisms of regulation will be discussed.

3.1. Cytoplasm-to-nucleus ratio

A number of observations support the model that the changes at the MBT are regulated by the cytoplasm-to-nucleus ratio, and not by the number of cleavages, by a count of rounds of DNA synthesis, or by elapsed time after fertilization (Newport and Kirschner, 1982a).

First, it has been observed that both in enocytic fertilized eggs, in which the cytosol is stratified by centrifugation and cleavages are abortive, and in cytochalasin B treated embryos, in which cytochalasin B inhibits the assembly of actin filaments required for cleavage, RNA synthesis starts at the same time as in normal embryos.

Second, partially constricted fertilized eggs were used, in which the nucleus is trapped at one side, and only this nucleated side cleaves until one of the daughter nuclei migrates through a narrow channel to the other half. This half then resumes cleaving and undergoes MBT-related cell cycle changes at a later time, suggesting it is not the number of rounds of DNA synthesis of individual nuclei, but the cytoplasm-to-DNA ratio which regulates the MBT.

B. In late blastula embryos, chromatin is less repressive toward transcription (symbolized by lower density of nucleosomes) and many genes are transcribed, which for a subset of genes is facilitated by the developmentally regulated translation of TBP mRNA. A number of genes are transcribed at low levels by the general transcription machinery, with constraints on activator function still in place. The general transcription machinery is depicted at "open" spots of the chromatin, with the transcription start site depicted with an arrow. C. During gastrulation chromatin becomes more repressive towards transcription due to incorporation of linker histone H1 into chromatin and a more prominent role of histone deacetylases. A more prominent role for targeted, gene-selective activation and repression events is observed, symbolized by the presence of additional proteins (circle and oval) in the vicinity of the general transcription machinery. As a consequence, some genes are induced to high levels whereas some of the genes that were transcribed initially are repressed by the time of gastrulation, symbolized in this panel by disappearance of one transcription complex (compare with panel B). (*For a colored version of this figure, see plate section, page 273.*)

Third, as judged from metabolic labeling of newly synthesized RNA, synthesis of the most abundant transcripts (mostly RNA polymerase III-dependent) starts earlier in polyspermic embryos than in control embryos (Newport and Kirschner, 1982a).

Fourth, RNA polymerase III-dependent transcription could be induced precociously by microinjection of 16–32 ng of DNA into fertilized eggs—an amount of DNA roughly corresponding to the 24 ng of genomic DNA accumulated at the MBT (Newport and Kirschner, 1982b). Although analysis of the earliest transcription in these reports has mainly focused on RNA polymerase III-dependent transcription, RNA polymerase II-dependent transcription could—at least to some extent—be subject to similar controls, as it has been shown that hyperphosphorylation of the large subunit of RNA polymerase II, which normally coincides with the onset of transcription, occurs precociously in polyspermic embryos (Palancade et al., 2001).

3.2. Regulatory complexity

A number of observations indicate that the regulation of the onset of transcription is more complicated than a simple model of a repressor titrated by genomic DNA would suggest. For example, experiments involving cleavage-arrested and cytoplasm-extracted embryos have indicated that the onset of rRNA transcription (RNA polymerase I-mediated) is independent of the cytoplasm-to-nucleus ratio, and is regulated by an independent cytoplasmic "clock" (Shiokawa et al., 1985; Takeichi et al., 1985).

In addition, tRNA synthesis (RNA polymerase III-dependent) was observed to start earlier in enocytic fertilized eggs injected with the protein translation inhibitor cycloheximide than in uninjected enocytic eggs, even though the DNA content in cycloheximide treated eggs was much lower due to cell cycle arrest and inhibition of DNA replication (Kimelman et al., 1987). Lengthening of the cell cycle could allow higher levels of transcription relative to circumstances of rapid succession of mitosis and DNA replication. As the process of DNA replication has the potential to erase transcription complexes (Wolffe and Brown, 1986), and mitotic chromatin displaces many transcription factors (Martinez-Balbas et al., 1995), the pre-MBT characteristics of the cell cycle could contribute to repression of transcription. Though this is a plausible explanation, cycloheximide treatment may also prevent the translation of other components that influence the onset of transcription. This possibility should not be ignored in light of the fact that many maternal gene products are stored in the egg as both protein and RNA, and the full complement of the postulated inhibitor of transcription could depend on ongoing RNA translation.

Other data add further to the complexity of the picture, both in terms of gene-selective effects of regulatory mechanisms, and the fact that a longer cell cycle is not sufficient for activation of transcription before the MBT. For example, repression of an injected (RNA polymerase II-dependent) c-myc promoter could not be prevented by arresting embryos before the MBT with either cycloheximide (G2 phase arrest) or with a proteolysis-resistant mutant of cyclin B (M phase arrest) (Prioleau et al., 1994). Therefore, although cell cycle lengthening at some level may contribute to the onset of transcription, it is not sufficient to induce transcription. In addition, although the timing of the onset of transcription is generally the same in enocytic eggs and normal embryos, the pattern of endogenous gene activation is different, as satellite I gene transcription (RNA polymerase III-dependent) occurs at much higher levels in enocytic eggs compared to normal embryos (Lund and Dahlberg, 1992). This observation highlights the difficulty in comparing and interpreting observations obtained using different experimental approaches.

Adding to the notion that not all genes share identical regulatory requirements, is the observation that injection of exogenous DNA is not sufficient for transcription from a number of RNA polymerase II-dependent promoters before the MBT (Lund and Dahlberg, 1992; Prioleau et al., 1994, 1995; Almouzni and Wolffe, 1995; Veenstra et al., 1999a), and the evidence indicates that changes in the composition of the general transcription machinery are required for the onset of transcription from a subset of promoters at the mid-blastula transition (reviewed below).

Taken together, these data suggest an important role for a repressor of transcription prior to the MBT, a repressor which presumably is titrated by the exponentially increasing amount of genomic DNA. The molecular nature of this putative suppressor is unknown, although a component or aspect of chromatin is likely to be involved (Newport and Kirschner, 1982b; Prioleau et al., 1994, 1995; Almouzni and Wolffe, 1995). This is reviewed below in the section on chromatin. However, it is clear that a simple model of a titratable repressor cannot explain all the observations, and we now know that multiple mechanisms play a role in the onset of transcription at the MBT, some of which may function in gene-specific ways. Mechanisms known to play a role include regulation of nuclear localization, general transcription factors, and of co-activator function.

4. Transcription factor subcellular localization

Some proteins that are nuclear in the germinal vesicle (oocyte nucleus), translocate to the nucleus directly after fertilization or during early cleavage stages. Other proteins are cytoplasmic for some time during embryogenesis before they translocate to the nucleus (Dreyer, 1987). Yet other proteins are cytoplasmic in the oocyte, but translocate to the nucleus at some stage during embryogenesis. The timed nuclear translocation of some of these proteins is known to coincide with the onset of embryonic transcription. Many of these factors play roles related to gene expression, and circumstantial evidence suggests that in many cases their nuclear translocation constitutes a regulatory switch.

4.1. Regulated nuclear translocation

For example, the *Xenopus* nuclear factor xnf7 is a maternally expressed, DNA binding zinc finger protein that plays a role in dorsal–ventral patterning of the embryo (El-Hodiri et al., 1997). The protein translocates to the nucleus around the mid-blastula transition. Before the MBT, xnf7 is retained in the cytoplasm through an anchoring mechanism that depends on phosphorylation of four threonine residues that define a cytoplasmic retention domain (Miller et al., 1991; Li et al., 1994; Shou et al., 1996). When these amino acids are mutated to alanine, rendering the residues at these sites constitutively unphosphorylated, the mutated xnf7 protein translocates to the nucleus before the MBT. Conversely, if the threonine residues are mutated to glutamic acid, mimicking the phosphorylated state of these residues, the mutated xnf7 protein does not translocate to the nucleus until gastrulation

(Li et al., 1994; Shou et al., 1996). The mutated xnf7 protein with glutamic acid residues, not only does not translocate to the nucleus, if overexpressed it also prevents endogenous wild type xnf7 from entering the nucleus. Using this dominant-negative protein, it was shown that embryos are dorsalized if nuclear translocation of xnf7 is inhibited in the ventral side of the embryo, whereas the embryo is ventralized if nuclear translocation of xnf7 is inhibited on the dorsal side of the embryo (El-Hodiri et al., 1997). Therefore, the dephosphorylation-regulated nuclear translocation of xnf7 at the mid-blastula transition is required for normal dorsal–ventral patterning of the embryo.

A number of other proteins that are translocated to the nucleus also play important roles during gastrulation, for example at the site of mesoderm induction in the marginal zone, as is the case for the myogenic transcription factor MyoD. Cytoplasmic retention is involved in keeping MyoD from inducing ectopic muscle. The two genes encoding this myogenic transcription factor are transcribed ubiquitously at the MBT; in addition, a relatively small amount of MyoD RNA is maternally derived (Rupp and Weintraub, 1991). However, MyoD is not competent for myogenesis at the MBT, due to sequestration of MyoD protein in the cytoplasm (Rupp et al., 1994). During gastrulation, MyoD translocates to the nucleus in the induced marginal zone (prospective muscle), but not in other regions of the embryo. This regulated nuclear translocation begins an autoregulatory loop that could commit cells to myogenesis (Rupp et al., 1994). Another example is that of β-catenin, a protein involved in Wnt-signaling and axis formation, which translocates to the nucleus in the dorsal marginal zone at the MBT (Schneider et al., 1996). In this case, the TCF3 and LEF1 transcription factors interact with β-catenin, cause it to translocate to the nucleus, and mediate its downstream effects on gene transcription and axis formation (Huber et al., 1996; Molenaar et al., 1996). The functional relevance of nuclear translocation of other proteins, for example FGF (Shiurba et al., 1991) and ER1 (Luchman et al., 1999), is less clear. ER1 is a transcription factor whose expression is induced by FGF as an immediate early effect during gastrulation. Maternally derived cytoplasmic ER1 translocates to the nucleus in the marginal zone at the onset of transcription between stages 8 and 8 1/2. ER1 translocates to the nucleus of other cells in the embryo at a later stage, between the MBT and the onset of gastrulation (Luchman et al., 1999). These examples, especially those of MyoD and β-catenin, suggest that regulated nuclear translocation is a mechanism that causes rapid signaling-induced changes in gene expression in a spatio-temporally controlled way in the early embryo. This may be particularly relevant for organizer function and axis induction.

Another protein regulated by nuclear translocation is the CCAAT box transcription factor CBTF, a factor capable of binding and presumably activating the GATA-2 promoter at the beginning of gastrulation (Brewer et al., 1995). CBTF[p122], a double-stranded RNA-binding subunit of CBTF, is perinuclear during cleavage stages but moves from the cytoplasm into the nucleus at stage 9 (Orford et al., 1998). CBTF[p122] is associated with the translationally masked mRNP complexes, and the timing of CBTF[p122] nuclear translocation depends on its RNA binding domain, in the absence of which CBTF[p122] translocates to the nucleus well before the MBT (Brzostowski et al., 2000). Translation and turnover of maternal RNA present in the early embryo might trigger the nuclear translocation of CBTF[p122], which then could contribute to CBTF activity and GATA-2 activation during gastrulation. The regulation of nuclear translocation of CBTF[p122] might be more complicated however. The human CBTF[p122] homolog NF90, also referred to as NFAR-1/2, interacts with and is phosphorylated in its RNA-binding domain by the double-stranded RNA-activated protein kinase PKR (Parker et al., 2001; Saunders et al., 2001). Activated PKR is known to inhibit translation by phosphorylating eIF2α, a potentially relevant fact since CBTF^{p122} was found to associate with translationally masked mRNPs. NFAR protein enhances mRNA synthesis and interacts with the SMN and FUS proteins, implicated in RNA splicing (Saunders et al., 2001). Interaction of *Xenopus* CBTF^{p122} and human NF90 with a sequence-specific DNA binding transcription factor could provide a targeting mechanism for gene-selective regulation of mRNA processing. As PKR phosphorylates NF90/NFAR in its RNA-binding domain (Parker et al., 2001), a domain implicated in cytoplasmic retention of *Xenopus* CBTF^{p122} (Brzostowski et al., 2000), and as PKR localizes to both the cytoplasm and the nucleus but inhibits translation primarily in the cytoplasm through phosphorylation of eIF2α, it is possible that a similar phosphorylation event inhibits nuclear translocation of CBTF^{p122} in *Xenopus*, contributing to regulation of translation and nuclear mRNA processing in a developmentally controlled fashion. It will be important to test this hypothesis.

S-Adenylhomocystein Hydrolase (SAHH) may play an important role in methylation and elongation of pre-mRNA. It is predominantly located in the cytoplasm of cleavage and early blastula embryos, but translocates to the nucleus in late blastula and early gastrula embryos (Radomski et al., 1999), coincident with a strong increase in embryonic RNA synthesis.

Nucleolin is a major component of nucleoli in oocytes and somatic cells. During early embryogenesis however, maternal nucleolin is found in the cytoplasm, and it gradually translocates back to the nucleus just before the MBT (Messmer and Dreyer, 1993). Localization to the nucleoli is not observed until the onset of gastrulation. The translocation back to the nucleus during early blastula stages is driven by dephosphorylation of cdc2 and CK2 phosphorylation sites found closely to the bipartite nuclear localization signal of nucleolin (Schwab and Dreyer, 1997).

The maternal POU homeodomain transcription factor Oct-1 localizes to the cytoplasm during cleavage and early blastula stages, then gradually translocates to the nucleus, a process completed at the onset of gastrulation (Veenstra et al., 1999b). This regulated nuclear translocation is mediated by the POU domain, which mediates sequence-specific binding to DNA. The POU domain is also critical for nuclear translocation of another POU factor, XLPOU-60, which accumulates during oogenesis and early embryogenesis, and reaches a peak of expression around the MBT, between stages 8 and 9 (Whitfield et al., 1995). Total XLPOU-60 protein levels drop precipitously subsequently, and whole-embryo XLPOU-60 protein levels are relatively low from gastrulation stages onwards. Interestingly, XLPOU-60 is cytoplasmic in the oocyte and translocates to the nucleus between stage 8 and 9, during the peak of its expression (Whitfield et al., 1995). Therefore, nuclear XLPOU-60 function is very narrowly restricted by localization and expression dynamics around the MBT, suggesting a transient but potentially important role in the onset of transcription of POU factor target genes. The POU domain, which mediates the regulated nuclear translocation of both Oct-1 and XLPOU-60, contains a nuclear localization signal that is juxtaposed to phosphorylation sites widely found in POU domain proteins (Sock et al., 1996), raising the possibility that the phosphorylation status of the POU domain regulates nuclear translocation—similar to how phosphorylation regulates nuclear translocation of xnf7 and nucleolin.

4.2. Function of cytoplasmic retention

The nuclear translocation of transcription factors at the MBT could suggest that this regulation plays a role in the onset of transcription at the MBT, and in the qualitative and quantitative changes in gene regulation between the MBT and the onset of gastrulation. However, although some of these proteins are known to be required in the nucleus after the MBT (for example xnf7), it is not known what role, if any, cytoplasmic retention of these proteins plays before the MBT. Suggestive as the coincidence of nuclear translocation of select transcriptional regulators around the time of onset of transcription may be, there is no published data yet to show that cytoplasmic localization of these proteins before the MBT is required for, or contributes to transcriptional quiescence before the MBT. To address the role of cytoplasmic retention, the transcriptional status of in vivo target genes of these transcriptional regulators need to be determined under experimental conditions that favor precocious nuclear translocation. Such experiments are complicated by the fact that the onset of transcription at the MBT is regulated not by a single mechanism, but by a variety of mechanisms that act in concert. Nevertheless, it is an important issue to address because the role of cytoplasmic sequestration of transcription factors is not clear. This concern is highlighted by the fact that not all transcription factors are retained in the cytoplasm before the MBT. For example, c-Myc translocates to the nucleus after fertilization, resulting in exceptionally high levels of protein in the nucleus during early development (Lemaitre et al., 1995). The amount of c-Myc per nucleus declines during cleavage, to reach levels comparable to those found in somatic cells around the MBT. The cell cycle transcriptional regulator p53 also translocates to the nucleus soon after fertilization, in this case within 20 min, coinciding with the first S-phase (Tchang and Méchali, 1999). It is not known what role, if any, p53 and c-Myc play during pre-MBT development. The exceptionally high nuclear levels of c-Myc during cleavage, well above levels observed in transcriptionally active somatic cells, raise the possibility that cytoplasmic retention does not serve a transcription regulatory role, but simply mediates storage of protein required later during development.

These two possibilities, cytoplasmic retention being important for transcriptional quiescence, or for storage of protein, are not mutually exclusive. Storage of protein in the cytoplasm may be necessary to prevent non-sequence-specific binding and non-specific transcription initiation complex-assembly in the nucleus, driven by excessively high protein concentrations if all protein would translocate to the nucleus before the MBT. Because the number of nuclei increases exponentially during early development, any maternally derived protein translocating to the nucleus would reach its highest levels in the one- or two-cell embryo, after which it is diluted by cell division until new protein synthesis takes place. From a quantitative perspective, the nuclear concentration of some transcriptional regulators will reach unphysiologically high levels if translocated to the nucleus in early embryos in a constitutive fashion. For example, the total amount of Oct-1 protein is more or less constant during development until late neurula stages (Hinkley and Perry, 1992; Hinkley et al., 1992), whereas the number of cells increases from 1 to 10^5. During this time, Oct-1 is cytoplasmic before the MBT (Veenstra et al., 1999b), and from gastrula to tadpole stages of development the Oct-1 nuclear concentration decreases in many cell types (Veenstra et al., 1995). If the fixed amount of Oct-1 protein were to be translocated to the nucleus in a constitutive way, the local concentration reached in the single nucleus of the zygote would exceed 100 µM (Veenstra et al., 1999b), well above its K_d for nonspecific DNA (Verrijzer et al., 1992). Therefore, constitutive nuclear translocation of Oct-1 could potentially drive nonspecific transcriptional activation before the MBT. Cytoplasmic sequestration could therefore provide a strategy to store maternal transcription factors without activating transcription in a nonspecific way.

5. Regulation of the general transcription machinery

Components of the general transcription machinery are present in the transcriptionally inactive egg and in the early embryo. In eggs, pre-MBT embryos, and in egg extract, low levels of transcription can be observed if chromatin assembly is impaired, for example by injection into embryos or addition to extract of relatively large amounts of exogenous DNA (Newport and Kirschner, 1982b; Toyoda and Wolffe, 1992; Prioleau et al., 1994, 1995; Almouzni and Wolffe, 1995; Veenstra et al., 1999a). This is true for both RNA polymerase II and RNA polymerase III-dependent transcription, showing that all proteins required for transcription initiation and elongation by these polymerases are present and functional before the MBT. Nevertheless, regulation of general transcription factors, which have recently been recognized to play gene-selective developmental roles (reviewed in Veenstra and Wolffe, 2001b), contributes to the onset of transcription in *Xenopus*. Whereas the amounts of TFIIB and the RAP75 subunit of TFIIF are invariant during early embryogenesis, the levels of RNA polymerase I, the large subunit of RNA polymerase II, and TATA binding protein (TBP) increase during early embryogenesis (Bell and Scheer, 1999; Veenstra et al., 1999a).

5.1. Regulated translation of maternal TBP mRNA

TBP is rate-limiting for transcription before the mid-blastula transition (Veenstra et al., 1999a). TBP is hardly detectable in the egg and in cleavage stage embryos. In contrast, oocytes and eggs contain relatively large stores of translationally masked TBP RNA, which is translated from early to late blastula stages, resulting in a peak of TBP expression coinciding with the MBT (Veenstra et al., 1999a). In concordance with TBP being rate-limiting for transcription before the MBT, transcription is greatly stimulated by precocious translation of TBP mRNA or microinjection of the protein if chromatin assembly is incomplete or impaired (Prioleau et al., 1994, 1995; Almouzni and Wolffe, 1995; Veenstra et al., 1999a). However, if chromatin assembly is normal, no transcription is observed in the presence or absence of exogenous TBP, suggesting that chromatin-mediated repression is a separate, dominant mechanism of repression independent of regulation of TBP. Experiments in which chromatin assembly was compromised uncovered the auxiliary, redundant nature of constraining transcription before the MBT by translational masking of maternal TBP mRNA (Veenstra et al., 1999a). As discussed below, the maternally derived nucleosomal ATPase ISWI is capable of antagonizing TBP function by actively removing TBP from chromatin in eggs (Kikyo et al., 2000), adding a new twist to the saga of TBP, chromatin and early development.

5.2. TBP-like factor

In the absence of TBP mRNA translation, the embryo develops past the MBT and starts gastrulation, as judged by the appearance of the blastopore, but fails to complete gastrulation (Veenstra et al., 2000). Some developmental regulator genes are transcribed under these circumstances, whereas other genes are not. At least some of the genes transcribed are dependent on TBP-like factor (TLF) rather than TBP. Targeting TLF mRNA with antisense oligonucleotides resulted in an embryonic arrest at the MBT (Veenstra et al., 2000), similar to what happens to *C. elegans* and zebrafish if TLF function is inhibited by RNAi or a dominant-negative mutation (Dantonel et al., 2000; Kaltenbach et al., 2000; Muller et al., 2001), but clearly different from the situation in the mouse where TLF is not required for embryonic viability (Martianov et al., 2001; Zhang et al., 2001). It should be noted that this observation does not necessarily preclude a function of mouse TLF in transcription of embryonic genes non-essential for viability, and a careful analysis of target genes using functional genomics approaches will shed more light on the role TLF plays in embryonic gene transcription in different species. In *Xenopus*, however, TLF seems to play an essential role in transcription of a subset of RNA polymerase II-dependent embryonic genes (Veenstra et al., 2000). The requirement of translation of TBP mRNA for transcription of a subset of genes, explains some of the observations that indicated that titration of a putative repressor with injected exogenous DNA is not sufficient for the onset of transcription of all genes before the MBT.

6. Regulation of activator function

Another level of regulation involves the ability of activators to effectively and functionally interact with the basal transcription machinery and stimulate transcription. Data are scarce on this mechanism, but the published body of evidence suggests that specific co-activators could play an important role in regulating transcription of subsets of genes.

At least some transcriptional activators seem to be impaired in their function during early embryogenesis. The artificial activator GAL4-VP16 was found to anti-repress chromatin and activate transcription under conditions endogenous activators could not (Almouzni and Wolffe, 1995), whereas in another study the same activator could bind its *cis*-acting element but failed to activate transcription (Prioleau et al., 1995), thus behaving more like endogenous activators under these circumstances.

One of the endogenous activators functionally constrained before the MBT is Oct-1. Histone H2B promoters contain an Oct-1 binding site directly upstream of the TATA box that contributes to the regulation of this promoter in vivo (Fletcher et al., 1987; LaBella et al., 1988; Heintz, 1991; Hinkley and Perry, 1992). Although more than 95% of Oct-1 protein is retained in the cytoplasm before the MBT, Oct-1 protein levels in pre-MBT nuclei are most likely sufficiently high to bind Oct-1 binding sites (Veenstra et al., 1999b). In egg extracts, however, Oct-1 is not capable of activating a histone H2B promoter, as immunodepletion of Oct-1 does not affect histone H2B promoter activity (Veenstra et al., 1999a). Moreover, the full length histone H2B promoter is as active as its core promoter, both in egg extract and in vivo in mid-to-late blastula stage embryos. The core promoter is a minimal promoter derived from the histone H2B promoter, consisting of the TATA box and initiator, which lacks additional upstream promoter elements. Between the MBT and the onset of gastrulation, the upstream promoter becomes functionally relevant, now driving levels of transcription well above those seen with the core promoter (Veenstra et al., 1999a). These data suggest that Oct-1, and other endogenous activators that can bind to the full length H2B promoter, are constrained in their activity, much like the Gal4-VP16 fusion in the experiments by Prioleau et al., (1995). In the case of the histone H2B promoter, the constraints on activator function are mediated—at least in part—by a limitation of co-activator function. Injection of synthetic mRNA encoding one particular co-activator restored activated transcription to the histone H2B promoter in late blastula stage embryos [GJCV, unpublished, cf. Veenstra et al., 1999a]. Co-activators, transcriptional regulators that do not bind to DNA directly, may therefore play important regulatory roles in constraining activator function during early embryogenesis. However, many different co-activators exist which often regulate subsets of *trans*-acting factors. More work is required to examine both the generality of co-activators restraining activator function, and the mechanisms underlying such regulation.

7. Chromatin dynamics in early development

As reviewed in the previous paragraphs, it is important for early embryonic transcription to have transcription factors translocate to the nucleus, to have an unconstrained general transcription machinery, and to allow sequence-specific transcription factors to interact functionally with the general transcription machinery. In this section we will consider the regulation of the transcription template itself, and it appears that the importance of the structure and accessibility of the DNA template for transcription cannot easily be overemphasized, as a major role has been attributed to chromatin in regulating transcription at both a global and a locus-specific level during *Xenopus* embryogenesis (reviewed in Landsberger and Wolffe, 1995; Patterton and Wolffe, 1996). Nucleosomal density, chromatin composition, histone post-translational modifications, and nuclear organization of chromatin are highly dynamic during embryogenesis.

7.1. Nucleosome assembly and remodeling

Upon hormonal stimulation, oocytes mature into eggs, a process accompanied by the progression of prophase of meiosis I to metaphase of meiosis II, germinal vesical breakdown, and transcriptional repression. This process of meiotic maturation was shown to be accompanied by the formation of more regular arrays of nucleosomes and an increased efficiency of nucleosome assembly on injected plasmid (Landsberger and Wolffe, 1997). Using injected promoter–reporter constructs as a model, it was found that as a result of increased nucleosome assembly, the nucleosome density increased, a DNase I hypersensitive site on the promoter was lost, and the DNA template became more resistant to enzymatic digestion (Landsberger and Wolffe, 1997). Similarly, nucleosome assembly is much more efficient in early embryos than it is in oocytes (Veenstra et al., 1999a), and before the MBT, transcription can only be observed if exogenous DNA is injected to reduce the efficiency of nucleosome assembly to levels observed in the oocyte (Prioleau et al., 1994, 1995; Almouzni and Wolffe, 1995; Veenstra et al., 1999a), implying a role for the changes during oocyte maturation (Landsberger and Wolffe, 1997) in early embryonic repression of transcription.

Chromatin-mediated repression before the MBT could be mediated by multiple molecular mechanisms, including those involving nucleosome remodeling enzymes. For example, the nucleosomal ATPase ISWI was found to be involved in remodeling of nuclei in egg cytoplasm (Kikyo et al., 2000), which is accompanied by repression of transcription. ISWI actively removes TATA binding protein (TBP) from chromatin (Kikyo et al., 2000), an activity which undoubtedly contributes to the transcription-repressive character of the egg. This finding sheds more light on earlier reports of the influence of chromatin and levels of TBP on transcription before the MBT (Prioleau et al., 1994, 1995; Almouzni and Wolffe, 1995; Veenstra et al., 1999a). Even though pre-MBT chromatin is repressive towards transcription by itself, and even though translational regulation of TBP mRNA represents a regulatory mechanism distinct from that which is mediated by chromatin, ISWI may connect these separate mechanisms in an interesting way, better explaining the effects of exogenous DNA, exogenous TBP, and combinations of exogenous DNA and TBP on the repression of transcription before the MBT.

7.2. DNA methylation

That pre-MBT chromatin is repressive towards transcription was also observed in experiments manipulating the levels of genomic DNA methylation. Methylated CpG dinucleotides recruit methyl CpG binding domain (MBD) proteins, most of which interact with histone deacetylases and repress transcription (reviewed in Ballestar and Wolffe, 2001; Wade, 2001). Both MeCP2, the founding member of the MBD protein family, and the MBD3-containing Mi2 complex are found in *Xenopus* oocytes and eggs (Jones et al., 1998; Wade et al., 1998, 1999). The Mi2 complex represents the most abundant source of histone deacetylase activity in oocytes and eggs (Wade et al., 1998). DNA methyltransferase I (Dnmt1) is one of the enzymes involved in maintaining DNA methylation levels in the genome following DNA replication. When Dnmt1 mRNA was targeted with antisense RNA, it was found that some genes are transcribed before the MBT, implicating methylation-mediated repression in maintaining transcriptional quiescence before the MBT (Stancheva and Meehan, 2000). Global levels of genomic DNA methylation do not change between early blastula and tadpole stages of embryonic development (Veenstra and Wolffe, 2001a), however, this does not preclude the possibility that relatively few local changes at specific loci—undetected at a global level—are of regulatory significance for the onset of transcription of specific genes at the MBT. Alternatively, disruption of DNA methylation by interfering with the activity of Dnmt1, could compromise the repressive nature of chromatin, even though the normal regulation of the onset of transcription involves relieving chromatin-mediated repression of transcription through mechanisms that do not involve changes in DNA methylation. These two interpretations are not mutually exclusive as some mechanisms may act in gene-selective ways, and additional experiments are required to address these issues. Regardless of what specific interpretation is found to be correct, it seems safe to conclude that DNA methylation and chromatin play important roles in maintaining transcriptional repression before the MBT.

Interestingly, experimental genomic hypomethylation also leads to apoptosis (Kaito et al., 2001; Stancheva et al., 2001) and an altered differentiation potential during gastrulation (Stancheva et al., 2001), implying a role for DNA methylation in the proper execution of embryonic signaling pathways.

7.3. Chromatin organization

Given the important role of chromatin in maintaining transcriptional repression before the MBT, it is important then to consider what is known about the developmental regulation of chromatin, its organization, composition and modification status. At the level of how genomic chromatin is organized in the nucleus, dramatic changes have been observed with regard to what sequences are associated with the nuclear matrix during early development (Vassetzky et al., 2000). Several genomic regions were examined, including those containing the rDNA, c-myc and somatic 5S genes, which are transcribed by RNA polymerase I, II and III, respectively. Whereas association with the nuclear matrix appeared random within each genomic region before the MBT, a preferred site of attachment for each region was observed by the time of gastrulation, when the corresponding genes are actively transcribed. In contrast, rearrangement of the nuclear matrix attachment sites of a keratin gene was not observed at the MBT, which fits with its lack of expression at these stages. In the case of the multi-copy somatic 5S gene family, actively transcribed somatic 5S genes preferentially associated with the post-MBT nuclear matrix, whereas transcriptionally quiescent somatic 5S genes did not, suggesting that regions attached to the nuclear matrix correspond to accessible and transcriptionally active chromatin domains (Vassetzky et al., 2000).

From these data it is not immediately clear whether the observed changes are causal for, or are the consequence of the gene-regulatory transition with which they coincide. Interestingly, specification of DNA replication origins also exhibits an increase in specificity during early development (Hyrien et al., 1995), similar to the transition in specificity of nuclear matrix attachment sites, and it is of interest to further explore the relationship between nuclear organization of chromatin, specification of DNA replication origins, and the onset of embryonic transcription.

7.4. Histone modifications

Not only the organization of chromatin in the nucleus is subject to change during early development, the composition of chromatin alters dramatically as well. In contrast to somatic cells in which histone synthesis is tightly regulated and tied to S-phase, histones accumulate during oogenesis, to be deposited during embryogenesis (Woodland and Adamson, 1977; Almouzni and Wolffe, 1993). Histone H4 is stored in the diacetylated form, and remains predominantly diacetylated upon incorporation into chromatin during cleavage stages. However, between cleavage stages and gastrulation histone H4 becomes progressively deacetylated (Dimitrov et al., 1993). Interestingly, the gradual deacetylation of histone H4 correlates with the ability of the histone deacetylase inhibitor sodium butyrate to cause hyperacetylation of histone H4, suggesting that the change in modification level of histone H4 is regulated at the level of histone deacetylases (Dimitrov et al., 1993). In addition to histone acetylation, other modifications, including histone phosphorylation and histone methylation, have been implicated in transcription regulatory control (reviewed in Jenuwein and Allis, 2001). It will be interesting and important to examine the extent to which histone phosphorylation and histone methylation vary in different stages of development, since these modifications may constitute layers of chromatin-mediated control of transcription that are important for the onset of transcription during early embryogenesis.

7.5. Linker histones

Another transition in chromatin composition involves linker histones. The embryonic linker histone B4 is predominantly found in early embryonic chromatin until gastrulation, whereas the H1 linker histones (H1A, H1B and H1C) progressively replace histone B4 from blastula stages onwards (Smith et al., 1988; Dimitrov et al., 1993). In addition, early embryonic chromatin is also enriched in the high mobility group proteins 1 and 2 (HMG1/2) (Dimitrov et al., 1994), which share linker histone-like structural properties with linker histones B4 and H1 (Nightingale et al., 1996).

The changes in linker histone composition have been found to correlate with a selective decrease of the permissiveness of chromatin for RNA polymerase III-dependent transcription, between the MBT and neurulation (Wormington and Brown, 1983; Wolffe, 1989; Andrews et al., 1991; Dimitrov et al., 1993). This decrease is selective as somatic 5S transcription is not affected, in contrast to oocyte 5S rRNA transcription, which is repressed. The progressive incorporation of histone H1 causes the selective repression of oocyte 5S rRNA transcription during early embryogenesis, both in vitro and in vivo (Wolffe, 1989; Bouvet et al., 1994). These global changes in chromatin composition not only have a profound impact on RNA polymerase III-dependent transcription, but on early embryonic transcription of protein-encoding genes as well. In particular, the replacement of linker histone B4 by linker histone H1 is causal for the loss of mesoderm competence (Steinbach et al., 1997). Apparently, the ability of activin to induce mesoderm in early embryonic cells is dependent on a permissive chromatin environment which is lost upon incorporation of histone H1 into chromatin. In addition, histone H1 was found to be important for the spatial limitation of mesoderm induction in the embryo (Steinbach et al., 1997). These results indicate that the B4 and H1 linker histones have differential effects on gene expression in vivo.

Although HMG1/2 and the linker histones B4 and H1 share a similar structural role in binding to linker DNA (Nightingale et al., 1996), their affinity for nucleosomal DNA and their ability to repress transcription are different (Ura et al., 1996). Among HMG1/2 and the linker histones B4 and H1, linker histone H1 binds to nucleosomal DNA most avidly and represses transcription most potently, observations which provide a mechanistic and structural basis for the in vivo effects of linker histone H1 on RNA polymerase III-dependent transcription and on mesoderm induction. This is consistent with a model in which global chromatin-mediated repression becomes more prominent during gastrulation, through deposition of linker histone H1 histone and a more prominent role for deacetylation (Fig. 1). Specific genes are not necessarily affected by a more prominent global repression because of targeted and locus-specific activation events.

In conclusion, a large body of evidence suggests a prominent role for chromatin in repressing transcription before the MBT. However, it is not known exactly which chromatin-mediated mechanism of repression is dynamically regulated in such a way as to cause the genome to become accessible to the transcription machinery at the MBT. A variety of properties and aspects of chromatin are subject to change during early embryogenesis, causing major changes in global accessibility and gene regulatory control during gastrulation.

8. Perspective

Multiple molecular mechanisms contribute to the timing and regulation of the onset of embryonic transcription in *Xenopus* at the mid-blastula transition. These mechanisms include regulation of subcellular localization of transcription factors, regulation of the general transcription machinery, regulation of activator function, and multiple chromatin-dependent mechanisms. The obvious implication is that redundancy plays a prominent role in the regulation of the onset of transcription. This redundancy may have a variety of functions (Thomas, 1993), including a role in enhancing the robustness and fidelity of the regulatory control of transcription. In addition to these functions, new properties of transcriptional regulation could emerge from multiple, temporally distinct mechanisms. For example, some proteins translocate to the nucleus during cleavage, some proteins during blastula stages, whereas other proteins translocate to the nucleus around the mid-blastula transition or even later. TATA binding protein starts to accumulate at the early blastula stage, just before the onset of transcription, whereas activated transcription becomes evident only well after the MBT for some promoters. A variety of aspects of chromatin, its composition, post-translational modifications and organization within the nucleus, are highly dynamic between the blastula and gastrula stages of development. Whereas all these mechanisms seem to cooperate to prevent transcription before the MBT, these constraints on transcription are relieved in temporally distinct ways, such that the nature and rate of transcription of individual genes changes as a function of time after the MBT (Fig. 1), which is a feat that none of the regulatory mechanisms could accomplish on its own.

Embryonic development in most species is characterized by a period during which the zygotic or embryonic genome is transcriptionally quiescent (Yasuda and Schubiger, 1992; Andéol, 1994). This early embryonic phenomenon is interesting to study in connection with the biology of the various organisms. However, the relevance of these dynamically regulated changes does not rely solely, or even primarily, on an interest in the biology of the organism. As with other model systems, experiments performed with *Xenopus* have served to provide powerful paradigms in molecular, cellular, developmental and cancer biology. Many of the molecular mechanisms reviewed in this chapter warrant further investigation, to add detail to known regulatory pathways and to explore new ones.

Acknowledgments

I thank Igor Dawid, Paul Wade and Matt Guille for helpful comments and suggestions.

References

Almouzni, G., Wolffe, A.P. 1993. Nuclear assembly, structure, and function: the use of *Xenopus* in vitro systems. Exp. Cell. Res. 205, 1–15.

Almouzni, G., Wolffe, A.P. 1995. Constraints on transcriptional activator function contribute to transcriptional quiescence during early *Xenopus* embryogenesis. EMBO J. 14, 1752–1765.

Andéol, Y. 1994. Early transcription in different animal species: implication for transition from maternal to zygotic control in development. Roux's Arch. Dev. Biol. 204, 3–10.

Andrews, M.T., Loo, S., Wilson. L.R. 1991. Coordinate inactivation of class III genes during the Gastrula-Neurula Transition in *Xenopus*. Dev. Biol. 146, 250–254.

Ballestar, E., Wolffe, A.P. 2001. Methyl-CpG-binding proteins: targeting specific gene repression. Eur. J. Biochem. 268, 1–6.

Bell, P., Scheer, U. 1999. Developmental changes in RNA polymerase I and TATA box-binding protein during early *Xenopus* embryogenesis. Exp. Cell. Res. 248, 122–135.

Bouvet, P., Dimitrov, S., Wolffe, A.P. 1994. Specific regulation of *Xenopus* chromosomal 5S rRNA gene transcription in vivo by histone H1. Genes Dev. 8, 1147–1159.

Brewer, A.C., Guille, M.J., Fear, D.J., Partington, G.A., Patient, R.K. 1995. Nuclear translocation of a maternal CCAAT factor at the start of gastrulation activates *Xenopus* GATA-2 transcription. EMBO J. 14, 757–766.

Brzostowski, J., Robinson, C., Orford, R., Elgar, S., Scarlett, G., Peterkin, T., Malartre, M., Kneale, G., Wormington M., Guille, M. 2000. RNA-dependent cytoplasmic anchoring of a transcription factor subunit during *Xenopus* development. EMBO J. 19, 3683–3693.

Dahmus, M.E. 1995. Phosphorylation of the C-terminal domain of RNA polymerase II. Biochim. Biophys. Acta 1261, 171–182.

Dantonel, J.C., Quintin, S., Lakatos, L., Labouesse M., Tora, L. 2000. TBP-like factor is required for embryonic RNA polymerase II transcription in *C. elegans*. Mol. Cell. 6, 715–722.

Dimitrov, S., Almouzni, G., Dasso M., Wolffe, A.P. 1993. Chromatin transitions during early *Xenopus* embryogenesis: changes in histone H4 acetylation and in linker histone type. Dev. Biol. 160, 214–227.

Dimitrov, S., Dasso, M.C., Wolffe, A.P. 1994. Remodelling sperm chromatin in *Xenopus laevis* egg extracts: the role of core histone phosphorylation and linker histone B4 in chromatin assembly. J. Cell. Biol. 126, 591–601.

Dreyer, C. 1987. Differential accumulation of oocyte nuclear proteins by embryonic nuclei of *Xenopus*. Development 101, 829–846.

El-Hodiri, H.M., Shou, W., Etkin, L.D. 1997. xnf7 functions in dorsal-ventral patterning of the *Xenopus* embryo. Dev. Biol. 190, 1–17.

Fletcher, C., Heintz N., Roeder, R.G. 1987. Purification and characterization of OTF-1, a transcription factor regulating cell cycle expression of a human histone H2b gene. Cell 51, 773–781.

Heintz, N. 1991. The regulation of histone gene expression during the cell cycle. Biochim. Biophys. Acta 1088, 327–339.

Hensey, C., Gautier, J. 1997. A developmental timer that regulates apoptosis at the onset of gastrulation. Mech. Dev. 69, 183–195.

Hensey, C., Gautier, J. 1998. Programmed cell death during *Xenopus* development: a spatio-temporal analysis. Dev. Biol. 203, 36–48.

Hinkley, C., Perry, M. 1992. Histone H2b gene transcription during *Xenopus* early development requires functional cooperation between proteins bound to the CCAAT and octamer motifs. Mol. Cell. Biol. 12, 4400–4411.

Hinkley, C.S., Martin, J.F., Leibham D., Perry, M. 1992. Sequential expression of multiple POU proteins during amphibian early development. Mol. Cell. Biol. 12, 638–649.

Howe, J.A., Newport, J.W. 1996. A developmental timer regulates degradation of cyclin E1 at the midblastula transition during *Xenopus* embryogenesis. Proc. Natl. Acad. Sci. USA 93, 2060–2064.

Howe, J.A., Howell, M., Hunt, T., Newport, J.W. 1995. Identification of a developmental timer regulating the stability of embryonic cyclin A and a new somatic A-type cyclin at gastrulation. Genes Dev. 9, 1164–1176.

Huber, O., Korn, R., McLaughlin, J., Ohsugi, M., Herrmann B.G., Kemler, R. 1996. Nuclear localization of beta-catenin by interaction with transcription factor LEF-1. Mech. Dev. 59, 3–10.

Hyrien, O., Maric, C., Méchali, M. 1995. Transition in specification of embryonic metazoan DNA replication origins. Science 270, 994–997.

Jenuwein, T., Allis, C.D. 2001. Translating the histone code. Science 293, 1074–1080.

Jones, P.L., Veenstra, G.J.C., Wade, P.A., Vermaak, D., Kass, S.U., Landsberger, N., Strouboulis, J., Wolffe, A.P. 1998. Methylated DNA and MeCP2 recruit histone deacetylase to repress transcription. Nature Genet. 19, 187–191.

Kaito, C., Kai, M., Higo, T., Takayama, E., Fukamachi, H., Sekimizu K., Shiokawa, K. 2001. Activation of the maternally preset program of apoptosis by microinjection of 5–aza-2'-deoxycytidine and 5–methyl-2'-deoxy-cytidine-5'-triphosphate in *Xenopus laevis* embryos. Dev. Growth. Differ., 43, 383–390.

Kaltenbach, L., Horner, M.A., Rothman, J.H., Mango, S.E. 2000. The TBP-like factor CeTLF is required to activate RNA polymerase II transcription during *C. elegans* embryogenesis. Mol. Cell 6, 705–713.

Kikyo, N., Wade, P.A., Guschin, D., Ge H., Wolffe, A.P. 2000. Active remodeling of somatic nuclei in egg cytoplasm by the nucleosomal ATPase ISWI. Science 289, 2360–2362.

Kimelman, D., Kirschner, M., Scherson, T. 1987. The events of the midblastula transition in *Xenopus* are regulated by changes in the cell cycle. Cell 48, 399–407.

LaBella, F., Sive, H.L., Roeder, R.G., Heintz, N. 1988. Cell-cycle regulation of a human histone H2b gene is mediated by the H2b subtype-specific consensus element. Genes Dev. 2, 32–39.

Landsberger, N., Wolffe, A.P. 1995. Chromatin and transcriptional activity in early *Xenopus* development. Seminars in Cell Biology 6, 191–199.

Landsberger, N., Wolffe, A.P. 1997. Remodeling of regulatory nucleoprotein complexes on the *Xenopus* hsp70 promoter during meiotic maturation of the *Xenopus* oocyte. EMBO J. 16, 4361–4373.

Lemaitre, J.-M., Bocquet, S., Buckle, R., Méchali, M. 1995. Selective and rapid nuclear translocation of a c-Myc-containing complex after fertilization of *Xenopus laevis* eggs. Mol. Cell. Biol. 15, 5054–5062.

Li, X., Shou, W., Kloc, M., Reddy, B.A., Etkin, L.D. 1994. Cytoplasmic retention of *Xenopus* Nuclear Factor 7 before the mid blastula transition uses a unique anchoring mechanism involving a retention domain and several phosphorylation sites. J. Cell. Biol. 124, 7–17.

Luchman, H.A., Paterno, G.D., Kao, K.R., Gillespie, L.L. 1999. Differential nuclear localization of ER1 protein during embryonic development in *Xenopus laevis*. Mech. Dev. 80, 111–114.

Lund, E., Dahlberg, J.E. 1992. Control of 4–8S RNA transcription at the midblastula transition in *Xenopus laevis* embryos. Genes Dev. 6, 1097–1106.

Martianov, I., Fimia, G.-M., Dierich, A., Parvinen, M., Sassone-Corsi P., Davidson I. 2001. Late arrest of spermiogenesis and germ cell apoptosis in mice lacking the TBP-like *TLF/TRF2* gene. Mol. Cell. 7, 509–515.

Martinez-Balbas, M.A., Dey, A., Rabindran, S.K., Ozato, K., Wu, C. 1995. Displacement of sequence-specific transcription factors from mitotic chromatin. Cell 83, 29–38.

Messmer, B., Dreyer, C. 1993. Requirements for nuclear translocation and nucleolar accumulation of nucleolin of *Xenopus laevis*. Eur. J. Cell. Biol. 61, 369–382.

Miller, M., Reddy, B.A., Kloc, M., Li, X.X., Dreyer, C., Etkin, L.D. 1991. The nuclear-cytoplasmic distribution of the *Xenopus* nuclear factor, xnf7, coincides with its state of phosphorylation during early development. Development, 113, 569–575.

Molenaar, M., van, de W.M., Oosterwegel, M., Peterson-Maduro, J., Godsave, S., Korinek, V., Roose, J., Destree, O., Clevers, H. 1996. XTcf-3 transcription factor mediates beta-catenin-induced axis formation in *Xenopus* embryos. Cell 86, 391–399.

Muller, F., Lakatos, L., Dantonel, J., Strahle, U., Tora, L. 2001. TBP is not universally required for zygotic RNA polymerase II transcription in zebrafish. Curr. Biol. 11, 282–287.

Nakakura, N., Miura, T., Yamana, K., Ito, A., Shiokawa, K. 1987. Synthesis of heterogeneous mRNA-like RNA and low-molecular-weight RNA before the midblastula transition in embryos of *Xenopus laevis*. Dev. Biol. 123, 421–429.

Newport, J., Dasso, M. 1989. On the coupling between DNA replication and mitosis. J. Cell. Sci. Suppl. 12, 149–160.

Newport, J., Kirschner, M. 1982a. A major developmental transition in early *Xenopus* embryos: I. Characterization and timing of cellular changes at the midblastula stage. Cell 30, 675–686.

Newport, J., Kirschner, M. 1982b. A major developmental transition in early *Xenopus* embryos: II. Control of the onset of transcription. Cell 30, 687–696.

Nightingale, K., Dimitrov, S., Reeves, R., Wolffe, A.P. 1996. Evidence for a shared structural role for HMG1 and linker histones B4 and H1 in organizing chromatin. EMBO J. 15, 548–561.

Orford, R.L., Robinson, C., Haydon, J.M., Patient, R.K., Guille, M.J. 1998. The maternal CCAAT box transcription factor which controls GATA-2 expression is novel and developmentally regulated and contains a double-stranded-RNA-binding subunit. Mol. Cell. Biol. 18, 5557–5566.

Palancade, B., Bellier, S., Almouzni, G., Bensaude, O. 2001. Incomplete RNA polymerase II phosphorylation in *Xenopus laevis* early embryos. J. Cell. Sci. 114(Jul Pt 13), 2483–2489.

Parker, L.M., Fierro-Monti, I., Mathews, M.B. 2001. Nuclear factor 90 is a substrate and regulator of the eukaryotic initiation factor 2 kinase double-stranded RNA-activated protein kinase. J. Biol. Chem. 276, 32522–32530.

Patterton, D., Wolffe, A.P. 1996. Developmental roles for chromatin and chromosomal structure. Dev. Biol. 173, 2–13.

Prioleau, M.-N., Huet, J., Sentenac, A., Méchali, M. 1994. Competition between chromatin and transcription complex assembly regulates gene expression during early development. Cell 77, 439–449.

Prioleau, M.-N., Buckle, R.S., Méchali, M. 1995. Programming of a repressed but committed chromatin structure during early development. EMBO J. 14, 5073–5084.

Radomski, N., Kaufmann, C., Dreyer, C. 1999. Nuclear accumulation of S-adenosylhomocysteine hydrolase in transcriptionally active cells during development of *Xenopus laevis*. Mol. Biol. Cell. 10, 4283–4298.

Rupp, R.A., Weintraub, W.H. 1991. Ubiquitous MyoD transcription at the midblastula transition precedes induction-dependent MyoD expression in presumptive mesoderm of *X. laevis*. Cell 65, 927–937.

Rupp, R.A., Snider, W.L., Weintraub, H. 1994. *Xenopus* embryos regulate the nuclear localization of XMyoD. Genes Dev. 8, 1311–1323.

Saunders, L.R., Perkins, D.J., Balachandran, S., Michaels, R., Ford, R., Mayeda, A., Barber, G.N. 2001. Characterization of two evolutionarily conserved, alternatively spliced nuclear phosphoproteins, NFAR-1 and -2, that function in mRNA processing and interact with the double-stranded RNA-dependent protein kinase, PKR. J. Biol. Chem. 276, 32300–32312.

Schneider, S., Steinbesser, H., Warga, R.M., Hausen, P. 1996. β-catenin translocation into nuclei demarcates the dorsalizing centers in frog and fish embryos. Mech. Dev. 57, 191–198.

Schwab, M.S., Dreyer, C. 1997. Protein phosphorylation sites regulate the function of the bipartite NLS of nucleolin. Eur. J. Cell. Biol. 73, 287–297.

Shilatifard, A. 1998. The RNA polymerase II general elongation complex. Biol. Chem. 379, 27–31.

Shiokawa, K., Takeichi, T., Miyata, S., Tashiro, K., Matsuda, K. 1985. Timing of the initiation of rRNA gene expression and nucleolar formation in cleavage embryos arrested by cytochalasin B and podophyllotoxin and in cytoplasm-extracted embryos of *Xenopus laevis*. Cytobios. 43(174S), 319–334.

Shiurba, R.A., Jing, N., Sakakura, T., Godsave, S.F. 1991. Nuclear translocation of fibroblast growth factor during *Xenopus* mesoderm induction. Development 113, 487–493.

Shou, W., Li, X., Wu, C., Cao, T., Kuang, J., Che, S., Etkin, L.D. 1996. Finely tuned regulation of cytoplasmic retention of *Xenopus* nuclear factor 7 by phosphorylation of individual threonine residues. Mol. Cell. Biol. 16, 990–997.

Sible, J.C., Anderson, J.A., Lewellyn, A.L., Maller, J.L. 1997. Zygotic transcription is required to block a maternal program of apoptosis in *Xenopus* embryos. Dev. Biol. 189, 335–346.

Smith, R.C., Dworkin-Rastl E., Dworkin, M.B. 1988. Expression of a histone H1-like protein is restricted to early *Xenopus* development. Genes Dev. 2, 1284–1295.

Sock, E., Enderich, J., Rosenfeld, M.G., Wegner, M. 1996. Identification of the nuclear localization signal of the POU domain protein Tst-1/Oct6. J. Biol. Chem. 271, 17512–17518.

Stack, J.H., Newport, J.W. 1997. Developmentally regulated activation of apoptosis early in *Xenopus* gastrulation results in cyclin A degradation during interphase of the cell cycle. Development 124, 3182–3195.

Stancheva, I., Meehan, R.R. 2000. Transient depletion of xDnmt1 leads to premature gene activation in *Xenopus* embryos. Genes Dev. 14, 313–327.

Stancheva, I., Hensey, C., Meehan, R.R. 2001. Loss of the maintenance methyltransferase, xDnmt1, induces apoptosis in *Xenopus* embryos. EMBO J. 20, 1963–1973.

Steinbach, O.C., Wolffe, A.P., Rupp, R.A.W. 1997. Somatic linker histones cause loss of mesodermal competence in *Xenopus*. Nature 389, 395–399.

Strouboulis, J., Damjanovski, S., Vermaak, D., Meric, F., Wolffe, A.P. 1999. Transcriptional repression by XPc1, a new polycomb homolog in *Xenopus laevis* embryos, is independent of histone deacetylase. Mol. Cell. Biol. 19, 3958–3968.

Takeichi, T., Satoh, N., Tashiro, K., Shiokawa, K. 1985. Temporal control of rRNA synthesis in cleavage-arrested embryos of *Xenopus laevis*. Dev. Biol. 112, 443–450.

Tchang, F., Méchali, M. 1999. Nuclear import of p53 during *Xenopus laevis* early development in relation to DNA replication and DNA repair. Exp. Cell. Res. 251, 46–56.

Thomas, J.H. 1993. Thinking about genetic redundancy. Trends Genet. 9, 395–399.

Toyoda, T., Wolffe, A.P. 1992. Characterization of RNA polymerase II-dependent transcription in *Xenopus* extracts. Dev. Biol. 153, 150–157.

Ura, K., Nightingale, K., Wolffe, A.P. 1996. Differential association of HMG1 and linker histones B4 and H1 with dinucleosomal DNA: structural transitions and transcriptional repression. EMBO J. 15, 4959–4969.

Vassetzky, Y., Hair, A., Mechali, M. 2000. Rearrangement of chromatin domains during development in *Xenopus*. Genes Dev. 14, 1541–1552.

Veenstra, G.J.C., Wolffe, A.P. 2001a. Constitutive genomic methylation during embryonic development of *Xenopus*. Biochim. Biophys. Acta 1521, 39–44.

Veenstra, G.J.C., Wolffe, A.P. 2001b. Gene-selective developmental roles of general transcription factors. Trends Biochem. Sci. 26, 665–671.

Veenstra, G.J.C., Beumer, T.L., Peterson-Maduro, J., Stegeman, B.I., Karg, H.A., Van der Vliet, P.C., Destrée, O.H.J. 1995. Dynamic and differential *Oct-1* expression during early *Xenopus* embryogenesis: persistence of Oct-1 protein following down-regulation of the RNA. Mech. Dev. 50, 103–117.

Veenstra, G.J.C., Peterson-Maduro, J., Mathu, M.T., Van der Vliet, P.C., Destrée, O.H.J. 1998. Non-cell autonomous induction of apoptosis and loss of posterior structures by activation domain-specific interactions of Oct-1 in the *Xenopus* embryo. Cell Death Differ. 5, 774–784.

Veenstra, G.J.C., Destrée, O.H.J., Wolffe, A.P. 1999a. Translation of maternal TBP mRNA potentiates basal but not activated transcription in *Xenopus* embryos at the midblastula transition. Mol. Cell. Biol. 19, 7972–7982.

Veenstra, G.J.C., Mathu, M.T., Destrée, O.H.J. 1999b. The Oct-1 POU domain directs developmentally regulated nuclear translocation in *Xenopus* embryos. Biol. Chem. 380, 253–257.

Veenstra, G.J.C., Weeks, D.L., Wolffe, A.P. 2000. Distinct roles for TBP and TBP-like factor in early embryonic gene transcription in *Xenopus*. Science 290, 2312–2315.

Verrijzer, C.P., Alkema, M.J., Van Weperen, W.W., Van Leeuwen, H.C., Strating, M.J.J., Van der Vliet, P.C. 1992. The DNA binding specificity of the bipartite POU domain and its subdomains. EMBO J. 11, 4993–5003.

Wade, P.A. 2001. Methyl CpG binding proteins: coupling chromatin architecture to gene regulation. Oncogene 20, 3166–3173.

Wade, P.A., Jones, P.L., Vermaak, D., Wolffe, A.P. 1998. A multiple subunit Mi-2 histone deacetylase from *Xenopus laevis* cofractionsates with an associated Snf2 superfamily ATPase. Curr. Biol. 8, 843–886.

Wade, P.A., Gegonne, A., Jones, P.L., Ballestar, E., Aubry, F., Wolffe, A.P. 1999. Mi-2 complex couples DNA methylation to chromatin remodelling and histone deacetylation. Nature Genet. 23, 62–66.

Whitfield, T.T., Heasman, J., Wylie, C.C. 1995. Early embryonic expression of XLPOU-60, a *Xenopus* POU-domain protein. Dev. Biol. 169, 759–769.

Wolffe, A.P. 1989. Dominant and specific repression of *Xenopus* oocyte 5S RNA genes and satellite I DNA by histone H1. EMBO J. 8, 527–537.

Wolffe, A.P., Brown, D.D. 1986. DNA replication in vitro erases a *Xenopus* 5S RNA gene transcription complex. Cell 47, 217–227.

Woodland, H.R., Adamson, E.D. 1977. The synthesis and storage of histones during the oogenesis of *Xenopus laevis*. Dev. Biol. 57, 118–135.

Wormington, W.M., Brown, D.D. 1983. Onset of 5S RNA gene regulation during *Xenopus* embryogenesis. Dev. Biol. 99, 248–257.

Yasuda, G.K., Schubiger, G. 1992. Temporal regulation in the early embryo: is MBT too good to be true? Trends Genet. 8, 124–127.

Zhang, D., Penttila, T.-L., Morris, P.L., Teichmann, M., Roeder, R.G. 2001. Spermiogenesis deficiency in mice lacking the *Trf2* gene. Science 292, 1153–1155.

Advances in Developmental Biology and Biochemistry, Vol. 12
M. DePamphilis (Editor)

The cell cycle during oogenesis and early embryogenesis in *Drosophila*

Giovanni Bosco[1] and Terry L. Orr-Weaver[1,2]

[1]Whitehead Institute for Biomedical Research, [2]Department of Biology, Massachusetts Institute of Technology, Nine Cambridge Center, Cambridge, MA 02142, USA

Summary

Although changes in gene expression play a large role in determination and differentiation, most organisms also employ variant cell cycles in order to achieve particular developmental goals. This strategy is used widely during *Drosophila* development. In this chapter we address how cell cycle regulation interfaces with development, focusing on *Drosophila* oogenesis and early embryogenesis. Five modified cell cycles occur during these developmental stages: meiosis to produce haploid gametes, endo cycles to produce polyploid cells, amplification to increase gene copy number, S/M cycles for rapid division, and S/G2/M cycles. In each case we discuss the coordination between developmental events and alteration of the cell cycle, and we review recent advances in defining the regulation of these variant cell cycles. The unique feature of meiosis during oogenesis is that progression through the meiotic cell cycle must be halted at specific points to permit oocyte differentiation. The endo cycle is used in specific tissues throughout plants and animals. Analysis of this cycle in *Drosophila* nurse cells and follicle cells has provided key insights into how endo cycles are initiated and terminated, as well as how parameters of this cycle can be adjusted to produce differential DNA replication and alterations in chromosome morphology. Amplification provides a powerful model for deciphering the control of initiation of DNA replication in metazoans. We describe specialized regulation required for S/M cycles and how developmental regulatory genes directly affect the cell cycle to insert a G2 phase during embryogenesis.

Contents

1. Introduction

An integral component of development is the control of cell proliferation to ensure the proper number of cells in each tissue and also to permit differentiation. In addition to this general control, all organisms utilize specialized cell cycles for particular developmental goals. The two universal cell cycle variants are meiosis and the endo cycle. Meiosis produces haploid gamete cells, and it also has the developmental constraint that the meiotic cycle must be coordinated with oocyte differentiation during oogenesis. The endo cycle is a cell cycle with S and G phases but no mitosis, and it thus produces polyploid or polytene cells found in some tissues throughout plants and animals. Two cell types in the *Drosophila* ovary undergo endo cycles, the nurse cells and the follicle cells. Analysis of the endo cycle control in these two tissues has provided pivotal insights into how endo cycles initiate and are shut off in response to developmental signals, particularly the demonstration that the Rb and E2F tumor suppressor genes control endo cycle exit. Recovery of mutants that affect nurse cell endo cycles illustrated control mechanisms that permit particular genomic regions to be shut down for replication during endo cycle S phases. Such mutants also revealed that mitotic functions introduced into endo cycles can alter chromosome morphology from polytene to polyploid.

In a more specialized mechanism described in several insect species, specific genomic regions are amplified during development to facilitate high levels of gene expression. Amplification of the eggshell protein genes in *Drosophila* follicle cells has been exploited as a model for the initiation of metazoan replication origins, permitting the delineation of *cis*-acting elements necessary for origin activity as well as the recruitment of replication proteins. Key insights into the regulation of initiation have emerged from this system.

Organisms that require rapid embryogenesis accomplish this by accelerated S/M cell cycles in which DNA replication oscillates with mitosis, but there are no gap phases for transcription or cell growth. This is made possible by the deposition of maternal pools of gene products, such as cell cycle regulators, during oogenesis. Because the S/M cycles are controlled posttranscriptionally, they require unique regulation not used during archetypal cell cycles. In addition, in *Drosophila* these cycles involve nuclear division within a shared cytoplasm, a feature that also might require specialized controls. A protein kinase complex has been recently identified that specifically controls these cycles by ensuring adequate accumulation of mitotic Cyclin proteins. Transcription of the embryo's own genome requires that the cell cycle slow down; this is accomplished by the addition of a G2 phase. The insertion of this gap phase is exquisitely coordinated with developmental events in the embryo, illustrating how regulatory genes can influence both pattern formation and cell cycle regulation.

2. Meiosis in oogenesis

In *Drosophila* oogenesis progression through meiosis is subjected to developmental regulation at two critical levels. The decision to enter meiosis is linked to determination of the oocyte. Once initiated, the meiotic cell cycle is interrupted to permit oocyte differentiation, and there are feedback controls between meiotic progression and oocyte differentiation.

2.1. Entry into meiosis

At the onset of oogenesis a stem-cell division produces a precursor cystoblast that then goes through four mitotic divisions to produce a 16-cell cyst (for review de Cuevas et al., 1997). These 16 cells will ultimately become 15 nurse cells that become polyploid and an oocyte that goes through meiosis. Cytokinesis is incomplete in these divisions, and the cells remain connected by ring canals joined by a cytoskeletal structure called the fusome that is likely to be important both in cell communication and determination. These events occur at the anterior end of the ovary, and within each of the 16 ovarioles comprising each ovary there is a progression of advancing stage eggchambers towards the posterior. Meiosis initiates within a region called the germarium, and the onset of meiosis is prior to the formation of a stage 1 eggchamber in which the nurse cell-oocyte cluster is surrounded by somatic follicle cells (Fig. 1).

The synaptonemal complex (SC) is a proteinaceous structure that forms as homologous chromosomes synapse during prophase I of meiosis, and thus it is a cytological marker for early meiosis. Until recently the *Drosophila* SC could be recognized only by electron microscopy, but the development of antibodies against SC proteins makes it possible to analyze the requirements for and timing of SC formation. One set of these antibodies is against an unknown protein (Huynh and St. Johnston, 2000), while another is against the C(3)G protein which has been shown to be required for homolog synapsis and recombination (Page and Hawley, 2001). Use of these cytological markers has yielded crucial insights into the initiation of meiosis, oocyte determination, and SC formation. The SC first appears in two cells of the 16-cell cyst in one specific region of the germarium designated region 2A. These two cells are distinctive because they are the cells containing four ring canals. SC also begins to form in two adjacent cells connected by three ring canals, but the staining is less bright and more punctate than the thread-like staining seen in the four ring canal cells (the prooocytes). Later in region 2A only the prooocytes contain SC, and in region 2B of the germarium a single cell, the oocyte, contains SC with the homologous chromosomes synapsed in pachytene of prophase I. The marker used by Huynh et al. is no longer visible by the time the eggchamber leaves the germarium (Huynh and St. Johnston, 2000). The C(3)G protein staining persists in the oocyte in stages after the germarium, appearing to gradually come off the chromosomes by stage 6 of oogenesis,

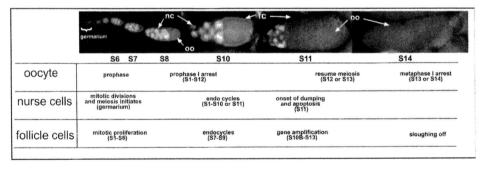

	S6 S7	S8	S10	S11	S14
oocyte	prophase	prophase I arrest (S1-S12)		resume meiosis (S12 or S13)	metaphase I arrest (S13 or S14)
nurse cells	mitotic divisions and meiosis initiates (germarium)		endo cycles (S1-S10 or S11)	onset of dumping and apoptosis (S11)	
follicle cells	mitotic proliferation (S1-S6)		endocycles (S7-S9)	gene amplification (S10B-S13)	sloughing off

Fig.1. A DAPI stained *Drosophila* ovariole showing various stages of eggchamber development. Germline and somatic stem cells reside in the germarium, the anterior tip of the ovariole. Nurse cells (nc) are germline derived and follicle cells (fc) are somatically derived cells. The oocyte (oo) enters meiosis while still in the germarium and proceeds to meiosis metaphase I arrest many days later in stage 14 oocytes (oocyte nucleus not visible). Nurse cells enter endo cycles in stage 1 and continue until stage 10 or 11. At stage 11 nurse cells initiate dumping of their proteins and mRNA into the oocyte, and then proceed to undergo apoptosis. Follicle cells proliferate mitotically until about stage 6. They stop proliferating and enter endo cycles at stage 7. At stage 9 or 10A they exit endo cycles and stop genomic replication. Amplification of the chorion gene cluster on chromosome III starts in stage 9, but the bulk of the amplification initiation occurs in stage 10B–11. The follicle cells then secrete chorion proteins and slough off in later stages.

while persisting in the nucleus (Page and Hawley, 2001). This correlates with the completion of pachytene and the formation of a compact karyosome containing the chromosomes within the oocyte nucleus.

These analyses reveal several important features about oocyte determination. First, SC appears in the cell that will become the oocyte prior to the localization of the earliest known oocyte proteins (Huynh and St. Johnston, 2000). Thus the onset of meiosis precedes known assays for oocyte determination. Second, genes such as *BicD* and *egalitarian* that affect oocyte-nurse cell determination cause all 16 cells to enter transiently into meiosis and form SC. Third, analyses of these mutants showed that the microtubule organizing center which forms in the oocyte early in determination is not needed for restriction of the SC to the oocyte. Proper microtubule function is required, however, for maintenance of the SC in the oocyte (Cox et al., 2001; Huynh et al., 2001).

Finally, these experiments uncover a recurrent theme that regulatory genes can function at different times of oogenesis in seemingly different processes. For example, the *mei-P26* gene is needed for proper distribution of recombination events and thus accurate chromosome segregation, but it is also necessary for localization of the C(3)G SC protein (Page and Hawley, 2001). This gene additionally restricts the germline mitotic divisions that precede the onset of meiosis (Page et al., 2000). As described in more detail below, the spindle genes, such as *spn B, C* and *D*, are needed both for meiosis and pattern formation in the oocyte. These mutants also inhibit determination of the oocyte, because restriction of the SC to a single cell is delayed in the mutants (Huynh et al., 2001). The distinction between nurse cell and oocyte fate is affected by cell cycle parameters, although the precise mechanism underlying this is not understood. Weak alleles of the *cyclin E* gene that is female sterile causes the formation of one or two extra oocytes in the 16 cell cluster (Lilly and Spradling, 1996).

2.2. Progression through prophase I

Two key parameters of prophase I emerged from analysis of yeast meiosis: meiotic recombination is initiated by double-strand breaks which are repaired by recombination and failure to repair double-strand breaks results in pachytene arrest in response to a checkpoint (Roeder, 1997; Roeder and Bailis, 2000). Characterization of *Drosophila* mutants has demonstrated that these meiotic properties are conserved in *Drosophila* oogenesis. The enzyme that produces meiotic double-strand breaks in yeast, Spo11, is conserved in flies, and is mutated in the *mei-W68* mutants (McKim and Hayashi-Hagihara, 1998). These mutants fail to undergo recombination and consequently have high levels of nondisjunction. The lack of recombination strongly suggests that in flies double-strand breaks initiate recombination. In addition, mutations have been recovered in two *Drosophila* homologs of genes needed for double-strand break repair, *okr* (*rad54*) and *spn-B* (*rad51/dmc1*) (Ghabrial et al., 1998). These mutants are sterile, but their sterility is suppressed by *mei-W68* mutation, presumably because double-strand breaks are not formed in the double mutant, and thus failure to repair them does not disrupt oogenesis (Ghabrial and Schupbach, 1999).

Mutant analyses show that progression through meiosis can be arrested in response to checkpoints as well. Although *Drosophila* oocytes do not have the diplotene and diakinesis stages of prophase I, exit from pachytene is marked by the formation of a karyosome in which the chromosomes are tightly compacted [for review see (Spradling, 1993)]. The karyosome persists until the oocyte nucleus breaks down and meiosis progresses to metaphase I, an event equivalent to oocyte maturation in vertebrates (Theurkauf and Hawley, 1992). The mutations in *okr* and *spn-B* which are likely to be defective in double-strand break repair delay formation of the karyosome and result in a karyosome with altered morphology, suggestive of a pachytene arrest. Strikingly this is suppressed by mutations in *mei-41*, the *Drosophila* homolog of the ATM checkpoint kinase (Ghabrial and Schupbach, 1999), implicating checkpoint signalling.

The *okr* and *spn-B* genes are part of a regulatory pathway that controls oocyte patterning (Ghabrial et al., 1998; Ghabrial and Schupbach, 1999). In these mutants, as well as *spn-C* and *D* and *vasa*, the oocyte fails to form dorsal structures and is ventralized. Dorsal differentiation is under the control of Egf signalling, including the key step of *gurken* mRNA and protein localization (for review see van Buskirk and Schupbach, 1999). The *okr* and *spn-B, C,* and *D* mutants initially were identified by their defects in oocyte patterning. This phenotype also is suppressed by mutations in *mei-W68* and *mei-41* (Ghabrial and Schupbach, 1999). Therefore, it appears that failure to repair double-strand breaks causes a meiotic arrest, in response to the MEI-41 checkpoint kinase, that also affects translation of the GURKEN protein, possibly by modification of the VASA protein (Ghabrial and Schupbach, 1999). If double-strand breaks are not initiated patterning is not affected. Thus patterning in the oocyte seems to be dependent on and to respond to progression through meiosis. It is not clear why these two processes need to be linked in their timing and whether there is an advantage to coordinating patterning with meiotic progression. One possibility is that the oocyte nucleus must be at the appropriate stages of meiosis in order for genes needed for patterning to be expressed. It also has been demonstrated that the E2F cell cycle regulator, a transcription factor that controls the G1-S transition, is needed for proper dorsal-ventral patterning in the oocyte (Myster et al., 2000). This molecular basis of this unexpected link is not understood.

The *encore* (*enc*) gene encodes a large novel protein that has multiple roles during oogenesis: it affects the number of cystoblast divisions, it localizes with *gurken* mRNA and affects protein levels, and it appears necessary for meiosis because in *enc* mutants, the karyosome morphology is abnormal (van Buskirk et al., 2000). Two kinds of karyosome defects are observed in that either the oocyte nucleus is polyploid like a nurse cell or the chromosomes are more diffuse in the mutant karyosome than in wild type. The *enc* mutants differ, however, from the *okr* and *spn* mutants because neither the karyosome or the ventralization phenotypes are rescued by *mei-41* mutations (van Buskirk et al., 2000). Thus ENC is similar to the OKR family in controlling multiple steps of oogenesis and it is independent of the meiotic checkpoint. One other gene, *fs(2)cup*, has been reported to control karyosome morphology, but this most likely reflects a role in chromosome structure rather than a specific effect on meiotic progression (Keyes and Spradling, 1997). Mutations in *fs(2)cup* affect nurse cell as well as oocyte chromosome morphology.

2.3. Metaphase I arrest

At stage 12 of eggchamber development the oocyte matures and resumes meiosis. The nuclear envelope breaks down, and the meiosis I chromosomes capture microtubules that are bundled into an anastral, acentriolar spindle (Theurkauf and Hawley, 1992). The oocyte then arrests in metaphase I. Chromosomes such as the fourth chromosome which have not undergone recombination move apart from each other towards the poles, away from the chromosome mass at the metaphase plate. The developmental signal responsible for the resumption of meiosis and progression to metaphase I is unknown, but some of the mechanisms leading to metaphase I arrest have been defined. It appears likely that active CDK1/mitotic cyclin kinase is needed for the metaphase arrest, because mutations in a meiosis-specific Cdc25 phosphatase called *twine* cause a failure to maintain the metaphase I arrest (Alphey et al., 1992; Courtot et al., 1992). The Cdc25 phosphatase is responsible for dephosphorylating an inhibitory tyrosine phosphorylation on CDK1.

Hawley and coworkers demonstrated that the metaphase I arrest requires the presence of chromosomes that have undergone exchange (McKim et al., 1993). Crossing over produces chiasmata that have been observed in many organisms to hold homologs together until they are released at the metaphase I/anaphase I transition. In *Drosophila* the requirement for recombination and presumably chiasmata is not explained solely by a physical block, however, with the homologs held together by chiasmata being restricted to the metaphase plate. This is evidenced by the fact that crossing over between a single pair of homologs is sufficient to permit a metaphase I arrest (Jang et al., 1995). This suggests that the presence of even a single chiasma produces a signal for the metaphase I arrest. Interestingly, in mutants for the *mei-41* checkpoint gene oocytes arrest at metaphase I even if they fail to undergo exchange (McKim et al., 2000). It is not obvious how failure of a checkpoint function could restore a normal arrest point, and the mechanism and rationale behind this control requires further elucidation.

2.4. Oocyte activation and the completion of meiosis

The oocyte is activated to exit the metaphase I arrest and complete both meiotic divisions as it progresses through the uterus. This occurs regardless of whether the oocyte becomes fertilized (Doane, 1960). The molecular basis of this activation is not understood, but it appears to involve hydration, as it can be mimicked by swelling oocytes in hypotonic buffers (Mahowald et al., 1983; Page and Orr-Weaver, 1997). Although there is an increase of translation at activation, the completion of the meiotic divisions does not require new translation (Page and Orr-Weaver, 1997). In an impressive set of observations, Heifetz et al. dissected oocytes from regions of the female reproductive tract to determine when and where activation occurs (Heifetz et al., 2001). Activation begins during ovulation and proceeds as the oocyte moves down the oviduct to the uterus. By the time the oocyte reaches the uterus, meiosis has resumed and it is in anaphase I. These observations raise the possibility that mechanical pressure from the oviducts as well as hydration may signal egg activation.

Few genes involved in egg activation have been identified. Mutants in the mitotic cyclin, *cyclin B3*, were reported to be defective in exiting metaphase I (Jacobs et al., 1998). Presumably exit from meiosis involves inactivation of CDK1/mitotic cyclin activity, making this observation puzzling. There are additional mitotic cyclins that could be involved in the maintenance of the metaphase I arrest if CycB3 participates in exit from meiosis (Jacobs et al., 1998). Two other genes, *cortex* (*cort*) and *grauzone* (*grau*), are required for egg activation and the completion of meiosis (Lieberfarb et al., 1996;

Page and Orr-Weaver, 1996). In *cort* or *grau* mutants microtubule reorganization that accompanies egg activation does not occur, there are defects in meiosis I segregation, and the eggs arrest at metaphase II. The *grau* gene encodes a transcription factor whose critical target is *cort*, as evidenced by the fact that *grau* sterility can be suppressed by induced expression of *cort* (Chen et al., 2000; Harms et al., 2000). The role of *cort* in the meiotic metaphase/anaphase transitions can be explained by its identity as a distant member of the Cdc20 protein family. These proteins promote the metaphase/anaphase transition of mitosis by activating the Anaphase Promoting Complex which targets mitotic cyclins and other mitotic regulators for destruction (Chu et al., 2001).

The four meiotic products decondense their chromosomes and go through a transient interphase before recondensing the chromosomes. Normally, the meiotic nucleus that lies most interior in the egg is captured by microtubules emanating from the male pronucleus, and this female pronucleus undergoes the first mitosis on a shared spindle with the male pronuclear chromosomes. The three unused meiotic products, the polar bodies, arrest in a metaphase-like state on the dorsal side of the egg, often fusing into a single nucleus. The protein kinase complex encoded by the *pan gu* and *plutonium* genes, together with the product of the *gnu* gene are required for recondensation of the meiotic products after their transient interphase (Fenger et al., 2000; Freeman and Glover, 1987; Freeman et al., 1986; Shamanski and Orr-Weaver, 1991). If these genes are mutated, the polar body and pronuclei remain in interphase and undergo extensive DNA replication but not mitosis. Condensation of the polar body chromosomes can be restored by increasing Cyclin B protein, implying that active CDK1/Cyc B kinase is needed for the postmeiotic condensation (Lee et al., 2001). Proper arrest of the polar bodies also requires the activity of the POLO kinase; in embryos from *polo* mutants mothers the polar bodies undergo inappropriate mitotic divisions (Riparbelli et al., 2000).

3. Endo cycles

3.1. Biological rationale for polyploidy

In most metazoa that have been studied endo cycles occur in some tissues during the normal course of development (for review see Edgar and Orr-Weaver, 2001). Endo cycles are modified cell cycles that allow cells to increase their ploidy. Endo cycles are developmentally controlled such that specific tissues enter endo cycle programs on cue at discrete developmental stages (Smith and Orr-Weaver, 1991). As a general rule the direct end-result of the endo cycle is an increase in genetic material from which mRNA and proteins can be made. High ploidy also allows cells to grow in size, and this seems to be critical for some cell types such as mammalian megakaryocytes and certain plant cells (for reviews see Zybina and Zybina, 1996; Zimmet and Ravid, 2000). The endo cycle results in cells that are factories, producing large amounts of gene products.

Polyploidy is usually achieved by repeated rounds of DNA replication in the absence of mitosis and cytokinesis. This process of increasing ploidy is sometimes referred to as endoreplication or endoreduplication. The S-phase of endo cycles, when DNA is replicated, is comprised of ordered steps and has discrete initiation and termination.

Furthermore, consecutive S-phases are always interrupted by a Gap phase when no DNA synthesis is occurring.

This endo cycle Gap phase can have both G1 and G2 attributes, including various aspects of mitosis (for review see Edgar and Orr-Weaver, 2001). Cell growth can take place during G1, while replication origins are reset for activation in G1 and blocked from re-firing in G2. Cells may enter mitosis after a G2-like phase, but then abort and exit a truncated mitosis. Which aspects of mitosis are included or bypassed is species specific as well as specific to tissue type. Two extreme examples of how mitosis is abbreviated can be found in mammals. The mammalian megakaryocytes that give rise to platelets achieve a ploidy of 128n (Zimmet and Ravid, 2000). Megakaryocytes chromosomes complete metaphase and enter anaphase and separate sister chromatids, but then abruptly abort this process at the anaphase-A/anaphase-B transition where the chromosomes are enveloped by one nucleus (reviewed in Edgar and Orr-Weaver, 2001). The giant cells of the mammalian placental trophectoderm also undergo endo cycles and reach greater than 1000 genome equivalents (Zybina and Zybina, 1996). In this case, however, the replicated chromosomes remain attached to one another, and many aspects of mitosis are averted (Zybina and Zybina, 1996; Edgar and Orr-Weaver, 2001). Thus, two tissue types in the same organism have evolved very different strategies for achieving polyploidy that utilize mitosis to different extents.

In this section we will focus primarily on developmental control of the endo cycle in *Drosophila*. Like other animals, *Drosophila* has evolved various strategies for controlling endo cycles so as to suit the specific developmental functions of those cells undergoing endo cycles. In contrast to many animals, most of the *Drosophila* larvae and adult tissues undergo endo cycles, thus providing the researcher with a myriad of regulatory mechanisms that customize endo cycles in a tissue-specific manner. Here we will discuss endo cycles that mainly occur in the nurse cells and follicle cells of the *Drosophila* ovary. Other cell types will be discussed only when comparisons are useful.

3.2. Polytene chromosomes versus polyploid cells

Polyploid cells can often have polytene chromosomes (recently reviewed in Edgar and Orr-Weaver, 2001). Historically, polytene chromosomes have been defined as cytologically visible chromosomes where the replicated sister chromatids have not separated and are maintained as physically attached structures (for review see Spradling and Orr-Weaver, 1987). These attachments are usually throughout the length of the chromatids, revealing highly reproducible banding patterns visible under light microscopy. For the purposes of this article we will confine our usage of "polytene" to this more historical definition. More modern techniques, such as fluorescence in situ hybridization, have shown that in some polyploid cells chromatid attachments can be restricted to short tracts of homologous sequences dispersed throughout the arms and centromeric regions. Although these chromosomes do not exhibit any visible banding patterns characteristic of polytene chromosomes, they also have been referred to as polytene structures in the literature (Dej and Spradling, 1999).

A beautiful example of polytene chromosomes exists in the giant cells of the *Drosophila* salivary glands. These polytene chromosomes have paired homologs and stereotypical polytene banding patterns. These are the most well studied system of polytene chromosomes, yet very little is known about how the replicated chromosomes are kept together, how the homologs are paired and how the banding pattern is established. Some models suggest that incomplete replication of chromosomes or incomplete decatenation leads to chromatids that cannot be separated (Lilly and Spradling, 1996). In addition, factors normally involved in sister chromatid cohesion during metaphase may serve to hold replicated chromatids together when S-phase is completed. However two mutants, *Drosophila ord* and *mei-S332*, that loose sister-chromatid cohesion have no visible polytene chromosome defect. This is likely due to the fact that these proteins are essential for cohesion only during meiosis (Kerrebrock et al., 1992); (Miyazaki and Orr-Weaver, 1992). Several mutants exist that enhance the banding patterns of nurse cell polytene chromosomes (Keyes and Spradling, 1997; King et al., 1981), but it is not yet known what role these gene products play in a molecular mechanism that holds chromatids together. As will be discussed below, early stage nurse cells have polytene chromosomes despite the fact that they are fully replicated (Dej and Spradling, 1999). Furthermore, neither incomplete replication nor decatenation can adequately explain how homologs are brought together and how their pairing is maintained through repeated rounds of DNA replication. Any model of how polytene structures arise and are maintained must take into account chromatid as well homolog pairing.

As we discussed above, the usefulness of polyploidization is almost certainly in achieving large amounts of mRNA and protein production. Having accommodated such high manufacturing demands, the obvious question arises as to what might be the biological utility of establishing and maintaining polytene structures. A possible clue lies in the actual structure of *Drosophila* polytene chromosomes. Many polytene chromosomes vary in DNA content along their lengths, and it has been well documented that the DNA content of heterochromatic regions in particular is under-represented (Gall et al., 1971; Hammond and Laird, 1985; Lilly and Spradling, 1996). How this is achieved is not yet understood, but certainly some sort of replication suppression in these regions leads to fewer copies of DNA (Lilly and Spradling, 1996). Perhaps then, for a cell whose function it is to streamline high production, it is economical not to replicate heterochromatic regions containing few if any genes. The polytene structure then may be a consequence of this economy, where chromosomes are held together at regions where replication has stalled or is otherwise suppressed (Fig. 2). Whether the polytene structure itself is required for a critical function is not known.

3.3. Nurse cells/endo cycle parameters

As described above (see Section 2), *Drosophila* nurse cells are derived from germ line stem cells (reviewed in Spradling, 1993). A progenitor cell undergoes four mitotic cell divisions and form a sixteen cell cyst, where all the sister cells are connected by intercellular bridges that arise from incomplete cytokinesis (de Cuevas et al., 1997). One cell is specified to become the oocyte while the remaining fifteen commence endo cycles and become nurse cells. These nurse cells undergo approximately 12 additional endo cycles. Their polyploid state is thought to be required for the critical function as nurse cells. As the name implies, they nurse the oocyte by serving as factory cells that produce large

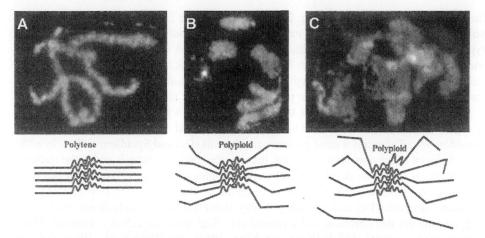

Fig. 2. DAPI stained nurse cells chromosome squash and schematic. A. Early stage (before endo cycle 5) nurse cells have very well defined polytene chromosomes where some simple banding patterns can be seen. These chromosomes are attached at their bright DAPI staining chromocenter. Associations between chromatids and homologues are very tight and they are perfectly aligned, as drawn schematically. B. Chromatids have begun to disperse and lose their polytene structure. C. A late stage nurse cell nucleus where the chromatids have completed dispersing and little or no chromosomal structure is visible.

quantities of mRNA and proteins that are then transported into the oocyte via intercellular bridges (for review see Spradling and Orr-Weaver, 1987).

Nurse cell endo cycles are noteworthy because they undergo several alterations that are clearly linked to developmental changes. Chromosome morphology also undergoes dramatic transformations. Thus nurse cells provide an excellent system in which to study the developmental control of cell cycle modifications as well as regulation of chromosome morphology.

3.3.1. Developmental onset of the endo cycle

Cyclin E, and presumably its partner CDK2 (*Dmcdc2c*), is required for the S/G oscillations observed during embryonic and larval endo cycles (Duronio and O'Farrell, 1995; Knoblich et al., 1994; Weiss et al., 1998). In the ovary, a female-sterile, loss-of-function *cycE* mutation allows more than one cell of the sixteen cell cyst to remain arrested and differentiate into an oocyte (Lilly and Spradling, 1996). This observation suggests that high levels of CycE/CDK2 activity is required for efficient entry into endo cycles. However, this same mutation has no effect on initiation of ovarian follicle cell endo cycles (discussed below), raising the question as to whether the primary defect in *cycE* mutants might be a failure to inhibit oocyte determination as opposed to a failure to initiate endo cycles. In either case, the link between CycE/CDK2, oocyte determination and entry into endo cycles is intriguing, but the molecular details of how endo cycles are inhibited in the oocyte while initiated in nurse cells remain to be deciphered.

3.3.2. Some polytene chromosomes have mitotic attributes

During the first four nurse cell endo cycles homologous chromosomes are paired and replication is not followed by chromatid separation, although the chromosomes are thought to be fully replicated (Fig. 2). These early nurse cell chromosomes are clearly polytene in nature, but they do not exhibit the striking banding patterns seen in other cell types. Following S-phase of the fifth endo cycle (64c DNA content) these polytene chromosomes separate and disperse into 32 pairs of chromatids, and it is thought that each 2c pair is held together by unreplicated heterochromatic regions (Dej and Spradling, 1999). Subsequent endo cycles continue to replicate euchromatic regions of these chromosomes.

Why this chromosome dispersal should occur at such a precise stage of development and what its utility is, are unclear. It has been postulated that since this phenomenon is conserved in other Dipterans it must present some benefit (Dej and Spradling, 1999). In addition, dispersal of the nurse cell nucleolus, the site of ribosome synthesis within the nucleus, also occurs at this same developmental stage (Dapples and King, 1970), and it has been proposed that chromosome dispersal is essential for nucleolar dispersion and efficient ribosome synthesis (Dej and Spradling, 1999). However, there is little direct evidence to suggest whether this chromosomal dispersal is critical for nurse cell development. Three female-sterile mutants, fs(2)B (Koch and King, 1964), otu (King et al., 1981) and cup (Keyes and Spradling, 1997) block the separation of the nurse cell polytene chromosomes at endo cycle five, and the nurse cell chromosomes go on to become giant polytene chromosomes with banding patterns similar to that observed in salivary gland cells. These mutants also block nucleolar dispersal (Khipple and King, 1976). It is not clear, however, whether chromosome and nucleolar dispersal is in fact essential for nurse cell development.

It has been proposed that nurse cell chromosomes experience a "mitotic-like" state precisely at the stage when chromosome dispersal has been observed (Dapples and King, 1970; Dej and Spradling, 1999; Reed and Orr-Weaver, 1997). No mitotic factors such as cyclins A and B nor any kind of spindle apparatus are detected at any stage of nurse cell endo cycles. However, in nurse cells carrying female-sterile alleles of the *morula* gene polytene chromosomes condense, CycB protein accumulates and formation of spindles can be seen (Reed and Orr-Weaver, 1997). *Morula* female-sterile mutants exhibit this phenotype only in nurse cells that are at the chromosome dispersal stage. Lethal alleles of *morula* affect mitotically dividing cells also, blocking the transition from metaphase to anaphase (Reed and Orr-Weaver, 1997). This strongly suggests that endo cycle five is unique and possibly has some mitotic character that must be kept in check by the function of the Morula protein (Dej and Spradling, 1999; Reed and Orr-Weaver, 1997). The phenotype observed in *morula* mutant nurse cells also suggests that these endo cycling cells are able to produce, and perhaps are even primed to express, mitotic factors that may play a role in chromosome condensation and separation of polytene chromosomes. This is another example of development controlling the parameters of the cell cycle where specific events, in this case mitotic events, are included or bypassed.

3.3.3. Heterochromatin is under-represented

An examination of satellite DNA copy number has revealed that many Dipterans underrepresent these sequences in their polyploid and polytene cells (Gall et al., 1971; Spradling and Orr-Weaver, 1987). Modulation of two S-phase parameters are thought to result in the underrepresentation of heterochromatin. First, the duration of S-phase in nurse cell endo cycles is thought to be abbreviated and thus does not allow time for complete replication of the genome. Second, the checkpoint surveillance that normally delays entry into mitosis so that DNA replication can be completed is bypassed.

Centric and pericentric satellite sequences make up the bulk of the heterochromatin found in *Drosophila* polytene chromosomes (Gall et al., 1971). The centromeric regions of polytene chromosomes are held together by an unknown mechanism to form a highly compacted structure known as the chromocenter (Gall et al., 1971). This chromocenter is easily detected because the DNA stain 4,6-diamino-2-phenylindole (DAPI) preferentially binds to it, and it is visualized under fluorescence microscopy as a very bright DAPI staining spot within the nucleus. Mutants that fail to under-replicate these satellite sequences are easily seen as having an unusually bright DAPI spot. The brightness of this chromocenter spot reflects the amount of heterochromatin replication (Lilly and Spradling, 1996).

Visualization of S-phase cells with BrdU labeling (a nucleotide analog) reveals two patterns: S-phase is said to have early and late replication patterns, where early replication occurs in euchromatic sequences and late replication of heterochromatin occurs during the latter part of S-phase (for review see Wintersberger, 2000). Endo cycle S-phases can lack any visible evidence of the late replication pattern (Lilly and Spradling, 1996). In nurse cell endo cycles after cycle 5, genomic replication does not go to completion, having omitted the latter part of S-phase (Dej and Spradling, 1999; Lilly and Spradling, 1996). As a consequence heterochromatic regions are not replicated whereas euchromatic regions are (Dej and Spradling, 1999; Hammond and Laird, 1985; Lilly and Spradling, 1996).

How does an abbreviated S-phase lead to incomplete replication? Replication initiates at chromosomal sites known as origins of replication (Dutta and Bell, 1997). After initiation has occurred, replication forks proceed to move along the chromosome bidirectionally away from the origin. In order to replicate chromosomes in a timely manner many origins fire and initiate forks that eventually meet and complete replication of the entire chromosome. Typically, not all replication origins need to be activated in order to completely replicate a genome (Dutta and Bell, 1997). With some exceptions, replication origins generally can be divided into two categories: Those that are activated early in S-phase and others that fire late in S-phase (Dutta and Bell, 1997; Wintersberger, 2000). The differential activation of replication origins is what allows for the temporal distinction between early and late replication. Early firing origins are abundant in euchromatin while late origins are enriched in heterochromatin. Thus, an abbreviated S-phase in which early origins are always activated but late origins are only sometimes activated could lead to underrepresentation of heterochromatin.

Normally, genome surveillance by checkpoint proteins delays exit from S-phase until all the genome is replicated. This delay allows time for sufficient numbers of late origins to activate or simply give time for replication forks originating in euchromatin to proceed through heterochromatin. Thus, a failure to activate late origins is normally inconsequential. It has been proposed that what causes incomplete replication during endo cycles is a combination of short S-phase duration and bypass of the replication checkpoint (Lilly and Spradling, 1996). This allows endo cycle cells to exit S-phase before heterochromatin has been replicated. Repeated rounds of "abbreviated" S-phases, where late origins are not allowed to fire, results in under-representation of heterochromatic sequences (Lilly and Spradling, 1996).

A critical regulator of S-phase length is Cyclin E (CycE). CycE, and its partner CDK2, are required for entry into S-phase (Richardson et al., 1993; Knoblich et al., 1994; Richardson et al., 1995). In *Drosophila* the loss-of-function *cycE* mutant that has low CycE protein levels (described above) results in female sterility (Lilly and Spradling, 1996).

One consequence is the development of an extra oocyte (**2 Meiosis**). Another effect is on replication of heterochromatin. This mutant exhibits very bright DAPI staining nurse cell chromocenters, and quantitation of satellite DNA contained within chromocenters confirms that CycE deficient nurse cells have replicated more of their genome than wild-type nurse cells (Lilly and Spradling, 1996). Labeling of cells with BrdU also has confirmed that these *cycE* nurse cells display both early and late replication patterns (Lilly and Spradling, 1996). In this *cycE* mutant the lower levels of CycE protein fail to oscillate with proper kinetics during nurse cell endo cycles, and this is presumed to allow extra time for late firing origins to activate and more efficiently replicate heterochromatin (Lilly and Spradling, 1996).

Since CycE is such a critical regulator of S-phase, any perturbation of CycE oscillations resulting in longer S-phases is expected to produce large, overreplicated chromocenters. Indeed, overexpression of either CycE or CycA (a mitotic cyclin that also plays a role in late S-phase) leads to larger chromocenters (Lilly and Spradling, 1996). In addition, a mutation in the *Drosophila effete* gene encoding a ubiquitin conjugating enzyme, thought to be important for cyclin proteolysis, results in bright chromocenters (Lilly and Spradling, 1996). Also consistent with this model is the observation that mutations in subunits of the E2F transcription factor, *dE2F1* and *dDP*, result in bright DAPI staining chromocenter (I. Royzman and T.L. Orr-Weaver, unpublished results). The *CycE* gene is transcriptionally regulated by E2F, and mutations in the dE2F1 subunit may affect CycE levels.

Although the duration of S-phase is a cell cycle parameter that can be modulated, it is difficult to measure directly how long S-phase lasts in vivo. Thus, in the absence of direct evidence of how S-phase duration is affected in *cycE* mutants it should be noted that other models of how heterochromatin is underreplicated are also consistent with the data discussed above. For example, the structure of heterochromatin itself may be disrupted in these mutants thereby leading to activation of late origins in early or middle S-phase. CycE/CDK2 are known to phosphorylate E2F and RBF (the RB *Drosophila* homolog) that potentially recruit histone modifying enzymes to heterochromatin and serve to keep replication origins silenced (for review see Dyson, 1998); (Trimarchi and Lees, 2002). dE2F1 itself is involved in regulating heterochromatin, as expression of heterochromatic genes is sensitive to dE2F1 dosage (Seum et al., 1996). Alternatively, CycE protein may act directly at replication origins through its interaction with a pre-replication complex component, the Cdc6 protein (Furstenthal et al., 2001), and thereby affect origin activity directly. Many other possibilities exist, and clearly more work is needed in order to ascertain the mechanism of how certain sequences are chosen for replication while others are not.

3.4. Follicle cells and the onset and exit from endo cycles

Follicle cells are somatically derived cells that form a monolayer of epithelial cells around the developing egg. They are important for many aspects of egg development as well as formation of the eggshell (Spradling, 1993; Dobens and Raftery, 2000). They originate from two or three non-clonal somatic stem cells and proliferate mitotically in order to reach approximately 1000 follicle cells per egg. Follicle cells stop proliferating and, like nurse cells, commence endo cycles at a precise point in their development (Fig. 1). However, many differences exist in the manner in which nurse and follicle cells regulate their endo cycles. Finally, follicle cells must exit endo cycles at a precise time in development in order to allow a gene amplification event to occur. This gene amplification event is critical for eggshell formation and will be discussed in detail in the following section.

Drosophila ovaries contain assembly lines of developing eggs, or eggchambers, where all the developmental stages can be found (Fig. 1). The various stages can be distinguished by gross eggchamber morphology as well as by nurse cell and follicle cell morphology (Spradling, 1993). Follicle cells undergo several mitotic cell divisions in stages 1–6 eggchambers. In the latter part of stage six, follicle cells stop proliferating and exit the mitotic cycle.

3.4.1. Onset of endo cycles is developmentally controlled

As the eggchambers transition from stage six to seven, the follicle cells enter endo cycles (Fig. 1). Although the developmental stage of an eggchamber when endo cycles commence is discrete, not all follicle cells within one eggchamber initiate endo cycles at precisely the same time. Consequently follicle endo cycles within a single eggchamber are asynchronous. Recent work has shown that follicle cells that are defective in the Notch and Delta signaling pathway fail to stop proliferating and do not enter endo cycles (Lopez-Schier and St Johnston, 2001; Deng et al., 2001). Notch signaling has been proposed to be the temporal signal that allows follicle cells to transition into endo cycles. However, downstream targets that cause follicle cells to exit mitotic cycles are not known. In general, follicle cells remain relatively unexplored as a model system for this developmentally controlled cell cycle switch.

3.4.2. Endo cycles occur with no apparent oscillations of regulators

Follicle cells undergo precisely three endo cycle S-phases during stages seven through nine to achieve 16n ploidy. This is in contrast to nurse cells that reach > 500n ploidy. During follicle cell endo cycles there is no obvious oscillation of cell cycle or replication proteins (Asano and Wharton, 1999; Calvi et al., 1998; Lilly and Spradling, 1996; Royzman et al., 1999). Cyclin E seems to be an exception in that although protein levels appear constant during endo cycles, its activity is presumed to oscillate based on the staining of a putative substrate, that is recognized by the MPM2 antibody (Calvi et al., 1998; Lilly and Spradling, 1996). Such CycE oscillations could lead to transient phosphorylation of CDK2 targets. The mitotic marker, MPM2, normally recognizes mitosis-specific phospho-proteins, but it also has been correlated with CycE activity, because it marks follicle cells that are in S-phase (Calvi et al., 1998). In addition, MPM2 staining increases when CycE is overexpressed and decreased when CycE/CDK2 activity is inhibited (Calvi et al., 1998). If MPM2 staining patterns indeed reflect CycE/CDK2 activity then it would appear that CDK2 activity does oscillate during follicle cell endo cycles. In nurse cell endo cycles CycE/CDK2 activity is also thought to oscillate, because levels of CycE protein change (Lilly and Spradling, 1996). However, MPM2 staining patterns do not reflect such oscillations in nurse cells (Calvi et al., 1998; I. Royzman and T.L. Orr-Weaver, unpublished results).

3.4.3. Under-replication of heterochromatin

Follicle cells almost completely suppress replication of centromeric and pericentromeric satellite DNA during endo cycles (Hammond and Laird, 1985). As detailed above, a loss-of-function *cycE* mutation restores late replication to nurse cells and allows for replication of heterochromatin, but this mutation has no such detectable effect on follicle cell endo cycle DNA replication (Lilly and Spradling, 1996). Furthermore, mutations in *effete*, *dDP* and *dE2F1* that also lead to heterochromatin replication in nurse cells have no effect on follicle cell heterochromatin replication ((Lilly and Spradling, 1996); I. Royzman and T.L. Orr-Weaver, unpublished). The observations that certain mutations can restore late replication patterns to nurse cells but not follicle cells suggests that some fundamental difference exists between nurse cells and follicle cell endo cycle regulation. One interesting exception exists. Overexpression of CycE can restore late replication to both nurse cells and follicle cells (Lilly and Spradling, 1996). It is proposed that high CycE levels leads to a prolonged S-phase, late origin firing and subsequent heterochromatin replication.

3.4.4. Mitotic character of follicle cell endo cycles

Follicle cells do not undergo the dramatic changes in chromosomal morphology observed in nurse cells. The chromosomes in follicle cells never appear polytene, rather they are dispersed and interphase-like. The *morula* mutation that reverts nurse cell endo cycles to a mitotic like state has no effect on follicle cell endo cycles (Reed and Orr-Weaver, 1997). An extra layer of follicle cell epithelium is observed in the *morula* mutant, however this is likely due to defects in the earlier mitotic cycles when follicle cells are still proliferating (Reed and Orr-Weaver, 1997). This is further evidence that nurse cell and follicle cell endo cycles are controlled differently.

Phosphorylation of histones H1 and H3 is normally associated with condensed mitotic chromosomes, and antibodies specific for these phospho-epitopes can be used to label chromosomes in mitosis. There are no detectable levels of histone H3 phosphorylation in either nurse cells or follicle cells in endo cycles. This is despite the fact that nurse cell polytene chromosomes undergo some condensation (Dej and Spradling, 1999; K. Dej, personal communication). Although follicle cells exhibit no visible change in chromosome morphology phosphorylation of histone H1 does occur during endo cycles (G. Bosco and T.L. Orr-Weaver, unpublished data) raising the possibility that follicle cell endo cycles may enter some abbreviated mitotic prophase. Alternatively, phosphorylation of histone H1 may merely reflect high CDK activity as it is postulated to peak in middle or late S-phase (Lilly and Spradling, 1996), and it may not necessarily be indicative of any mitotic kinase activity. In either case, it will be of interest to determine whether phosphorylation of histone H1 plays a critical role in endo cycle regulation.

3.4.5. Exit from endo cycles

The transition from stage nine to stage 10A eggchambers is when follicle cells exit endo cycles and shut off genomic replication. Only rarely does one see BrdU incorporation in stage 10A follicle cells, and even then BrdU incorporation is minimal (Calvi et al., 1998). This prolonged time marked by a lack of DNA replication can last up to 6 h (Spradling, 1993), and it may be a G1-like phase during which nuclear factors are reorganized for the next step of development. We speculate that there are at least two mechanisms that might account for this exit from endo cycles. Developmental signals and/or achieving appropriate ploidy could cause termination of endo cycles in follicle cells.

The stage 9–10A transition is the stage during which the oocyte nucleus begins signaling the follicle cells, instructing them to differentiate and produce dorsal/anterior patterns (reviewed in Spradling, 1993; Dobens and Raftery, 2000). Thus, it is conceivable that exit from endo cycles is specified by one or more developmental programs impinging on follicle cell cycle control, and that there is a developmental signal that is directly responsible for terminating endo cycles. However, different follicle cell fates are specified at different times, but exit from endo cycles occurs at the same time for all follicle cells of a given egg chamber. This suggests that signals specifying cell fates may be indirectly involved in terminating endo cycles.

Alternatively, follicle cells may be programmed early in development "to count" their ploidy, and shut off endo cycle DNA replication by simply running out of one or more limiting replication factors. Although there is no direct evidence to support such a model, some experimental data are at least consistent with this idea. Overexpression of a DNA replication initiation factor, origin recognition complex 1 (ORC1), is sufficient for initiation of an extra S-phase in stage 10 (Asano and Wharton, 1999). Induction of high amounts of ORC1 protein are not, however, sufficient for the completion of an extra S-phase, as cells overexpressing ORC1 do not double their genome (Asano and Wharton, 1999). This suggests that these cells are capable of at least entering S-phase at this time of development, and perhaps are kept from doing so by insufficient quantities of limiting replication factors.

3.4.6. Rb/E2f pathway genes and exit from endo cycles

In most eukaryotic cells the cell cycle transition from G1-phase to S-phase requires induction of many S-phase specific genes whose expression is sufficient for S-phase initiation (Lam and La Thangue, 1994; Reed, 1997). This restriction point is controlled in large part by the retinoblastoma (RB) family of tumor suppressor proteins (for reviews see Dyson, 1998; Trimarchi and Lees, 2002). This is accomplished when RB is recruited to promoter regions by the E2F transcription factor, and the complex actively represses transcription. Therefore, it might be predicted that genes in the RB pathway would be involved in keeping cells from re-entering S-phase once they have exited endo cycles, a prediction verified by mutant effects in follicle cells (Bosco et al., 2001).

Drosophila female-sterile mutants have been recovered in which there is a failure to keep DNA replication off in stage 10 follicle cells. The genes encoding two subunits of the E2F transcription factor, *dDP* (Royzman et al., 1999) and *dE2F2* (Cayirlioglu et al., 2001), when mutated initiate genomic replication in stage ten follicle cells that should have exited endo cycles and shut off replication. Failure to repress transcription of S-phase specific E2F target genes could explain why these mutants initiate an extra round of DNA replication. Transcript levels of E2F controlled genes were analyzed by in situ hybridizations and were found not to be affected in the *dDP* mutant (Royzman et al., 1999). A more sensitive assay, quantitative RT-PCR, was used for the *dE2F2* mutant, and of five potential E2F target genes that were tested only one of these, *DmORC5*, was found reproducibly to have elevated transcript levels (Cayirlioglu et al., 2001). Furthermore, misexpression of E2F target genes cannot be sufficient for completing an extra S-phase in follicle cells, because FACS analysis and quantitation of DNA content shows that genomic replication is incomplete in the *dE2F2* mutant (Cayirlioglu et al., 2001). DNA content has not been analyzed for the *dDP* mutant, and thus it is not known whether in these mutant follicle cells S-phase is completed (Royzman et al., 1999).

Repression of E2F target genes is predicted to require one of two RB homologs, *Rbf* and *Rbf2* (Du et al., 1996; Du and Dyson, 1999; Frolov et al., 2001). Thus mutations in either of these genes would be expected to have similar phenotypes to *dDP* and *dE2F2*, a failure to keep replication off once follicle cells have exited the endo cycle. Mutations in the *Rbf2* gene have not been reported, however a transposon insertion at the *Rbf* locus reveals an interesting phenotype (Bosco et al., 2001). In this mutant the RBF protein is unaltered but the levels are approximately one half that of normal levels (Du and Dyson, 1999), and this *Rbf* mutant results in sterile females in which follicle cells fail to keep DNA replication off after stage ten (Bosco et al., 2001). In contrast to the *dE2F2* mutants and the overexpression of *DmORC1* that can initiate but not complete an extra S-phase, the *Rbf* mutant can complete a whole round of genomic replication (Bosco et al., 2001). Interestingly, mutations in the *dE2F1* gene exhibit no defect in exiting endo cycles nor in maintaining a G-phase in which DNA replication is off (Royzman et al., 1999). These data suggest that mere derepression of E2F target genes could be sufficient to drive follicle cells to initiate an extra S-phase, but additional lowering of RBF levels is required for completing S-phase.

4. Gene amplification: co-opting a cell cycle event to achieve a developmental end

4.1. A model replication origin

Amplification of the *Drosophila* chorion gene clusters provides a unique metazoan model system for genetic, cytological and biochemical analysis of replication origin structure and function. It is also a wonderful example of how development can co-opt and modify specific aspects of the cell cycle, in this case DNA replication, in order to achieve the demands of differentiated cells. A major role of the *Drosophila* follicle cells is to secrete proteins that make up the chorion (the eggshell) at the proper developmental stage (Spradling, 1993). Most of the chorion genes are arranged into two clusters, one on the X-chromosome and the other on the third chromosome (Spradling, 1981). Construction of the chorion requires synthesis of large amounts of the chorion proteins in a short developmental time, and to accommodate this high demand follicle cells amplify these two chorion gene clusters (Calvi and Spradling, 1999). The X-chromosome cluster is amplified approximately 16-fold whereas the third chromosome cluster is amplified about 60-fold to 80-fold, and the onset as well as termination of amplification of both gene clusters is strictly controlled by development (Calvi and Spradling, 1999; Orr-Weaver, 1991; Spradling, 1999).

Chorion gene amplification employs a replicative mechanism. This requires DNA synthesis to occur in a controlled manner. Most if not all of the DNA replication machinery that is normally used during S-phase is involved in this process (Spradling, 1999). In addition, specific DNA sequences behave as replication origins, where replication proteins are nucleated and initiate replication. Thus, amplification of the chorion genes in *Drosophila* follicle cells provides researchers with a model system for investigating how metazoan replication origins are structured as well as how replication factors activate them.

Gene amplification is known to occur in insect and mammalian cells, including many human tumor cells (for reviews see Calvi and Spradling, 1999; Spradling, 1999; Hamlin et al., 1991; Windle et al., 1991). The mechanisms by which cells can amplify genes vary, but in general two mechanisms have been described. One is through a DNA rearrangement that usually leads to tandem repeats, inverted repeats or extrachromosomal copies (Singer et al., 2000; reviewed in Hamlin et al., 1991; Windle et al., 1991). We will not be discussing this mechanism, because it is not a developmentally controlled mechanism.

In *Drosophila* and other insects gene amplification occurs by repeated rounds of replication initiation. Repeated initiation events results in an "onion skin" structure, first coined by Michael Botchan and colleagues (Botchan et al., 1979). In this structure the maximum copy number is achieved at the origin and tapers off with distance from the origin (Fig. 3). Direct evidence for the onion skin amplification model has been obtained (Osheim and Miller, 1983; Osheim et al., 1988). Electron micrographs of chromatin spreads from *Drosophila* follicle cells undergoing amplification revealed replication forks within replication forks (Osheim and Miller, 1983; Osheim et al., 1988).

Replication origins have been defined most precisely in the budding yeast. In *S. cerevisiae* small specific sequences have been identified that both recruit replication proteins and act as sites for replication initiation (for review see Diffley, 2001; Mechali, 2001). In other systems such as *S. pombe, Xenopus, Drosophila* and mammalian cell lines replication origins are more complex than those found in the budding yeast, and discrete DNA sequence delineating an origin consensus sequence has not been forthcoming (for review see Mechali, 2001; DePamphilis, 1999; Dutta and Bell, 1997). In yeast a powerful test for origin function has been the ability of DNA fragments to permit plasmid replication, hence the name autonomously replicating sequences (ARS) (Stinchcomb et al., 1979). Such a plasmid assay has not been possible in metazoa, and analyses have been restricted to genomic sites. Mapping of replication initiation sites in metazoa reveals that origin regions actually have several sites at which DNA synthesis initiates, spread throughout many kilobases (DePamphilis, 1999; Hamlin and Dijkwel, 1995; Linskens and Huberman, 1990). This is in contrast to yeast origins where one initiation site exists per origin (Bielinsky and Gerbi, 1999). In addition, sequences far from an origin can strongly influence origin activity (DePamphilis, 1999), making it difficult to ascertain the function of integrated constructs at ectopic sites.

Is follicle cell chorion gene amplification a good model system for studying metazoan replication? Two facts, detailed below, argue that it is: First, the *cis*-regulatory elements that control amplification have structural and functional similarities to budding yeast replication origins (Swimmer et al., 1990). Second, many of the *Drosophila* replication initiation factor homologs have been shown to be important for proper onset and levels of chorion gene amplification (Table 1) (for review see (Spradling, 1999). Moreover, these factors appear to play analogous roles as in other systems. Amplification, however, does differ from *bona fide* metazoan DNA replication, because eukaryotic replication origins normally fire once and only once per cell cycle (Diffley, 1996; Diffley, 2001). This is thought to be true even during endo cycles. By contrast, during gene amplification the origins activated at the chorion loci fire repeatedly within a small window of time (Calvi et al., 1998). This indicates that chorion loci origins may be immune to mechanisms that usually inhibit origins from firing more than once. Consequently, amplification provides

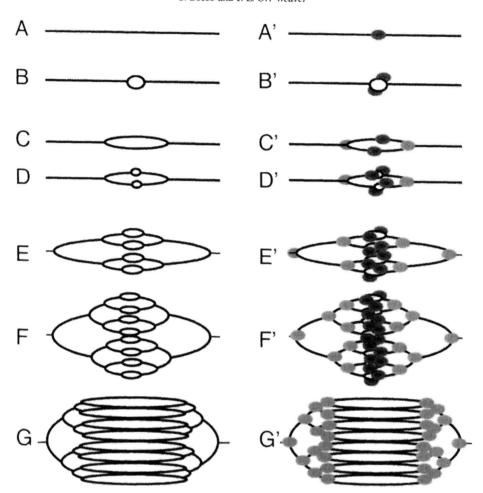

Fig. 3. Schematic and cytological illustration of gene amplification. A. A linear chromosome before replication initiates loads initiator proteins, such as ORC (shown in red, A′), and is first detected in stage 10A. At this stage there is no detectable BrdU incorporation (DNA synthesis) and it is not clear whether initiation events are taking place. B. Replication initiation is first detected cytologically at stage 10B and initiator proteins are still bound (B′). C–F. Many initiation events occur within a short window of time, forming an "onion skin" structure of bubbles within bubbles (see text). C′–F′. Initiator proteins (red) are still bound to chromatin as intiation events continue. Replication elongation moves replication forks away from the origin and replication factors involved in elongation (e.g. PCNA and MCMs, depicted in green) travel with the replication forks. G. Late stage 10B and stages 11–13 continue to synthesize DNA as replication forks proceed away from the origin, but initiation events have ceased. The maximum level of amplification has occurred at the origin. G′. Initiator proteins are no longer localized, but elongation factors (green) are still present at replication forks. (*For a colored version of this figure, see plate section, page 274.*)

Table 1
Drosophila cell cycle regulatory genes

Gene	Protein product, function	Mutations	Evidence for role in:	Reference
Broad-Complex	Broad-Complex, zinc finger transcription factor	Y	amplification	Tzolovsky et al., 1999
c(3)G	C(3)G, synaptonemal complex component	Y*	meiosis	Page and Hawley, 2001
cdc2	CDK1, protein kinase	Y	S/M, S/G2/M, G1/S/G2/M	Stern et al., 1993
cdc2c	CDK2, protein kinase	Y	? all types	Lane, 2000
cdc45	CDC45, replication factor	N	? all types	Loebel et al., 2000
cdc6	CDC6, replication factor	N	? all types	
cdc7	CDC7, CDK-like protein kinase with Chiffon/Dbf4	N	? all types	
cdk4	CDK4, protein kinase	Y	cell growth	Sauer et al., 1996; Jacobs et al., 2001
cfo	CFO	Y	S/M	Wakefield et al., 2000
chiffon	CHIF/Dbf4, binds Cdc7 kinase	Y	amplification	Landis and Tower, 1999
cortex	CORT, Cdc20-like	Y	meiosis	Lieberfarb et al., 1996; Page and Orr-Weaver, 1996; Chu et al., 2001
cup	CUP	Y	nurse cell endo cycle	Keyes and Spradling, 1997
cyclin A	CycA	Y	S/M, S/G2/M, G1/S/G2/M	Lehner and O'Farrell, 1989; Lehner and O'Farrell, 1990
cyclin B	CycB	Y	S/M, S/G2/M, G1/S/G2/M	Sigrist et al., 1995
cyclin B3	CycB3	Y	S/M, S/G2/M, meiosis	Jacobs et al., 1998
cyclin D	CycD	N		Datar et al., 2000; Meyer et al., 2000
cyclin E	CycE	Y	all types	Knoblich et al., 1994; Richardson et al., 1993
dacapo	DAP, CDK2 inhibitor	Y	G1/S/G2/M	de Nooij et al., 1996; Lane et al., 1996
DmORC1	DmORC1, replication initiator	N	all types?, amplification	Asano and Wharton, 1999; Chesnokov et al., 2001
DmORC2	DmORC2, replication initiator	Y	all types, amplification	Landis et al., 1997; Royzman et al., 1999
DmORC5	DmORC5, replication initiator	N	all types?, amplification	Loebel et al., 2000
double parked	DUP/Cdt1, replication factor	Y	all types, amplification	Whittaker et al., 2000
dDP	dDP, transcription factor	Y	endo cycle, G1/S/G2/M amplification	Royzman et al., 1997; Duronio et al., 1998; Royzman et al., 1999
Dwee1	DWEE1/Wee1, protein kinase	Y	S/G2/M	Price et al., 2000
dE2F1	dE2F1, transcription factor	Y	endo cycle, G1/S/G2/M amplification	Duronio et al., 1998; Royzman et al., 1999

(*continued on next page*)

Table 1 (*continued*)

Gene	Protein product, function	Mutations	Evidence for role in:	Reference
dE2F2	dE2F2, transcription factor	Y	follicle cell endo cycle	Cayirlioglu et al., 2001; Frolov et al., 2001; Sawado et al., 1998
effete	EFF, ubiquitin conjugating enzyme	Y	nurse cell endo cycle	Lilly and Spradling, 1996
fs(2)B	?	Y	nurse cell endo cycle	Koch and King, 1964
furhstart	?	Y	S/G2/M	Grosshans and Wieschaus, 2000
geminin	GEMININ, inihibitor of DNA replication	Y	amplification, S/G2/M, endo cycle	Quinn et al., 2001
gnu	GNU	Y	S/M	Freeman and Glover, 1987
grapes	GRP/Chk1, protein kinase, DNA checkpoint	Y	S/G2/M	Sibon et al., 1997; Yu et al., 2000
grauzone	GRAU, transcription factor	Y	meiosis	Lieberfarb et al., 1996; Page and Orr-Weaver, 1996; Chen et al., 2000
mei-41	MEI-41/ATM, protein kinase, DNA checkpoint	Y	meiosis, S/G2/M	Hari et al., 1995; Sibon et al., 1999
mei-P26	MEI-P26	Y	meiosis	Page et al., 2000
mei-S332	MEI-S332, sister-chromatid cohesion at the centromere	Y	meiosis	Kerrebrock et al., 1992
mei-W68	MEI-W68/Spo11, meiotic recombination initiation	Y	meiosis	McKim and Hayashi-Hagihara, 1998
morula	?	Y	endo cycle, S/M, G1/S/G2/M	Reed and Orr-Weaver, 1997
mus101	MUS101, DNA repair	Y	amplification	Yamamoto et al., 2000
ord	ORD, sister-chromatid cohesion	Y	meiosis	Miyazaki and Orr-Weaver, 1992
okra	OKR, DNA repair	Y	meiosis	Ghabrial et al., 1998
otu	OTU	Y	nurse cell endo cycle	King et al., 1981
pan gu	PNG, protein kinase	Y	S/M	Shamanski and Orr-Weaver, 1991; Fenger et al., 2000
mus209	PCNA, DNA pol. processivity	Y	all types, amplification	Henderson et al., 1994, 2000
plutonium	PLU	Y	S/M	Shamanski and Orr-Weaver, 1991
rbf	RBF, pRB-like transcriptional repressor	Y	follicle cell endo cycle, G1/S/G2/M amplification	Du et al., 1996; Bosco et al., 2001
rbf2	RBF2, pRB-like transcriptional repressor	N	?	
spn-B	SPN-B, recombinase	Y	meiosis	Ghabrial et al., 1998
spn-C	SPN-C, recombinase	Y	meiosis	Ghabrial et al., 1998
spn-D	SPN-D, recombinase	Y	meiosis	Ghabrial et al., 1998
string	STG/Cdc25, protein phosphatase	Y	S/G2/M	Edgar and O'Farrell, 1989
tribbles	TRB, inhibitor of STG/Cdc25	Y	S/G2/M	Grosshans and Wieschaus, 2000; Mata et al., 2000; Seher and Leptin, 2000
twine	TWE/Cdc25, protein phosphatase	Y	meiosis	Alphey et al., 1992; Courtot et al., 1992

the opportunity to analyze *cis* elements of an origin and proteins acting there, but it is also a model system for understanding the regulatory mechanisms affecting origin activation.

4.2. Cis-acting amplification controls

Since chorion gene amplification in follicle cells can be measured quantitatively and temporally it has been possible to determine what DNA sequences are necessary to support both the timing and levels of amplification. Transgenic flies can be made by using P-element transposons that carry genes of interest (Spradling, 1986). These integrate at ectopic sites in the genome of germ line cells to produce stably transformed lines. Using this method allows for integration of various deletion constructs that can be assayed for their ability to amplify in vivo, thus delineating the chorion sequences involved in controlling amplification. Such studies with third chromosome sequences determined that several regions were important: A 320 basepair amplification control element (ACE3) was found to be important for high levels of amplification (Orr-Weaver and Spradling, 1986), whereas four amplification enhancing regions (AER-A,B,C and D) were defined as stimulating amplification (Delidakis and Kafatos, 1987; 1988). Replication was observed to initiate within more than one of these regions, both at the endogenous locus as well as at an ectopic transgene (Delidakis and Kafatos, 1989; Heck and Spradling, 1990).

ACE3 and AER sequences seem to have redundant functions (Swimmer et al., 1989); (Carminati et al., 1992). Constructs with few AER sequences require ACE3 for amplification, whereas constructs containing all the AER sequences and some flanking DNA do not require ACE3 DNA (Swimmer et al., 1989). Multimers of ACE3 alone are sufficient for low levels of amplification that remarkably retained tissue and temporal specificity (Carminati et al., 1992). Interestingly, these ACE3 multimers exhibited peak amplification levels in flanking regions outside the ACE3 multimer, suggesting that replication may have initiated at some origin(s) flanking ACE3 and not within ACE3 itself (Carminati et al., 1992). Normally the highest level of amplification is observed at the site of replication initiation (Fig. 3).

Because each transformant line has the transposon inserted at a different site, position effects were observed to greatly influence amplification levels (de Cicco and Spradling, 1984). This necessitated analysis of a large number of independent transformants to obtain an accurate measurement of amplification levels for any given construct. The observation by Lu and Tower that transcriptional insulator elements (suppressor of Hairy-wing protein binding sites) would protect amplification transposons from position effects provided a powerful method to delineate amplification regulatory elements (Lu and Tower, 1997). Insulated transgenes are able to amplify to levels comparable to the endogenous locus at most insertion sites (Lu and Tower, 1997). Both ACE3 and a region containing AER-D, called ori-β, are necessary and sufficient for supporting the proper levels of gene amplification (Lu et al., 2001). Two-dimensional gel mapping studies demonstrated that initiation occurs in ori-β DNA but not in ACE3 sequences, indicating that ACE3 function is to enhance origin activity at the adjacent site (Lu et al., 2001). This observation is consistent with previous results where multimers of ACE3 stimulated origin activity outside the multimer array (Carminati et al., 1992). Further evidence that ACE3 controls adjacent origin activity is evidenced by the fact that insulator elements placed between ACE3 and ori-β abolishes origin activity at ori-β (Lu et al., 2001).The fact that ACE3 is not essential when large fragments in the transposon contain all AERs suggests that other sequences in the region are able to similarly stimulate origin activity within ori-β (Orr-Weaver and Spradling, 1986; Swimmer et al., 1989).

Both the 320 bp ACE3 and the 884 bp ori-β regions contain A/T rich sequences that resemble the yeast ARS consensus that recruit replication proteins, such as the origin recognition complex (ORC) (reviewed in Calvi and Spradling, 1999; Spradling, 1999; Bell and Stillman, 1992). The ORC complex contains a set of six conserved proteins that bind to origin DNA in vivo and in vitro. Austin et al. (Austin et al., 1999) showed that both ACE3 and AER-D (contained within ori-β) can be bound by *Drosophila* ORC in vitro. Chromatin immunoprecipitations demonstrated that ORC also binds these regions in vivo (Austin et al., 1999). Immunolocalization experiments localized ORC to endogenous amplifying loci as well as some ectopic amplifying transposons (Royzman et al., 1999; Austin et al., 1999). A transposon containing a multimer of ACE3, but not ori-β, can localize ORC, whereas transposons with ACE3 deleted cannot localize ORC (Austin et al., 1999). Interestingly, ChIP analysis suggested that ORC binds DNA outside the ACE3 multimer array, but when this same sequence was tested in the absence of ACE3 ORC was not bound to it (Austin et al., 1999). Taken together these data suggests that ACE3 is necessary and sufficient to localize ORC in vivo. These data also indicate that ACE3 nucleates ORC, and that ORC may move or spread to adjacent sequences and activate origins. This view is consistent with data indicating that ACE3 multimers could stimulate replication initiation within flanking DNA (Carminati et al., 1992).

One curious exception to ACE3 being sufficient for ORC localization comes from the study done by Lu et al. (Lu et al., 2001). As described above, transposons carrying both ACE3 and ori-β regions flanked by transcriptional insulator elements are protected from position effects and were able to amplify to levels comparable to the endogenous locus (Lu and Tower, 1997). However, the authors failed to immunolocalize ORC to this transposon even though it contained both ACE3 and ori-β and amplification levels were high (Lu et al., 2001). One possible explanation for this is that nucleation of ORC to ACE3 may be sufficient for activation of nearby origins but accumulation and movement or spreading of ORC to adjacent sites (in this case beyond the insulator sites) may be necessary for visualization by immunolocalization. ORC movement may be blocked by the insulator elements. Consistent with the idea that insulator elements may block ORC accumulation is the observation that insulator elements inhibit ACE3 from acting on flanking DNA origins and on ori-β itself when the insulator element was placed between ACE3 and ori-β (Lu et al., 2001).

4.3. Replication factors important for amplification

Identification of the trans-acting factors that bind to the *cis*-regulatory amplification elements has been facilitated by three aspects of *Drosophila* follicle cell biology. First, reduction of amplification causes thin eggshells, a phenotype associated with female sterility and is easy to identify. A curious aspect of chorion gene amplification is that it places extremely high demand on the replication process itself. Therefore, weak mutant alleles of essential replication factors allow for adult viability but result in female sterility

(Spradling, 1999; Calvi and Spradling, 1999). It is possible that very high replication factor activity is necessary for the rapid bursts of replication initiation occurring during amplification, and factors that may be even moderately compromised will lead to defects in chorion gene amplification. Second, immunolocalization of trans-acting factors has made it possible to ascertain the regulation of protein loading onto the origins of replication. Third, follicle cells remain alive for several hours in vitro making it possible to detect replication foci by BrdU labeling and treatment with various inhibitors. When combined these techniques make chorion gene amplification a very powerful model metazoan system for studying the regulation of replication.

ORC is the first of many replication proteins that are recruited to origins (Bell and Stillman, 1992). ORC loading allows further recruitment of other proteins and establishment of a pre-replicative complex (for review see Dutta and Bell, 1997; Diffley, 2001). The pre-replicative complex allows the origin to fire and initiate DNA synthesis. In follicle cells undergoing amplification, ORC has been shown to localize to specific subnuclear foci (Royzman et al., 1999; Austin et al., 1999; Asano and Wharton, 1999). *Drosophila* ORC has been shown to be necessary for an in vitro replication assay using embryonic extracts, and this is dependent on the ATPase activity of ORC1 (Chesnokov et al., 1999; Chesnokov et al., 2001). One of the first thin eggshell mutants (*fs(3)293*) described is an allele of the *k43* gene, which has been cloned and shown to be the *Drosophila* ORC2 homolog, now referred to as *DmORC2* (Snyder et al., 1986; Landis et al., 1997). A strong loss-of-function allele of DmORC2 is recessive lethal. Thus, it is clear that *Drosophila* ORC is important for DNA replication in general and chorion gene amplification in particular.

At least 11 proteins (excluding the six ORC proteins, cyclin dependent kinases and polymerases) are required for activating an origin and initiating replication (Diffley, 1996; 2001; Dutta and Bell, 1997). Homologs for all these proteins have been identified in *Drosophila* and some have been shown to be important for gene amplification (Spradling, 1999). A list of factors thought to be involved in chorion gene amplification is shown in Table 1. Here we will discuss only some of those factors that are known to have a critical role in amplification.

The Double Parked protein (*Drosophila* DUP/Cdt1) has been shown to localize to the chorion genes during amplification, and interestingly DUP loads on the chorion loci after ORC and persists there after ORC is removed (Whittaker et al., 2000). A female-sterile allele of the *DUP* gene (*fs(2)PA77*) exhibits a thin chorion phenotype, delays the onset of amplification and decreases the levels of gene amplification (Underwood et al., 1990; Whittaker et al., 2000). DUP is essential for DNA replication at other times in development as well, and studies in Xenopus and *S. pombe* have shown that Cdt1 (with Cdc6) is needed to load the MCM proteins at origins (Nishitani et al., 2000; Whittaker et al., 2000; Maiorano et al., 2000). These data indicate that DUP/Cdt1 is an essential component for proper amplification, and consistent with yeast and *Xenopus* experiments DUP loads onto chromatin after ORC (Whittaker et al., 2000).

The MCMs (a hexamer of MCM2-7) must be loaded for an origin to be competent for activation, but they also move with the replication forks and have helicase activity (Aparicio et al., 1997; Ishimi, 1997). They have been immunolocalized in *Drosophila* and shown to associate and dissociate from chromatin in a CycE/CDK2 dependent manner (Su et al., 1996, 1997; Su and O'Farrell, 1997, 1998). Recently, MCMs have also been localized to the chorion loci during amplification (D. MacAlpine and S.P. Bell, personal communication). Although lethal *Drosophila* mutants of some of the MCM genes exist (Feger et al., 1995) female-sterile alleles that specifically affect amplification have not been reported. A Cdc6 homolog also exists in flies, but it has not been studied genetically.

One of the last activating steps in replication initiation is the phosphorylation of origin components by the Cdc7-Dbf4 complex (for review see Dutta and Bell, 1997; Diffley, 2001). Dbf4 interacts directly with origin components (Dowell et al., 1994). In budding yeast the Cdc7 protein has been shown to be a Cdk-like serine/threonine protein kinase that is recruited to origins of replication by virtue of its tight association to the Dbf4 protein (Dowell et al., 1994). The *Drosophila* Cdc7 homolog has not been studied. However, there are female-sterile mutations in *chiffon*, a *Drosophila* gene that shows homology to Dbf4 (Landis and Tower, 1999). These mutant females produce embryos with thin eggshells that result from decreased levels of gene amplification (Landis and Tower, 1999). The Chiffon protein has not been tested for localization to chorion loci, but given the prediction that Cdc7-Dbf4 phosphorylation of origin components is a late step in origin activation it would be of interest if one could visualize a temporal difference in loading of Chiffon protein and early initiator proteins such as ORC.

4.4. Regulation of initiation timing and levels of amplification

Two reported mutants result in levels of amplification that are higher than normal. One mutation is in the *dE2F1* gene and the other is in the *Rbf* gene (the fly RB homolog) (Royzman et al., 1999; Bosco et al., 2001). Both are loss-of-function mutations, suggesting that some mechanism must exist that normally limits the number of origin initiation events. There are two ways in which the number of initiation events can be increased. First, the window of time during which amplification initiations occur could be lengthened. Second, the frequency of initiation events can be increased. These are not mutually exclusive possibilities that can lead to higher amplification levels.

The *dE2F1^{i2}* mutation causes a truncated E2F1 protein that cannot transactivate transcription or bind the RBF repressor (Royzman et al., 1999; Bosco et al., 2001). Mutants in which this truncated form is the only form of dE2F1 are female sterile with higher than normal levels of chorion gene amplification (Royzman et al., 1999). Careful staging of eggchambers and quantitative PCR analysis revealed that this mutant commences amplification earlier than their sibling control flies (Royzman et al., 1999).

RBF is usually recruited to chromatin through its interaction with dE2F. Therefore, an *Rbf* loss-of-function mutation might be expected to phenocopy the *dE2F1^{i2}* truncation mutation that produces a protein that is unable to interact with RBF. Deletion of the *Rbf* gene leads to lethality, whereas a mutation that produces low levels of wild-type protein, *Rbf^{120a}*, leads to female sterility (Du et al., 1996; Du and Dyson, 1999). This *Rbf^{120a}* mutation also leads to abnormally high levels of gene amplification (Bosco et al., 2001). Analysis of various E2F target gene transcripts, by in situ hybridization, failed to detect any visible change in either *dE2F1^{i2}* or the *Rbf^{120a}* follicle cells (Royzman et al., 1999; Bosco et al., 2001). Moreover, in vitro culture of eggchambers in the presence of α-amanatin blocked general transcription but had no effect on the high levels of gene amplification in the *Rbf^120a* mutant follicle cells (Bosco et al., 2001). These data suggest that dE2F1 and RBF proteins act directly to control amplification.

Both dE2F1 and RBF proteins can be immunoprecipitated with DmORC from ovarian extracts, indicating that dE2F1/RBF complexes may directly effect amplification levels through their interaction with ORC (Bosco et al., 2001). Furthermore, chromatin immuno-precipitations showed that in vivo E2F1 and ORC can bind sequences at or near the amplification control element, and this binding occurs at the same developmental time when amplification occurs (Austin et al., 1999; Bosco et al., 2001). Thus, genetic, cytological and biochemical data all suggest that the *Drosophila* E2F1 and RBF proteins function to limit amplification levels directly. The molecular mechanism through which E2F1/RBF repress amplification is not known. The data are consistent, however, with a general model in which E2F1/RBF complexes bind to sites adjacent to origins and directly repress origin activity, possibly through their association with ORC (Bosco et al., 2001). An alternative, but not mutually exclusive, model is that E2F1/RBF complexes serve to recruit histone modifying enzymes and control origin activity by modifying histones and/or replication factors. Whatever the mechanism, control of replication by retinoblastoma-family proteins may be a general phenomenon. Human pRB has been shown to bind MCM7 and inhibit replication in vitro (Sterner et al., 1998). In addition, in early S-phase pRB and associated histone deacetylases has been shown to localize to replication foci in primary human cells, suggesting that pRB may be involved in regulating replication origins (Kennedy et al., 2000).

4.5. Other regulators of amplification

Cyclin E is a major S-phase cyclin and is critical for the G1/S transition. Cyclin E protein levels have been shown to oscillate in proliferating cells as well as endo cycles, and these oscillations are important for cells entering and exiting S-phase (Follette et al., 1998; Weiss et al., 1998; Su and O'Farrell, 1998; Lilly and Spradling, 1996). CycE/CDK2 kinase is thought to phosphorylate components of the pre-replication complex, and in yeast and *Xenopus* CycE can associate directly with ORC2 and Cdc6 proteins (Leatherwood et al., 1996; Furstenthal et al., 2001). Inhibition of CycE/CDK2 activity by in vivo mis-expression of the CDK-inhibitor Dacapo reduced amplification levels, as did in vitro incubation of ovaries with the CDK inhibitor 6-DMAP (Calvi et al., 1998). In stage ten follicle cells undergoing amplification, CycE protein levels accumulate and are not seen to oscillate (Calvi et al., 1998). Taken together, these data indicate that high CycE/CDK2 activity is crucial for proper amplification levels. However, mis-expression of CycE was not sufficient for inducing amplification in stage 10A, a stage where amplification normally does not occur (Calvi et al., 1998).

The Geminin protein is an inhibitor of DNA replication (for review see Diffley, 2001; McGarry and Kirschner, 1998; Tada et al., 2000; Wohlschlegel et al., 2000). Recently, a *Drosophila* Geminin homolog has been reported. Mutations in the *geminin* gene causing female-sterility exhibit increased chorion amplification (Quinn et al., 2001). Geminin is thought to be important for preventing re-initiation of replication origins possibly by inhibiting replication factor recruitment by DUP/Cdt1 in late S-phase, G2 and mitosis (Tada et al., 2001; Wohlschlegel et al. 2000). Geminin is then targeted for proteolysis at the metaphase/anaphase transition by the anaphase promoting complex

(McGarry and Kirschner, 1998). Interestingly, Geminin protein in late follicles cells undergoing amplification has been shown to accumulate and not to oscillate (Quinn et al., 2001). This indicates that although nuclear levels of Geminin are high, the chorion origins are able to escape, at least for some time, the inhibitory effects of Geminin. Ultimately Geminin is required to limit the amount of amplification. Since nuclear Geminin protein levels remain high while amplification is occurring, Geminin either must be activated before it can inhibit amplification or a change at the chorion loci must make the origins vulnerable to Geminin inhibition. Further analysis of Geminin activity at the chorion origins will further elucidate how this replication inhibitor functions in vivo.

Many of the important transitions during *Drosophila* development are controlled by the activity of the hormone ecdysone, including transitions in eggchamber development when follicle cells undergo amplification (Buszczak et al., 1999). Ecdysone pulses lead to dramatic changes in transcriptional programs, and the *Broad-Complex* is an early ecdysone responsive gene that may play an important role in regulating gene amplification (Tzolovsky et al., 1999). The *Broad-Complex* locus encodes several isoforms from a family of zinc finger transcription factors. Ectopic expression of the *Broad-Complex* leads to overamplification of the chorion loci as well as initiation of amplification of other unknown loci (Tzolovsky et al., 1999). In addition, transcriptional activity of the chorion genes is advanced developmentally, and chorion is synthesized several hours earlier than normal (Tzolovsky et al., 1999). Several putative *Broad-Complex* binding sites are present on the third chromosome chorion cluster (G. Bosco and T.L. Orr-Weaver, unpublished). Interestingly, the *fs(1)k10* mutant was first isolated in a screen for thin chorions (Wieschaus, 1978), and this mutant causes mis-expression of another early ecdysone responsive gene, *E75*, in the anterior follicle cells (Buszczak et al., 1999). However, it is not known whether one or more ecdysone responsive genes affect amplification directly at the chorion loci or by indirectly mis-regulating other target genes. Nevertheless, it is intriguing to speculate that ecdysone may play a role in the developmental signal that switches follicle cells from an endo cycle program to gene amplification.

5. S/M embryonic cycles

In many animals early embryogenesis is supported by high levels of maternally supplied factors present in the egg that promote rapid cell cycles without requiring growth or transcription of zygotic genes. In *Drosophila* embryogenesis the first 13 cycles occur in a syncytium, where DNA replication and nuclear divisions occur in a common cytoplasm (for review see Foe et al., 1993). The formation of cell membranes and cellularization of the syncytial nuclei takes place at cycle 14 (for review see Foe et al., 1993). In this section we define "early embryogenesis" as the first 13 nuclear divisions. We will further distinguish between the first nine cycles that are rapid and synchronous, and cycles 10–13 where the cortical nuclear divisions slow down and a gap phase (G2) is slowly added as feedback controls contribute to cell cycle regulation. These cycles 10–13 are called cortical divisions because the nuclei have moved to the cortex (surface) of the embryo (for review see Foe et al., 1993).

Although sharing a common cytoplasm with a stockpile of maternal factors facilitates rapid and nearly synchronous nuclear divisions, it also presents the *Drosophila* embryo with challenges: (1) each nucleus must replicate its genome in a very short period of time; (2) the fast paced S-phase must be coordinated with mitosis and faithful chromosome segregation; (3) the embryo must eventually slow down its cycles, add gap phases and cellularize; (4) there is little time in the first nine cycles to repair DNA damage or chromosome segregation failures and thus abnormal nuclei must be discarded without delaying or perturbing the remaining nuclei; (5) a common cytoplasm means that factors must be activated and inactivated locally at the site of their function; and (6) finally, these early cycles cannot rely on transcriptional control of cell cycle regulators to accomplish these feats, and thus must invoke post-transcriptional regulation of cell cycle factors.

5.1. S phase

5.1.1. Replication origin usage

It has long been observed that the early embryonic cell cycles are quite rapid as compared to later cell cycles or those of *Drosophila* cells in tissue culture (for review see Carminati and Orr-Weaver, 1996). S-phase in early *Drosophila* embryos can be as short as 3.8 min, whereas it can last for up to 600 min in tissue culture cells (Rabinowitz, 1941; Blumenthal et al., 1973). Therefore, some fundamental parameters of S-phase must be different in order to allow embryonic cycles to replicate the genome in such a short period. Using electron microscopy to analyze replicating DNA from early embryos, distances between "eyes" or bubbles that represent activated replication origins were measured (Blumenthal et al., 1973; Kriegstein and Hogness, 1974). They determined that the mean distance from one origin to another was approximately 8 kilobasepairs (Kb), and they also inferred replication fork movement at a rate of 2.6 Kb per minute (Blumenthal et al., 1973). When tissue culture cells were examined the fork movement rate was measured to be approximately the same as that observed in embryonic nuclei. However, the mean "eye-to-eye" distance was measured to be about 40 Kb (Blumenthal et al., 1973). This difference in origin spacing, although dramatic, could not account completely for the vast differences in S-phase duration between embryonic and tissue culture cells. If these origins in tissue culture cells were assumed to be evenly spaced, then S-phase should proceed faster than the observed 600 min. However, if origins in tissue culture cells were not evenly spaced, but were clustered this could account for the long time that is required for tissue culture cells to complete genomic replication. Conversely, origins in the early embryonic cycles are thought to be evenly spaced (Blumenthal et al., 1973). The mechanism by which origin usage is specified in embryos remains a mystery, but the frequency with which sequences are used raises the possibility that no specific sequence motifs are required for origin function at this stage.

In addition to uniform spacing of replication origins, it is likely that in early embryos all origins fire synchronously. In differentiated cells some origins are activated early in S phase, whereas other origins replicate late. In contrast, in early *Drosophila* embryos there is no cytological evidence of heterochromatin, and this would account for the lack of late replication patterns or any sort of suppression of origin activity within the centric and pericentric satellite sequences (McKnight and Miller, 1976, 1977).

5.1.2. Maternal stockpiles

During oogenesis the polyploid nurse cells act as factories that produce large quantities of proteins and mRNA that are then dumped into the oocyte (for reviews see Spradling, 1993; Foe et al., 1993). These maternally supplied stockpiles prime the oocyte for starting the rapid embryonic cell cycles upon fertilization. In these early cell cycles gene transcription does not occur (with the exception of pattern formation genes), thus many of the maternally supplied factors must be regulated post-transcriptionally. As will be discussed below, local degradation of certain regulators, such as Cyclin B, is critical for these rapid cycles to proceed, and the progressive diminishing levels of these regulators give way to slower cycles (cycles 11–13).

5.1.3. Entering and exiting S phase

The cell cycle machinery necessary for the onset of S-phase during the S/M cycles is not well understood. In normally proliferating cells the levels of CycE/CDK2 activity oscillate, peaking at the G1/S transition, and CycE/CDK2 activity drives the onset of S-phase (for review see, Nakayama et al., 2001). In early embryonic cell cycles high levels of maternally supplied CycE/CDK2 are constitutively active and no apparent oscillation is observed (Richardson et al., 1993; Knoblich et al., 1994). Thus it is presumed that local oscillations of CycE/CDK2 activity are responsible for regulating S-phase onset in the early embryonic nuclear divisions. We emphasize, however, that early embryonic nuclear divisions (cycles 1–9) do not have gap phases (i.e. no G1 and G2 phases), and the cortical divisions (cycles 10–13) only have modest G2 delays. As soon as chromosomes decondense in telophase DNA replication initiates. Therefore, it is not clear whether factors normally involved in the G1/S transition are as important in an M-to-S transition.

In cellularized embryos that have incorporated a G1 phase, CycA overexpression can drive cells into S-phase, even in the absence of CycE protein (Sprenger et al., 1997). This observation raises the possibility that Cyclin A may be important for S-phase onset, especially in the early cycles that do not have a G1 phase. This possibility has not been tested directly. A recent report has shown that CycA functions in mitosis, however these studies do not exclude an S-phase function for CycA (Jacobs et al., 2001).

At least three genes are needed for the exit from S-phase and the onset of M during the *Drosophila* S/M cycles. These genes, *pan gu* (*png*), *plutonium* (*plu*) and *giant nuclei* (*gnu*), are required for condensation of the meiotic products and therefore inhibition of S-phase in the unfertilized egg, as described in section 2.4 (Fig. 4) (Freeman and Glover, 1987; Freeman et al., 1986; Shamanski and Orr-Weaver, 1991). In addition, these genes specifically regulate the S/M cycles and are not required at any other time in development (Axton et al., 1994; Elfring et al., 1997; Fenger et al., 2000). In fertilized embryos from null mutant mothers DNA replication occurs but not mitosis (Freeman et al., 1986; Shamanski and Orr-Weaver, 1991). In fertilized embryos from mothers carrying weak alleles a few mitotic nuclear divisions can occur (Shamanski and Orr-Weaver, 1991). However, after only a few mitotic divisions the nuclei are locked into repeated rounds of DNA synthesis without mitosis, leading to giant polyploid nuclei (Shamanski and Orr-Weaver, 1991).

Interestingly, the levels of two mitotic cyclins, CycA and CycB, were found to be reduced in proportion to the severity of the mutant allele (Fenger et al., 2000). This observation led

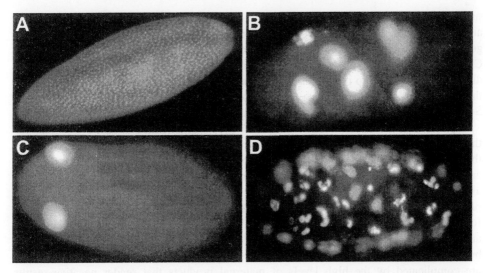

Fig. 4. DNA staining of wild-type and *png* mutant embryo. A. Wild-type syncytial embryo. B. Embryo from homozygous *png³³¹⁸* mothers. This is a weak allele that allows some mitotic divisions before the nuclei are stuck in S-phase and become giant polyploid nuclei. C. Weak *png* mutants are enhanced by reducing the dosage of *Cyclin B*. D. Introducing extra copies of the *Cyclin B* gene suppresses the *png* mutant defect and allows for many more mitotic divisions and condensation of chromosomes.

to the suggestion that normal mitotic cyclin activity is required for S-phase cessation in these early cycles. Indeed, the giant nuclei phenotype in these mutants was enhanced by lowering the CycB levels, whereas the phenotype was suppressed when extra copies of the CycB gene were introduced (Lee et al., 2001). Thus the *png*, *plu* and *gnu* gene products play two important roles in regulating S-phase in the early embryo. First, they serve to inhibit the very onset of S-phase until the egg can be fertilized. Second, once the embryonic cycles have initiated they serve to link S-phase completion with efficient entry into mitosis. Both of these functions are mediated by the same mechanism: ensuring adequate levels of mitotic Cyclin proteins for activation of Cyc/CDK1 kinase activity to inhibit rereplication of origins and to promote chromosome condensation.

5.1.4. Maternal stockpiles likely drive the early cycles

Many replication factors and other proteins important for S-phase are supplied to the embryo in maternal stockpiles, but in the early cycles a requirement for any one replication factor has not been directly demonstrated. It is likely that many of the factors that are critical for replication in other tissues are also important for early embryonic S-phase. However, the significant differences between these rapid S-phases and those that occur later in development raise the possibility that there could be some factors that are less important or dispensable for these special early S-phases. For example, the observation that replication origins must occur at a frequency of approximately one in every 8 Kb and are fairly evenly spaced suggests that during these rapid S-phases origins are loosely defined (Blumenthal et al., 1973; Kriegstein and Hogness, 1974). The initiator proteins, such as ORC, therefore may be indiscriminate in selecting binding sites or some mechanism independent of ORC could select and initiate replication. Recent studies have suggested that *Drosophila* ORC will bind to any A/T rich sequence (Austin et al., 1999; Chesnokov et al., 2001). An additional distinction is that these early cycles transition directly from S-phase to mitosis and from mitosis directly to S-phase. Replication factors that normally load onto chromatin in G1 do not have this luxury of time, thus these rapid S-phases may be different from the more orthodox S-phase that is preceded by G1.

Experiments that might seek to ascertain whether a particular factor is necessary for these early S-phases are complicated for two reasons. First, null mutations in the genes for these factors are expected to be lethal and germ line clones of null mutants are likely to arrest too early in oogenesis to produce an embryo. Second, although many hypomorphic alleles of replication factor genes exist, these are often too weak to exhibit an embryonic phenotype, and high levels of maternal products in the embryo potentially compensate for decreased activity of any one protein (Foe et al., 1993). Conceivably, injection of antibodies or inhibitory double stranded RNA (RNAi) may lead to informative results, however this has not been systematically done for replication factors. Again, such studies are complicated by the high levels of maternal products that render negative results meaningless.

5.2. M phase

The promotion of mitosis during the S/M cycles most likely utilizes the same cell cycle trigger as in archetypal cycles, activation of mitotic Cyclin/CDK1 complexes. The distinctions are that changes in the pools of mitotic Cyclin proteins are restricted to the vicinity of each nucleus. The increases responsible for the onset of mitosis and the degradation leading to exit from mitosis are likely to be localized around each nucleus. In addition, inhibitory phosphorylation of CDK1 does not occur prior to the addition of a G2 phase.

The mitotic cyclins, Cyclin A, Cyclin B and Cyclin B3, complex with a catalytic subunit, Cyclin Dependent Kinase 1 (CDK1), and serve to phosphorylate key cell cycle regulators (Sigrist et al., 1995). CDK1 activity itself is regulated by phosphorylation status, where phosphorylation of tyrosine-15 inactivates CDK1 kinase activity (for review see Murray and Hunt, 1993). All three mitotic cyclins and CDK1 are abundantly supplied to the embryo as maternal transcripts (Lehner and O'Farrell, 1990; Sigrist et al., 1995; Whitfield et al., 1990). *CycA* and *cycB3* mRNA are uniformly distributed throughout the embryo, whereas *cycB* mRNA is more concentrated at the posterior end of the embryo (Lehner and O'Farrell, 1990; Dalby and Glover, 1992). Analysis of protein levels with antibodies to these cyclins revealed that they were all evenly distributed throughout the embryo, and the apparent concentration of *cycB* mRNA in the posterior pole was not reflected by its protein levels (Lehner and O'Farrell, 1990). Embryos deficient for either *cycA* or *cycB* exhibited no cell cycle defects in cycles 1–9, but embryos from mothers heterozygous for mutations in these genes did slow down their cell cycle progression in cycles 10–13 when some cyclin degradation can be detected (Edgar et al., 1994). Embryos that are homozygous mutant for *cycA* arrest after the syncytial divisions, during cycle 15 (Lehner and O'Farrell, 1990). Cyclin B3 contains sequence similarity to both CycA and CycB, and all three cyclins are thought to share some functional redundancy

(Knoblich and Lehner, 1993; Jacobs et al., 1998). Surprisingly, mitosis can proceed in either *cycB* or *cycB3* null mutants, and neither gene is necessary for viability (Jacobs et al., 1998).

The phosphatase that is normally responsible for dephosphorylating, and thus activating, CDK1 is Cdc25, String (STG) in *Drosophila*. STG protein levels are very low before fertilization, but rise in the first eight cycles and then fall again (Edgar et al., 1994). STG protein levels change with no apparent alterations in *stg* mRNA levels (Edgar et al., 1994). The STG protein itself is regulated by phosphorylation, and its phosphorylation state begins to fluctuate as early as cycle 5 (Edgar et al., 1994). However, a similar fluctuation in CDK1 phosphorylation is not observed until much later (cycle 14) at cellularization (Edgar et al., 1994).

More recent data demonstrated that destruction of Cyclins is important for the control of the early embryonic cell cycles and that this Cyclin degradation must occur in a very localized manner. Injection of non-degradable forms of either CycA or CycB into these early embryos (before cycle 7) led to mitotic arrest in all stages of embryogenesis, as did injection of CycB amino terminal peptides (Su et al., 1998). This indicates that Cyclin degradation must be required in order for mitosis to proceed, even though it was not visibly detectable. Moreover, CycB protein degradation has been shown to be dependent on the presence of a mitotic spindle apparatus, and CycB itself localizes to the spindle. By using a GFP-tagged CycB protein, Huang and Raff (Huang and Raff, 1999) were able to show in live embryos during the cortical divisions that CycB protein accumulates at centrosomes during interphase and then moves along the mitotic spindle in prometaphase. In metaphase CycB moves to the middle of the spindle, and at the end of metaphase CycB-GFP disappears in a wave, starting at the spindle poles and ending at the spindle equator (Huang and Raff, 1999). This was observed in embryos after cycle 11, however it is likely that the same mechanism applies in the earlier precortical cycles (1–9). Endogenous CycB associated with spindles and centrosomes was found to behave the same in syncytial and cellularized embryos (Huang and Raff, 1999). However, the bulk cytoplasmic (non-microtubule associated) CycB was observed not to fluctuate in the syncytial embryos, consistent with previous reports.

Because CycB removal initiates at the spindle poles, where the centrosomes reside, the role of the centrosome in CycB degradation was examined. Mutation of the *Drosophila centrosome fall off* (*cfo*) gene leads to a coordinate detachment of centrosomes from the mitotic spindles and subsequent anaphase arrest of the centrosomeless spindle, with most embryos arresting in cycles 1–7 (Wakefield et al., 2000). In *cfo* mutant embryos CycB disappears at the detached centrosomes, but remains visibly associated with the arrested anaphase spindle and is enriched at the anaphase spindle equator (Wakefield et al., 2000). This is strong evidence that CycB removal not only initiates at the centrosome, but also that physical association of the centrosome with the spindle is required for removal of CycB from the spindle. Expression of a non-degradable form of CycB did not result in centrosome detachment (Huang and Raff, 1999). Thus, stabilization of CycB on the spindle is an effect (not a cause) of the centrosome detachment. CycB removal from the spindle is thought to reflect its degradation. The anaphase promoting complex/cyclosome (APC/C), a complex that targets proteins for proteolysis, can associate with microtubules from *Drosophila* embryonic extracts, and it has been postulated that spatial regulation of the APC/C at the centrosome is what regulates CycB degradation on the spindle

(Huang and Raff, 1999; Wakefield et al., 2000). Although CycB is known to be targeted by the APC/C for destruction, this has not been shown to occur directly at the centrosome.

Further evidence that CycA and CycB levels are important for entry into mitosis comes from the analysis of giant nuclei mutants. As described above, three *Drosophila* genes *pan gu* (*png*), *plutonium* (*plu*) and *giant nuclei* (*gnu*) are required for mitotic progression in the very early embryo (Freeman and Glover, 1987; Freeman et al., 1986; Shamanski and Orr-Weaver, 1991). Embryos from homozygous mutant mothers have reduced levels of CycA and CycB, and these mutants are defective in the onset of mitosis. The severity of these mutant phenotypes was suppressed when extra copies of the *cycB* gene were introduced and mitotic divisions were restored (Fig. 4) (Fenger et al., 2000; Lee et al., 2001). Thus, proper CycA and CycB levels are also necessary to link S-phase completion with efficient entry into mitosis.

5.3. Addition of G2-phase to the embryonic cycles

The detectable oscillations in cyclin levels in cycles 11–13 coincide with the slowing down of the cortical nuclear divisions (Edgar et al., 1994; Foe and Alberts, 1983). The duration of interphase increases starting after the tenth mitosis. Cycles 2–9 each are approximately 9 min in duration, whereas cycle 11 is 10 min, cycle 12 is 13 min and cycle 13 is 18 min long (Foe and Alberts, 1983). A gap phase (G2) is gradually introduced, and the embryo arrests in a prolonged G2 phase in cycle 14 at cellularization (Foe and Alberts, 1983). The mechanism controlling this transition is not clearly understood, but several models have been proposed. The timing during development of this transition can be altered by changing the nuclear/cytoplasmic ratio, indicating that titration of some maternal factor during the nuclear divisions may control the timing of the transition (Edgar et al., 1986). Interestingly, transcription of zygotic genes is also linked to changes in the nuclear/cytoplasmic ratio (Pritchard and Schubiger, 1996).

The cell cycle slow-down starting at cycle 10 and subsequent introduction of a G2 phase has been shown to be dependent on the ATM/Chk1 DNA damage/replication checkpoint genes. DNA damage/replication is normally monitored and any damage or incomplete replication activates the Chk1 and Rad3/ATM checkpoint pathway. In *Drosophila* the *grapes* (*grp*) gene encodes a homolog of Chk1 and *mei-41* is a homolog of the Rad3/ATM tumor suppressor gene (Fogarty et al., 1997). Chk1/GRP is a serine/threonine kinase and its activity is dependent on ATM/Mei-41 and detection of DNA damage or incomplete replication (for review see Westphal, 1997; Sibon et al, 1999). The Chk1/GRP kinase phosphorylates the Cdc25/String protein (Sigrist et al., 1995). This inactivates the Cdc25/String protein phosphatase. Cdc25/String dephosphorylates CDK1 and activates CycA/CDK1 and CycB/CDK1 kinase activity that drives cells into mitosis. In the absence of an active Cdc25/String, CDK1 is phosphorylated by the Wee1 kinase (Dwee1 in *Drosophila*), and remains inactive (Lundgren et al., 1991; Price et., al, 2000). Therefore an S-phase checkpoint leads to an inactive phosphorylated CDK1 which delays cells from entering mitosis. This delay gives cells the time to repair the problem or undergo cell death.

Drosophila mei-41 mutants (the ATM homolog) and *grp* mutants (the Chk1 homolog) have abnormally short mitoses during the cortical nuclear divisions, they do not stop their divisions after mitosis 13 and fail to cellularize (Sibon et al., 1999, 1997). The mitotic defects also include mis-aligned metaphase chromosomes in cycles 12 and 13, and mutant embryos exhibit structures similar to radiation damaged chromosomes (Fogarty et al., 1997). These mutations result in a dephosphorylated, active form of CDK1 that is thought to promote inappropriate entry into extra mitotic divisions (Fogarty et al., 1997; Sibon et al., 1999). Similarly, mutations in the *Dwee1* gene result in a failure to properly exit the syncytial nuclear divisions and introduce a G2 phase like the *grp* and *mei-41* mutations (Price et al., 2000). Together, these data suggest that triggering of the ATM/Chk1 DNA damage/replication checkpoint is critical for terminating the syncytial divisions and transitioning the embryos to a cell cycle program controlled by zygotic gene expression.

What triggers the Grp/Mei-41 checkpoint that leads to the slow-down of the cortical nuclear divisions? It has been suggested that a key replication factor may be limiting in the embryo. As the syncytial divisions progress, and the number of nuclei increase, this increasingly limiting factor slows down progression of S-phase in the cortical divisions. Thus the normal function of the Grp/Mei-41 S-phase checkpoint is to monitor DNA synthesis and prevent the onset of mitosis until replication is completed (Sibon et al., 1997, 1999). This progressive lengthening of S-phase and delaying of the onset of mitosis allows for an introduction of a G2-phase.

The Grp/Chk1 pathway and the Dwee1 kinase ultimately converge to keep CDK1 in a phosphorylated, inactive state. This raises the possibility that these checkpoint proteins may control CDK1 function in ways other than altering its phosphorylation, since a mitotic defect can be observed in *grp* and *mei-41* mutants as early as cycles 4–8 when CDK1 activity is constitutive (Su et al., 1999). In these mutant embryos CycA was found to be stabilized in cyclohexamide arrested interphase nuclei as well as colchicine arrested mitotic nuclei (Su et al., 1999). It had been previously shown that wild-type embryos normally degrade CycA in both interphase and mitotic arrest points (Edgar et al., 1994). This result has been interpreted to mean that the Grp/Mei-41 checkpoint normally promotes CycA turnover in interphase and mitosis, and that stabilization of CycA in the mutant embryos causes the failure to exit mitosis (Su et al., 1999). This is consistent with previous findings that non-degradable forms of CycA can also lead to a mitotic arrest (Su et al., 1998). It has also been suggested that the Dwee1 kinase may similarly regulate cyclin levels, since mutations in Dwee1 do not result in detectable changes in CDK1 phosphorylation (Price et al., 2000). This possibility has not been tested directly.

How then is a G2-phase introduced? A likely scenario is as follows: Limiting DNA replication factors are titrated out as the cycles progress and lead to slower and slower S-phases. This slow progression through S-phase triggers the Grp/Chk1 checkpoint as slow S-phase is registered as incomplete replication. Checkpoint activation promotes greater and greater maternal pools of CycA to be degraded, below a threshold where entry into mitosis is blocked. In cycle 14, once S-phase is complete and the checkpoint is released CycA levels begin to rise again. Thus, G2 is this period of time, after S-phase completion, when CycA levels are too low for entry into mitosis.

Recent reports suggest that the Grp/Chk1 pathway also is important for proper chromosome condensation (Yu et al., 2000). Embryos treated with a DNA synthesis inhibitor, aphidicolin, block in S-phase, but eventually go on to enter mitosis and condense their incompletely replicated chromosomes normally (Yu et al., 2000). Many mitotic defects ensue as a consequence of the incomplete replicated DNA. Despite the fact that DNA replication is not completed in *grp* mutant embryos, they also progress into mitosis and condense their chromosomes. However, chromosome condensation is abnormal in *grp* mutants, and the onset of nuclear envelope breakdown (NEB) occurs prematurely and the nuclei enter mitosis early (Yu et al., 2000). Why would a DNA-checkpoint mutation lead to a chromosome condensation defect in mitosis? One interpretation is that aphidocolin treated embryos, with an intact checkpoint, can detect incomplete replication and consequently delay NEB. This delay in NEB prevents cytoplasmic CycB/CDK1 access to the chromosomes and subsequent initiation of mitosis and chromosome condensation. In checkpoint deficient mutants, NEB is not delayed and mitosis initiates prematurely. This does not allow enough time for proper condensation (Yu et al., 2000). This model suggests that the Grp/Mei-41 checkpoint specifically controls NEB and exclusion from the nucleus of cytoplasmic pools of active CDK1.

It has also been observed that both *grp* and *mei-41* mutant embryos with nuclei arrested in mitosis have centrosomes that are detached from the nuclei (Sibon et al., 2000). Surprisingly, treatment of embryos with replication inhibitors or induction of DNA damage also leads to mitotic arrest and detachment of centrosomes in cycle 13. This observation is interpreted to mean that DNA damage and/or premature entry into mitosis triggers nuclear/centrosome dissociation, and this is what causes the mitotic arrest (Sibon et al., 2000). Since mitotic arrest and centrosome detachment occurs with or without functional Grp and Mei-41 proteins it has been suggested that centrosome detachment is checkpoint independent (Sibon et al., 2000). This mechanism of detaching centrosomes can be useful because it prevents nuclei with damaged chromosomes from further participation in mitotic divisions.

5.4. Pole cells

The cells that ultimately form in the posterior end of the embryos, the pole cells, are destined to become the germline cells of the adult fly. Whereas the cortical nuclei cellularize at cycle 14, pole cells form after mitosis of cycle 10. The newly formed pole cells continue to divide until the syncytial nuclear cycle 14, and then stop proliferating while cellularization proceeds in the rest of the embryo. Completion of cellularization allows the somatic cells to resume their cell cycles, but the pole cells remain withdrawn in a G2 phase (Su et al., 1998a). This withdrawal from the cell cycle coincides with pole cell migration to the interior of the embryo. It has been suggested that the pole cell withdrawal from the cell cycle is mediated through inhibition of CDK1 (Su et al., 1998a). Recent evidence also suggests that low CycB levels in pole cells are important for maintaining this G2 arrest (Asaoka-Taguchi et al., 1999). Maternally supplied Cyclin B mRNA is concentrated in the posterior pole of the embryo, and high levels of this transcript are incorporated into the pole cells as they form at the posterior end (Lehner and O'Farrell, 1990; Raff et al., 1990; Dalby and Glover, 1992). Apparently translation of this CycB mRNA must be suppressed until the appropriate time comes for pole cells to resume proliferation. This translational suppression is mediated by at least two proteins, Nanos and Pumilio, that interact with the 3'-UTR of the CycB mRNA. Mutations in the *nanos* and *pumilio* genes lead to translation of Cyclin B mRNA and induce inappropriate proliferation of the migrating pole cells

(Asaoka-Taguchi et al., 1999). Thus, germ line cells exert yet another level of control on cell cycle factors by specifically inhibiting CycB translation.

5.5. Nuclear discard mechanisms at cycles 11–13

We have discussed scenarios where alterations in the levels of cell cycle regulators and/or mutations cause global defects in the syncytial nuclear divisions. This results in embryonic arrest or cell cycle delays. However, it would seem inefficient to arrest all the nuclei in the embryo if only one or a few nuclei experienced a catastrophic and irreversible event. *Drosophila* syncytial embryos have evolved a mechanism for discarding individual aberrant or defective nuclei. For example, individual nuclei that have failed to segregate chromatids, two nuclei that have fused into one, or nuclei with centrosome defects tend to arrest in anaphase and then sink into the interior of the embryo (Sullivan et al., 1990; Sullivan et al., 1993). Thus, like checkpoints, this nuclear discard mechanism can detect aberrant nuclei, but unlike checkpoints this mechanism permanently withdraws nuclei from the nuclear divisions and spatially separates them from the rest of the cycling nuclei.

6. S/G2/M cycles

6.1. Control of string expression

Completion of S-phase in cycle 14 requires approximately 40 min. This is in stark contrast to the short (about 4 min) S-phases from earlier cycles. Once S-phase is complete the nuclei enter a prolonged G2 phase (minimum of 30 min). Cellularization initiates in the latter part of S-phase in cycle 14 and is completed about 65 min into cycle 14. Cellularization is a dynamic process of orchestrating membrane movements and cytoskeletal reorganization [for review see (Foe et al., 1993)].

Mitosis of cycle 14 can start as earlier as 70 min into cycle 14 for some cells while other cells initiate mitosis much later (Foe et al., 1993). Subsequent cell division must be coordinated with the various developmental demands to accommodate cell specific changes in morphology and cell migration. One way in which the embryo achieves this is by coordinating clusters of cells so that they enter mitosis together. These so called mitotic domains are useful since it allows groups of cells to coordinate cell division with developmental programs unique to a particular group of cells.

A critical regulator of mitosis at this point of embryonic development is the Cdc25 [String (STG) in *Drosophila*] protein phosphatase (Foe et al., 1993). This phosphatase acts directly on CDK1 and activates it. Active CycA/CDK1 and CycB/CDK1 drive cells into mitosis. A mutant embryo deficient for *string* (*stg*) arrests in interphase of cycle 14, while overexpression of *stg* can cause extra mitotic divisions in cycles 14–16 (Edgar and O'Farrell, 1989; 1990). Moreover, the expression pattern of *string* is quite telling in that it exactly anticipates the timing of mitosis and the specific domains that will initiate mitosis (Edgar and O'Farrell, 1989).

One way in which the *Drosophila* STG protein coordinates developmental processes with cell cycle progression is by being differentially expressed. Complete degradation of maternal *stg* mRNA is achieved by cycle 14. Transcriptional regulation of *stg* in subsequent cycles falls under the control of developmental patterning genes (Edgar et al., 1994). Moreover, it was shown that in cycles 15–17 cells that are arrested by various cyclin mutants *stg* expression is dictated by developmental patterns and timing and not by a cell cycle arrest point (Edgar et al., 1994). This supports the idea that zygotic transcription of the *stg* gene is controlled mainly by developmental factors rather than being cell cycle regulated.

Targeted destruction of the *stg* mRNA is also important in regulating entry into mitosis (Edgar et al., 1994). The transition from maternally controlled cell cycles to those controlled by zygotic gene transcription is complete by cycle 14, and by then most if not all of the maternal *stg* mRNA is degraded (Edgar et al., 1994). However, when the onset of zygotic transcription was blocked by injection of α-amanitin into early embryos (before cycle 6) the transcript was stabilized for two hours longer than normal in cycle 14. The injected embryos failed to cellularize and underwent an extra nuclear division (Edgar and Datar, 1996). Thus, the onset of zygotic transcription leads to the rapid turnover of *stg* mRNA, and this is thought to facilitate a switch to zygotic control of *stg* expression and regulation of mitosis.

6.2. Tribbles: a unique mitotic regulator

After cellularization, groups of adjacent cells, or domains, coordinately enter and exit mitosis in order to facilitate their differentiation and cell migration [for review see (Foe et al., 1993)]. As discussed above, transcriptional control of the *stg* gene is regulated by transcription factors that control development. In all mitotic domains, except one, expression of the *stg* gene immediately precedes the onset of mitosis. In mitotic domain 10, initiation of mitosis is delayed for some time after expression of *stg* can be seen. The cells of domain 10 are found in the ventral portion of the embryo and will go on to form the mesoderm. In order to do so these cells must first invaginate to form a ventral furrow, a process also known as gastrulation, and a failure to delay mitosis in these cells causes lethal defects in gastrulation. The question arises as to how mitosis is blocked in domain 10 even though *stg* expression is clearly on. Three independent reports have revealed that the *tribbles* gene is responsible for inactivating *stg* activity in the embryo (Mata et al., 2000; Seher and Leptin, 2000; Grosshans and Wieschaus, 2000); for review see (Johnston, 2000).

Embryos deficient for the *tribbles* gene product fail to delay mitosis in mitotic domain 10. In *tribbles*, *stg* double mutants gastrulation proceeds normally and the proliferation defect observed in *tribbles* is suppressed. This indicates that the mitotic defect in *tribbles* mutants is *stg* dependent. Furthermore, injection of *tribbles* mRNA into cleavage stage embryos could arrest mitosis at the site of injection, while overexpression of the *tribbles* gene could delay entry into mitosis in larval tissues (Grosshans and Wieschaus, 2000; Mata et al., 2000). In *tribbles* mutant embryos *stg* mRNA accumulates normally but protein levels are abnormally high (Mata et al., 2000). Taken together, these data indicate that the *tribbles* gene product specifically acts to promote String protein turnover in mitotic domain 10, and thus is able to delay mitosis while these cells invaginate. STG protein turnover does not occur in the *tribbles* mutant. Accumulation of STG activates CDK1 which in turn promotes mitosis, and invagination is disrupted because cell proliferation and changes in cell morphology necessary for gastrulation are incompatible.

The *tribbles* gene encodes a protein with homology to serine/threonine kinases (Grosshans and Wieschaus, 2000; Mata et al., 2000; Seher and Leptin, 2000). Although the Tribbles protein appears not to have kinase activity, and it is not clear how it might promote STG protein turnover in the embryo. It has been suggested that Tribbles promotes STG proteolysis through the proteosome pathway (Mata et al., 2000). A mutation in another gene, *furhstart*, has also been identified to have a very similar phenotype as the *tribbles* mutant embryos (Grosshans and Wieschaus, 2000). Exactly how these gene products regulate STG activity remains to be seen. What is clear is that the String/Cdc25 protein phosphatase is an important mitotic regulator, and that *stg* expression and STG activity are regulated at many different levels. The mechanisms(s) by which it is regulated seems to depend on the cell type and developmental timing.

7. Concluding remarks

An extensive foundation has been laid for understanding how the cell cycle and cell proliferation are linked to developmental events from research on this question in *Drosophila*. The variants of the cell cycle utilized at different developmental times have been determined, and the action of known cell cycle regulators within these modified cycles has been described. In the case of S/M cycles, novel cell cycle control pathways were identified. We now have a clear picture of when during the differentiation of several tissues the cell cycle is altered. This provides the framework for future studies to decipher which developmental regulators impact the cell cycle to cause these changes.

There are several areas in which identification of regulatory genes promises to yield insights into the molecular mechanisms responsible for the interface between development and cell cycle control. A linkage between cell cycle progression and oocyte versus nurse cell determination has been demonstrated, but the primary cause of this needs to be delineated. The regulatory signals for prophase I arrest, prophase I activation, metaphase I arrest, and activation to complete meiosis are unknown. They are likely to be most expeditiously identified by the recovery of mutants that affect these steps of meiosis. Similarly, isolation of mutants defective in the onset of the endo cycle in nurse cells and follicle cells will uncover control mechanisms governing the transition from mitotic proliferation to endo cycles. The elucidation of the role of Rb in endo cycle exit makes it feasible to identify the responsible targets. We now have examples of how parameters of the endo cycle can be changed, such as the differing extents to which mitotic functions are utilized. Additional analyses of the endo cycle in other developmental contexts will be important for determining differences between polyploid and polytene tissues as well as the use of underreplication of heterochromatin. This information will provide a framework for defining the cell cycle regulatory changes underlying these distinct endo cycle parameters.

Recent advances have revealed the molecular basis of other developmental cell cycle controls, yet further research will decipher these mechanisms with additional resolution. Amplification has proven a powerful model replication system that will permit the roles of additional replication proteins to be defined. Identification of these proteins and their binding sites in turn provides a means to recover additional replication origins. Insights into how origin firing is restricted and the role of Rb and E2F in this inhibition will emerge in the near future from this experimental system. It will be exciting to uncover the mechanism by which the PNG kinase complex controls Cyclin protein levels, and also to see whether there are additional regulators specific to the S/M cycles. The developmental control of the addition of G2 to the cell cycle serves as a paradigm for the joint activity of regulators in controlling differentiation and the cell cycle. This precise molecular mechanisms underlying this regulatory network are now poised to be solved.

Acknowledgments

We thank Astrid Clarke and David MacAlpine for helpful comments on the manuscript and Kim Dej and Laurie Lee for providing stained embryos for Fig. 4. We are grateful to Michael Botchan, Brian Calvi, Bob Duronio, Nick Dyson, Helena Richardson, and Mariana Wolfner for providing information prior to publication, and we thank David MacAlpine, Steve Bell, and Kim Dej for permission to cite unpublished results. G.B. was supported by a postdoctoral fellowship from the Damon Runyon-Walter Winchell Cancer Fund. T.O.-W. was supported by NIH grants GM39341 and GM57960.

References

Alphey, L., Jimenez, J., White-Cooper, H., Dawson, I., Nurse, P., Glover, D. 1992. *twine*, a *cdc25* homolog that functions in the male and female germline of Drosophila. Cell 69, 977–988.

Aparicio, O.M., Weinstein, D.M., Bell, S.P. 1997. Components and dynamics of DNA replication complexes in S. cerevisiae: redistribution of MCM proteins and Cdc45p during S phase. Cell 91, 59–69.

Asano, M., Wharton, R.P. 1999. E2F mediates developmental and cell cycle regulation of ORC1 in Drosophila. EMBO J. 18, 2435–2448.

Asaoka-Taguchi, M., Yamada, M., Nakamura, A., Hanyu, K., Kobayashi, S., Fogarty, P., Campbell, S.D., Abu-Shumays, R., Phalle, B.S., Yu, K.R., Uy, G.L., Goldberg, M.L., Sullivan, W., Seher, T.C., Leptin, M. 1999. Maternal Pumilio acts together with Nanos in germline development in *Drosophila* embryos. Nat. Cell Biol. 1, 431–437.

Austin, R.J., Orr-Weaver, T.L., Bell, S.P. 1999. *Drosophila* ORC specifically binds to *ACE3*, an origin of DNA replication control element. Genes and Dev. 13, 2639–2649.

Axton, J.M., Shamanski, F.L., Young, L.M., Henderson, D.S., Boyd, J.B., Orr-Weaver, T.L. 1994. The inhibitor of DNA replication encoded by the *Drosophila* gene *plutonium* is a small, ankyrin repeat protein. EMBO J. 13, 462–470.

Bell, S.P., Stillman, B. 1992. ATP-dependent recognition of eukaryotic origins of DNA replication by a multi-protein complex. Nature 357, 128–134.

Bielinsky, A.K., Gerbi, S.A. 1999. Chromosomal ARS1 has a single leading strand start site. Mol. Cell 3, 477–86.

Blumenthal, A., Kriegstein, H., Hogness, D. 1973. The units of DNA replication in *Drosophila melanogaster* chromosomes. Cold Spring Harbor Symp. Quant. Biol. 38, 205–223.

Bosco, G., Du, W., Orr-Weaver, T.L. 2001. DNA replication control through interaction of E2F-RB and the origin recognition complex. Nature Cell Biol. 3, 289–295.

Botchan, M., Topp, W., Sambrook, J. 1979. Studies on simian virus 40 excision from cellular chromosomes. Cold Spring Harb. Symp. Quant. Biol. 43 Pt 2, 709–719.

Buszczak, M., Freeman, M.R., Carlson, J.R., Bender, M., Cooley, L., Segraves, W.A. 1999. Ecdysone response genes govern egg chamber development during mid-oogenesis in Drosophila. Development 126, 4581–4589.

Calvi, B.R., Lilly, M.A., Spradling, A.C. 1998. Cell cycle control of chorion gene amplification. Genes and Dev. 12, 734–744.

Calvi, B.R., Spradling, A.C. 1999. Chorion gene amplification in Drosophila: A model for origins of DNA replication and S phase control. In: *Genetic Approaches to Eukaryotic Replication and Repair*, (P. Fisher, Ed.), New York: Academic Press, pp. 407–417.

Carminati, J., Orr-Weaver, T.L. 1996. Changes in DNA replication in animal development. In: *Eukaryotic DNA Replication*, (M. DePamphilis, Ed.), New York: Cold Spring Harbor Laboratory Press, pp. 409–434.

Carminati, J.L., Johnston, C.J., Orr-Weaver, T.L. 1992. The *Drosophila* ACE3 chorion element autonomously induces amplification. Mol. Cell Biol. 12, 2444–2453.

Cayirlioglu, P., Bonnette, P.C., Dickson, M.R., Duronio, R.J. 2001. *Drosophila* E2f2 promotes the conversion from genomic DNA replication to gene amplification in ovarian follicle cells. Development 128, 5085–5098.

Chen, B., Harms, E., Chu, T., Henrion, G., Strickland, S. 2000. Completion of meiosis in *Drosophila* oocytes requires transcriptional control by Grauzone, a new zinc finger protein. Development 127, 1243–1251.

Chesnokov, I., Gossen, M., Remus, D., Botchan, M. 1999. Assembly of functionally active *Drosophila* origin recognition complex from recombinant proteins. Genes and Dev. 13, 1289–1296.

Chesnokov, I., Remus, D., Botchan, M. 2001. Functional analysis of mutant and wild-type *Drosophila* origin recognition complex. Proc. Natl. Acad. Sci. USA 98, 11997–2002.

Chu, T., Henrion, G., Haegeli, V., Strickland, S. 2001. Cortex, a *Drosophila* gene required to complete oocyte meiosis, is a member of the Cdc20/fizzy protein family. Genesis 29, 141–152.

Courtot, C., Fankhauser, C., Simanis, V., Lehner, C.F. 1992. The *Drosophila cdc25* homolog *twine* is required for meiosis. Development 116, 405–416.

Cox, D.N., Lu, B., Sun, T.-Q., Williams, L.T., Jan, Y.N. 2001. *Drosophila par-1* is required for oocyte differentiation and microtubule organization. Curr. Biol. 11, 75–87.

Dalby, B., Glover, D.M. 1992. 3′ non-translated sequences in *Drosophila* cyclin B transcripts direct posterior pole accumulation late in oogenesis and peri-nuclear association in syncytial embryos. Development 115, 989–997.

Datar, S.A., Jacobs, H.W., de la Cruz, A.F., Lehner C.F., Edgar, B.A. 2000. The *Drosophila* cyclin D-Cdk4 complex promotes cellular growth. EMBO J. 19, 4543–4554.

Dapples, C.C., King, R.C. 1970. The development of the nucleolus of the ovarian nurse cell of *Drosophila melanogaster*. Z. Zellforsch Mikrosk. Anat. 103, 34–47.

de Cicco, D., Spradling, A. 1984. Localization of a cis-acting element responsible for the developmentally regulated amplification of *Drosophila* chorion genes. Cell 38, 45–54.

de Cuevas, M., Lilly, M.A., Spradling, A.C. 1997. Germline cyst formation in Drosophila. Annu. Rev. Genet. 31, 405–428.

de Nooij, J.C., Letendre, M.A., Hariharan, I.K. 1996. A cyclin-dependent kinase inhibitor, Dacapo, is necessary for timely exit from the cell cycle during *Drosophila* embryogenesis. Cell 87, 1237–1247.

Dej, K.J., Spradling, A.C. 1999. The endocycle controls nurse cell polytene chromosome structure during *Drosophila* oogenesis. Development 126, 293–303.

Delidakis, C., Kafatos, F.C. 1989. Amplification enhancers and replication origins in the autosomal chorion gene cluster of Drosophila. EMBO J. 8, 891–901.

Delidakis, C., Kafatos, F.C. 1987. Amplification of a chorion gene cluster in *Drosophila* is subject to multiple cis-regulatory elements and to long range position effects. J. Mol. Biol. 197, 11–26.

Delidakis, C., Kafatos, F.C. 1988. Deletion analysis of cis-acting elements for chorion gene amplification in *Drosophila* melanogaster. In: E*ukaryotic DNA Replication*, (T. J. Kelly and B. Stillman, Eds.), Cold Spring Harbor: Cold Spring Harbor Press, pp. 311–315.

Deng, W.M., Althauser, C., Ruohola-Baker, H. 2001. Notch-Delta signaling induces a transition from mitotic cell cycle to endocycle in *Drosophila* follicle cells. Development 128, 4737–4746.

DePamphilis, M.L. 1999. Replication origins in metazoan chromosomes: fact or fiction. BioEssays 21, 5–16.

Diffley, J.F. 2001. DNA replication: building the perfect switch. Curr. Biol. 11, R367–370.

Diffley, J.F. 1996. Once and only once upon a time: specifying and regulating origins of DNA replication in eukaryotic cells. Genes and Dev. 10, 2819–2830.

Doane, W. 1960. Completion of meiosis in uninseminated eggs of *Drosophila* melanogaster. Science 132, 677–678.

Dobens, L.L., Raftery, L.A. 2000. Integration of epithelial patterning and morphogenesis in *Drosophila* ovarian follicle cells. Dev. Dyn. 218, 80–93.

Dowell, S.J., Romanowski, P., Diffley, J.F. 1994. Interaction of Dbf4, the Cdc7 protein kinase regulatory subunit, with yeast replication origins in vivo. Science 265, 1243–6.

Du, W., Dyson, N. 1999. The role of RBF in the introduction of G1 regulation during *Drosophila* embryogenesis. EMBO J. 18, 916–925.

Du, W., Vidal, M., Xie, J.-E., Dyson, N. 1996. *RBF*, a novel RB-related gene that regulates E2F activity and interacts with *cyclin E* in *Drosophila*. Genes and Dev. 10, 1206–1218.

Duronio, R.J., Bonnette, P.C., O'Farrell, P.H. 1998. Mutations of the *Drosophila* dDP, dE2F, and cyclin E genes reveal distinct roles for the E2F-DP transcription factor and cyclin E during the G1-S transition. Mol. Cell. Biol. 18, 141–151.

Duronio, R.J., O'Farrell, P.H. 1995. Developmental control of the G_1 to S transition in *Drosophila*: cyclin E is a limiting downstream target of E2F. Genes and Dev. 9, 1456–1468.

Dutta, A., Bell, S.P. 1997. Initiation of DNA replication in eukaryotic cells. Annu. Rev. Cell Dev. Biol. 13, 293–332.

Dyson, N. 1998. The regulation of E2F by pRB-family proteins. Genes and Dev. 12, 2245–2262.

Edgar, B., Datar, S. 1996. Zygotic degradation of two maternal Cdc25 mRNAs terminates Drosophila's early cell cycle program. Genes and Dev. 10, 1966–77.

Edgar, B., Kiehle, C.P., Schubiger, G. 1986. Cell cycle control by the nucleocytoplasmic ratio in early *Drosophila* development. Cell 44, 365–372.

Edgar, B.A., Lehman, D.A., O'Farrell, P.H. 1994. Transcriptional regulation of *string* (*cdc25*): a link between developmental programming and the cell cycle. Development 120, 3131–3143.

Edgar, B.A., O'Farrell, P.H. 1989. Genetic control of cell division patterns in the *Drosophila* embryo. Cell 57, 177–187.

Edgar, B.A., O'Farrell, P.H. 1990. The three postblastoderm cell cycles of *Drosophila* embryogenesis are regulated in G2 by *string*. Cell 62, 469–480.

Edgar, B.A., Orr-Weaver, T.L. 2001. Endoreplication cell cycles: more for less. Cell 105, 297–306.

Edgar, B.A., Sprenger, F., Duronio, R.J., Leopold, P., O'Farrell, P.H. 1994. Distinct molecular mechanisms regulate cell cycle timing at successive stages of *Drosophila* embryogenesis. Genes and Dev. 8, 440–452.

Elfring, L.K., Axton, J.M., Fenger, D.D., Page, A.W., Carminati, J., Orr-Weaver, T.L. 1997. The *Drosophila* PLUTONIUM protein is a specialized cell cycle regulator required at the onset of development. Mol. Biol. Cell 8, 583–593.

Feger, G.V., H., Su, T.T., Wolff, E., Jan, L.Y., Jan, Y.N. 1995. dpa, a member of the MCM family, is required for mitotic DNA replication but not endoreplication in Drosopphila. EMBO J. 14, 5387–5398.

Fenger, D.D., Carminati, J.L., Burney-Sigman, D.L., Kashevsky, H., Dines, J.L., Elfring, L.K., Orr-Weaver, T.L. 2000. PAN GU: a protein kinase that inhibits S phase and promotes mitosis in early *Drosophila* development. Development 127, 4763–4774.

Foe, V.E., Alberts, B.M. 1983. Studies of nuclear and cytoplasmic behaviour during the five mitotic cycles that precede gastrulation in *Drosophila* embryogenesis. J. Cell Sci. 61, 31–70.

Foe, V.E., Odell, G.M., Edgar, B.A. 1993. Mitosis and morphogenesis in the *Drosophila* embryo: Point and counterpoint. In: *The Development of Drosophila Melanogaster*, (M. Bate and A. Martinez Arias, Eds.), Cold Spring Harbor, NY: Cold Spring Harbor Laboratory Press, pp. 149–300.

Fogarty, P., Campbell, S., Abu-Shumays, R., Phalle, B., Yu, K., Uy, G., Goldberg, M., Sullivan, W. 1997. The *Drosophila* grapes gene is related to checkpoint gene chk1/rad27 and is required for late syncytial division fidelity. Curr. Biol. 7, 418–426.

Follette, P.J., Duronio, R.J., O'Farrell, P.H. 1998. Fluctuations in cyclin E levels are required for multiple rounds of endocycle S phase in Drosophila. Curr. Biol. 8, 235–238.

Freeman, M., Glover, D. 1987. The *gnu* mutation of *Drosophila* causes inappropriate DNA synthesis in unfertilized and fertilized eggs. Genes and Dev. 1, 924–930.

Freeman, M., Nusslein-Volhard, C., Glover, D. 1986. The dissociation of nuclear and centrosomal division in *gnu*, a mutation causing giant nuclei in Drosophila. Cell 46, 457–468.

Frolov, M.V., Huen, D.S., Stevaux, O., Dimova, D., Balczarek-Strang, K., Elsdon, M., Dyson, N.J. 2001. Functional antagonism between E2F family members. Genes and Dev. 15, 2146–2160.

Furstenthal, L., Kaiser, B.K., Swanson, C., Jackson, P.K. 2001. Cyclin E uses Cdc6 as a chromatin-associated receptor required for DNA replication. J. Cell Biol. 152, 1267–1278.

Gall, J.G., Cohen, E.H., Polan, M.L. 1971. Reptitive DNA sequences in drosophila. Chromosoma 33, 319–344.

Ghabrial, A., Ray, R.P., Schupbach, T. 1998. okra and spindle-B encode components of the RAD52 DNA repair pathway and affect meiosis and patterning in *Drosophila* oogenesis. Genes and Dev. 12, 2711–2723.

Ghabrial, A., Schupbach, T. 1999. Activation of a meiotic checkpoint regulates translation of Gurken during *Drosophila* oogenesis. Nature Cell Biol. 1, 354–357.

Grosshans, J., Wieschaus, E. 2000. A genetic link between morphogenesis and cell division during formation of the ventral furrow in Drosophila. Cell 101, 523–531.

Hamlin, J.L., Dijkwel, P.A. 1995. On the nature of replication origins in higher eukaryotes. Curr. Opin. Genet. Dev. 5, 153–161.

Hamlin, J.L., Leu, T.H., Vaughn, J.P., Ma, C., Dijkwel, P.A. 1991. Amplification of DNA sequences in mammalian cells. Prog. Nuc. Acid Res. Mol. Biol. 41, 203–239.

Hammond, M.P., Laird, C.D. 1985. Chromosome structure and DNA replication in nurse and follicle cells of *Drosophila melanogaster*. Chromosoma 91, 267–278.

Hari, K.L., Santerre, A., Sekelsky, J.J., McKim, K.S., Boyd, J.B., Hawley, R.S. 1995. The mei-41 gene of *D. melanogaster* is a structural and functional homolog of the human ataxia telangiectasia gene. Cell 82, 815–821.

Harms, E., Chu, T., Henrion, G., Strickland, S. 2000. The only function of Grauzone required for *Drosophila* oocyte meiosis is transcriptional control of the *cortex* gene. Genetics 155, 1831–1839.

Heck, M., Spradling, A. 1990. Multiple replication origins are used during *Drosophila* chorion gene amplification. J. Cell Biol. 110, 903–914.

Heifetz, Y., Yu, J., Wolfner, M.F. 2001. Ovulation triggers activation of *Drosophila* oocytes. Dev. Biol. 234, 416–424.

Henderson, D.S., Banga, S.S., Grigliatti, T.A., Boyd, J.B. 1994. Mutagen sensitivity and suppression of position-effect variegation result from mutations in *mus209*, the *Drosophila* gene encoding PCNA. EMBO J. 13, 1450–1459.

Henderson, D.S., Wiegand, U.K., Norman, D.G., Glover, D.M. 2000. Mutual correction of faulty PCNA subunits in temperature-sensitive lethal *mus209* mutants of *Drosophila melanogaster*. Genetics 154, 1721–1733.

Huang, J., Raff, J.W. 1999. The disappearance of cyclin B at the end of mitosis is regulated spatially in *Drosophila* cells. EMBO J. 18, 2184–2195.

Huynh, J.-R., Shulman, J.M., Benton, R., St Johnston, D. 2001. PAR-1 is required for the maintenance of oocyte fate in Drosophila. Development 128, 1201–1209.

Huynh, J.-R., St. Johnston, D. 2000. The role of BicD, Egl, Orb and the microtubules in the restriction of meiosis to the *Drosophila* oocyte. Development 127, 2785–2794.

Ishimi, Y. 1997. A DNA helicase activity is associated with an MCM4, -6, and -7 protein complex. J. Biol. Chem. 272, 24508–24513.

Jacobs, H., Knoblich, J., Lehner, C. 1998. *Drosophila* Cyclin B3 is required for female fertility and is dispensable for mitosis like Cyclin B. Genes and Dev. 12, 3741–3751.

Jacobs, H.W., Keidel, E., Lehner, C.F. 2001. A complex degradation signal in Cyclin A required for G1 arrest, and a C-terminal region for mitosis. EMBO J. 20, 2376–2386.

Jang, J.K., Messina, L., Erdman, M.B., Arbel, T., Hawley, R.S. 1995. Induction of metaphase arrest in *Drosophila* oocytes by chiasma-based kinetochore tension. Science 268, 1917–1919.

Johnston, L.A. 2000. The trouble with tribbles. Curr Biol 10, R502–4.

Kennedy, B.K., Barbie, D.A., Classon, M., Dyson, N., Harlow, E. 2000. Nuclear organization of DNA replication in primary mammalian cells. Genes and Dev. 14, 2855–2868.

Kerrebrock, A.W., Miyazaki, W.Y., Birnby, D., Orr-Weaver, T.L. 1992. The *Drosophila mei-S332* gene promotes sister-chromatid cohesion in meiosis following kinetochore differentiation. Genetics 130, 827–841.

Keyes, L.N., Spradling, A.C. 1997. The *Drosophila* gene fs(2)cup interacts with otu to define a cytoplasmic pathway required for the structure and function of germ-line chromosomes. Development 124, 1419–1431.

Khipple, P., King, R.C. 1976. Oogenesis in the *female sterile(1)1304* mutant of *Drosophila* melanogaster. Int. J. Insect Morph. Embryol. 5, 127–135.

King, R.C., Riley, S.F., Cassidy, J.D., White, P.E., Paik, Y.K. 1981. Giant polytene chromosomes from the ovaries of a *Drosophila* mutant. Science 212, 441–443.

Knoblich, J., Lehner, C. 1993. Synergistic action of *Drosophila* cyclins A and B during the G2–M transition. EMBO J. 12, 65–74.

Knoblich, J.A., Sauer, K., Jones, L., Richardson, H., Saint, R., Lehner, C.F. 1994. Cyclin E controls S phase progression and its down-regulation during *Drosophila* embryogenesis is required for the arrest of cell proliferation. Cell 77, 107–120.

Koch, E.A., King, R.C. 1964. Studies on the *fes* mutant of *Drosophila melanogaster*. Growth 28, 325–369.

Kriegstein, H.J., Hogness, D.S. 1974. Mechanism of DNA replication in *Drosophila* chromosomes: Structure of replication forks and evidence for bidirectionality. Proc. Natl. Acad. Sci. USA 71, 135–139.

Lam, E.W., La Thangue, N.B. 1994. DP and E2F proteins: coordinating transcription with cell cycle progression. Curr. Opin. Cell Biol. 6, 859–866.

Landis, G., Kelley, R., Spradling, A.C., Tower, J. 1997. The *k43* gene, required for chorion gene amplification and diploid cell chromosome replication, encodes the *Drosophila* homolog of yeast origin recognition complex subunit 2. Proc. Natl. Acad. Sci. USA 94, 3888–3892.

Landis, G., Tower, J. 1999. The *Drosophila* chiffon gene is required for chorion gene amplification, and is related to the yeast dbf4 regulator of DNA replication and cell cycle. Development 126, 4281–4293.

Lane, M.E., Sauer, K., Wallace, K., Jan, Y.N., Lehner, C.F., Vaessin, H. 1996. Dacapo, a cyclin-dependent kinase inhibitor, stops cell proliferation during *Drosophila* development. Cell 87, 1225–1235.

Leatherwood, J., Lopez-Girona, A., Russell, P. 1996. Interaction of Cdc2 and Cdc18 with a fission yeast ORC2-like protein. Nature 379, 360–363.

Lee, L.A., Elfring, L.K., Bosco, G., Orr-Weaver, T.L. 2001. A genetic screen for suppressors and enhancers of the *Drosophila* PAN GU cell cycle kinase identifies Cyclin B as a target. Genetics 158, 1545–1556.

Lehner, C.F., O'Farrell, P.H. 1989. Expression and function of *Drosophila* cyclin A during embryonic cell cycle progression. Cell 56, 957–968.

Lehner, C.F., O'Farrell, P.H. 1990. The roles of *Drosophila* cyclins A and B in mitotic control. Cell 61, 535–547.

Lieberfarb, M.E., Chu, T., Wreden, C., Theurkauf, W., Gergen, J.P., Strickland, S. 1996. Mutations that perturb poly(A)-dependent maternal mRNA activation block the initiation of development. Development 122, 579–588.

Lilly, M.A., Spradling, A.C. 1996. The *Drosophila* endocycle is controlled by Cyclin E and lacks a checkpoint ensuring S-phase completion. Genes and Dev. 10, 2514–2526.

Linskens, M., Huberman, J.A. 1990. The two faces of higher eukaryotic DNA replication origins. Cell 62, 845–847.

Loebel, D., Huikeshoven, H., Cotterill, S. 2000. Localisation of the DmCdc45 DNA replication factor in the mitotic cycle and during chorion gene amplicfication. Nucleic Acids Res. 28, 3897–3903.

Lopez-Schier, H., St Johnston, D. 2001. Delta signaling from the germ line controls the proliferation and differentiation of the somatic follicle cells during *Drosophila* oogenesis. Genes and Dev. 15, 1393–1405.

Lu, L., Hongjun, Z., Tower, J. 2001. Functionally distinct, sequence specific replicator and origin elements are required for *Drosophila* chorion gene amplification. Genes and Dev. 15, 134–146.

Lu, L., Tower, J. 1997. A transcriptional insulator element, the su(Hw) binding site, protects a chromosomal DNA replication origin from position effects. Mol. Cell. Biol. 17, 2202–2206.

Lundgren, K., Walworth, N., Booher R., Dembski M., Beach, D. 1991. mik1 and wee1 cooperate in the inhibitory tyrosine phosphorylation on cdc2. Cell 22, 1111–1122

Mahowald, A.P., Goralski, T.J., Caulton, J.H. 1983. *In vitro* activation of *Drosophila* eggs. Dev. Biol. 98, 437–445.

Maiorano, D., Moreau, J., Mechali, M. 2000. XCDT1 is required for the assembly of pre-replicative complexes in *Xenopus laevis*. Nature 404, 622–625.

Mata, J., Curado, S., Ephrussi, A., Rorth, P. 2000. Tribbles coordinates mitosis and morphogenesis in *Drosophila* by regulating string/CDC25 proteolysis. Cell 101, 511–522.

McGarry, T.J., Kirschner, M.W. 1998. Geminin, an inhibitor of DNA replication, is degraded during mitosis. Cell 93, 1043–53.

McKim, K.S., Hayashi-Hagihara, A. 1998. *mei-W68* in *Drosophila melanogaster* encodes a Spo11 homolog: evidence that the mechanism for initiating meiotic recombination is conserved. Genes and Dev. 12, 2932–2942.

McKim, K.S., Jang, J.K., Sekelsky, J.J., Laurencon, A., Hawley, R.S. 2000. *mei-41* is required for precocious anaphase in *Drosophila* females. Chromosoma 109, 44–49.

McKim, K.S., Jang, J.K., Theurkauf, W.E., Hawley, R.S. 1993. Mechanical basis of meiotic metaphase arrest. Nature 362, 364–366.

McKnight, S., Miller, O. 1976. Ultrastructural patterns of RNA synthesis during early embryogenesis of *Drosophila melanogaster*. Cell 8, 305–319.

McKnight, S.L., Miller, O.L. 1977. Electron microscopic analysis of chromatin replication in the cellular blastoderm embryo. Cell 12, 795–804.

Mechali, M. 2001. DNA replication origins: from sequence specificity to epigenetics. Nat. Rev. Genet. 2, 640–645.

Meyer, C.A., Jacobs, H.W., Datar, S.A., Du, W., Edgar, B.A., Lehner, C.F., Grosshans, J., Wieschaus, E. 2000. *Drosophila* Cdk4 is required for normal growth and is dispensable for cell cycle progression. EMBO J. 19, 4533–4542.

Miyazaki, W.Y., Orr-Weaver, T.L. 1992. Sister-chromatid misbehavior in *Drosophila ord* mutants. Genetics 132, 1047–1061.

Murray, A., Hunt, T. 1993. *The Cell Cycle: An Introduction*, New York: Freeman.

Myster, D.L., Bonnette, P.C., Duronio, R.J. 2000. A role for the DP subunit of the E2F transcription factor in axis determination during *Drosophila* oogenesis. Development 127, 3249–3261.

Nakayama, K., Hatakeyama, S., Nakayama, K. 2001. Regulation of the cell cycle at the G1–S transition by proteolysis of Cyclin E and p27[Kip1]. Biochem. Biophys. Res. Comm. 282:852–860.

Nishitani, H., Lygerou, Z., Nishimoto, T., Nurse, P. 2000. The Cdt1 protein is required to license DNA for replication in fission yeast. Nature 404, 625–628.

Orr-Weaver, T., Spradling, A. 1986. *Drosophila* chorion gene amplification requires an upstream region regulating *s18* transcription. Mol. Cell Biol. 6, 4624–4633.

Orr-Weaver, T.L. 1991. *Drosophila* chorion genes: cracking the eggshell's secrets. BioEssays 13, 97–105.

Osheim, Y.N., Miller, O.L. 1983. Novel amplification and transcriptional activity of chorion genes in *Drosophila melanogaster* follicle cells. Cell 33, 543–553.

Osheim, Y.N., Miller, O.L., Beyer, A.L. 1988. Visualization of *Drosophila melanogaster* chorion genes undergoing amplification. Mol. Cell Biol. 8, 2811–2821.

Page, A., Orr-Weaver, T. 1997. Activation of the meiotic divisions in *Drosophila* oocytes. Dev. Biol. 183, 195–207.

Page, A.W., Orr-Weaver, T.L. 1996. The *Drosophila* genes *grauzone* and *cortex* are necessary for proper female meiosis. J. Cell Sci. 109, 1707–1715.

Page, S.L., Hawley, R.S. 2001. *c(3)G* encodes a *Drosophila* synaptonemal complex protein. Genes and Dev. 15, 3130–3143.

Page, S.L., McKim, K.S., Deneen, B., Van Hook, T.L., Hawley, R.S. 2000. Genetic studies of *mei-P26* reveal a link between the processes that control germ cell proliferation in both sexes and those that control meiotic exchange in Drosophila. Genetics 155, 1757–1772.

Price, D., Rabinovitch, S., O'Farrell, P.H., Campbell, S.D. 2000. *Drosophila* wee1 has an essential role in the nuclear divisions of early embryogenesis. Genetics 155, 159–166.

Pritchard, D.K., Schubiger, G. 1996. Activation of transcription in *Drosophila* embryos is a gradual process mediated by the nucleocytoplasmic ratio. Genes and Dev. 10, 1131–1142.

Quinn, L.M., Herr, A., McGarry, T.J., Richardson, H. 2001. The *Drosophila* Geminin homolog: roles for Geminin in limiting DNA replication, in anaphase and in neurogenesis. Genes and Dev. 15, 2741–2754.

Rabinowitz, M. 1941. Studies on the cytology and early embryology of the egg of *Drosophila melanogaster*. J. Morph. 69, 1–49.

Raff, J.W., Whitfield, W.G., Glover, D.M. 1990. Two distinct mechanisms localise cyclin B transcripts in syncytial *Drosophila* embryos. Development 110, 1249–1261.

Reed, B.H., Orr-Weaver, T.L. 1997. The *Drosophila* gene *morula* inhibits mitotic functions in the endo cell cycle and the mitotic cell cycle. Development 124, 3543–3553.

Reed, S.I. 1997. Control of the G1/S transition. Cancer Surv. 29, 7–23.

Richardson, H., O'Keefe, L.V., Marty, T., Saint, R. 1995. Ectopic cyclin E expression induces premature entry into S phase and disrupts pattern formation in the *Drosophila* eye imaginal disc. Development 121, 3371–3379.

Richardson, H.E., O'Keefe, L.V., Reed, S.I., Saint, R. 1993. A *Drosophila* G1-specific cyclin E homolog exhibits different modes of expression during embryogenesis. Development 119, 673–690.

Riparbelli, M.G., Callaini, G., Glover, D.M. 2000. Failure of pronuclear migration and repeated divisions of polar body nuclei associated with MTOC defects in *polo* eggs of Drosophila. J. Cell Sci. 113, 3341–3350.

Roeder, G.S. 1997. Meiotic chromosomes: it takes two to tango. Genes and Dev. 11, 2600–2621.

Roeder, G.S., Bailis, J.M. 2000. The pachytene checkpoint. Trends in Genet. 16, 395–403.

Royzman, I., Austin, R.J., Bosco, G., Bell, S.P., Orr-Weaver, T.L. 1999. ORC localization in *Drosophila* follicle cells and the effects of mutations in *dE2F* and *dDP*. Genes and Dev. 13, 827–840.

Royzman, I., Whittaker, A.J., Orr-Weaver, T.L. 1997. Mutations in *Drosophila DP* and *E2F* distinguish G1-S progression from an associated transcriptional program. Genes Dev. 11, 1999–2011.

Sauer, K., Weigmann, K., Sigrist, S., Lehner, C.F. 1996. Novel members of the cdc2-related kinase family in *Drosophila*: cdk4/6, cdk5, PFTAIRE, and PITSLRE kinase. Mol. Biol. Cell 7, 1759–1769.

Sawado, T., Yamaguchi, M., Nishimoto, Y., Ohno, K., Sakaguchi, K., Matsukage, A. 1998. dE2F2, a novel E2F-family transcription factor in *Drosophila melanogaster*. Biochem. Biophys. Res. Commun. 251, 409–415.

Seher, T.C., Leptin, M. 2000. Tribbles, a cell-cycle brake that coordinates proliferation and morphogenesis during *Drosophila* gastrulation. Curr. Biol. 10, 623–629.

Seum, C., Spierer, A., Pauli, D., Szidonya, J., Reuter, G., Spierer, P. 1996. Position-effect variegation in *Drosophila* depends on dose of the gene encoding the E2F transcriptional activator and cell cycle regulator. Development 122, 1949–1956.

Shamanski, F., Orr-Weaver, T. 1991. The *Drosophila plutonium* and *pan gu* genes regulate entry into S phase at fertilization. Cell 66, 1289–1300.

Sibon, O.C., Kelkar, A., Lemstra, W., Theurkauf, W.E. 2000. DNA-replication/DNA-damage-dependent centrosome inactivation in *Drosophila* embryos. Nat. Cell. Biol. 2, 90–95.

Sibon, O.C., Laurencon, A., Hawley, R., Theurkauf, W.E. 1999. The *Drosophila* ATM homologue Mei-41 has an essential checkpoint function at the midblastula transition. Curr. Biol. 9, 302–312.

Sibon, O.C., Stevenson, V.A., Theurkauf, W.E. 1997. DNA-replication checkpoint control at the *Drosophila* midblastula transition. Nature 388, 93–97.

Sigrist, S., Jacobs, H., Stratmann, R., Lehner, C.F. 1995. Exit from mitosis is regulated by *Drosophila fizzy* and the sequential destruction of cyclins A, B, and B3. EMBO J. 14, 4827–4838.

Singer, M.J., Mesner, L.D., Friedman, C.L., Trask, B.J., Hamlin, J.L. 2000. Amplification of the human dihydrofolate reductase gene via double minutes is initiated by chromosome breaks. Proc. Natl. Acad. Sci. USA 97, 7921–7926.

Smith, A.V., Orr-Weaver, T.L. 1991. The regulation of the cell cycle during *Drosophila* embryogenesis: the transition to polyteny. Development 112, 997–1008.

Snyder, P., Galanopoulos, V., Kafatos, F.C. 1986. *trans*-acting amplification mutants and other eggshell mutants of the third chromosme in *Drosophila melanogaster*. Proc. Natl. Acad. Sci. USA 83, 3341–3345.

Spradling, A. 1981. The organization and amplification of two clusters of *Drosophila* chorion genes. Cell 27, 193–202.

Spradling, A. 1986. P element mediated transformation. In: Drosophila: A practical approach, D.B. Roberts, Ed. Oxford, England: IRL Press, pp. 175–197.

Spradling, A., Orr-Weaver, T. 1987. Regulation of DNA replication during *Drosophila* development. Ann. Rev. Genetics 21, 373–403.

Spradling, A.C. 1993. Developmental genetics of oogenesis. In: *The Development of Drosophila melanogaster*, (M. Bate, A. Martinez Arias, Eds.), Cold Spring Harbor, NY: Cold Spring Harbor Laboratory Press, pp. 1–70.

Spradling, A.C. 1999. ORC binding, gene amplification, and the nature of metazoan replication origins. Genes and Dev. 13, 2619–2623.

Sprenger, F., Yakubovich, N., O'Farrell, P.H. 1997. S-phase function of *Drosophila* cyclin A and its downregulation in G1 phase. Current Biology 7, 488–499.

Stern, B., Ried, G., Clegg, N., Grigliatti, T., Lehner, C. 1993. Genetic analysis of the *Drosophila cdc2* homolog. Development 117, 219–232.

Sterner, J.M., Dew-Knight, S., Musahl, C., Kornbluth, S., Horowitz, J.M. 1998. Negative regulation of DNA replication by the retinoblastoma protein is mediated by its association with MCM7. Mol. Cell. Biol. 18, 2748–2757.

Stinchcomb, D.T., Struhl, K., Davis, R.W. 1979. Isolation and characterisation of a yeast chromosomal replicator. Nature 282, 39–43.

Su, T.T., Campbell, S.D., O'Farrell, P.H. 1998a. The cell cycle program in germ cells of the *Drosophila* embryo. Dev. Biol. 196, 160–170.

Su, T.T., Campbell, S.D., O'Farrell, P.H. 1999. *Drosophila* grapes/CHK1 mutants are defective in cyclin proteolysis and coordination of mitotic events. Curr. Biol. 9, 919–922.

Su, T.T., Feger, G., O'Farrell, P.H. 1996. *Drosophila* MCM protein complexes. Mol. Biol. Cell 7, 319–29.

Su, T.T., O'Farrell, P.H. 1998. Chromosome association of minichromosome maintenance proteins in *Drosophila* endoreplication cycles. J. Cell. Biol. 140, 451–460.

Su, T.T., O'Farrell, P.H. 1997. Chromosome association of minichromosome maintenance proteins in *Drosophila* mitotic cycles. J. Cell. Biol. 139, 13–21.

Su, T.T., Sprenger, F., DiGregorio, P.J., Campbell, S.D., O'Farrell, P.H. 1998. Exit from mitosis in *Drosophila* syncytial embryos requires proteolysis and cyclin degradation, and is associated with localized dephosphorylation. Genes and Dev. 12, 1495–1503.

Su, T.T., Yakubovich, N., O'Farrell, P.H. 1997. Cloning of *Drosophila* MCM homologs and analysis of their requirement during embryogenesis. Gene 192, 283–289.

Sullivan, W., Daily, D.R., Fogarty, P., Yook, K.J., Pimpinelli, S. 1993. Delays in anaphase initiation occur in individual nuclei of the syncytial *Drosophila* embryo. Mol. Biol. Cell 4, 885–896.

Sullivan, W., Minden, J.S., Alberts, B.M. 1990. daughterless-abo-like, a *Drosophila* maternal-effect mutation that exhibits abnormal centrosome separation during the late blastoderm divisions. Development 110, 311–323.

Swimmer, C., Delidakis, C., Kafatos, F.C. 1989. Amplification-control element ACE-3 is important but not essential for autosomal chorion gene amplification. Proc. Natl. Acad. Sci. USA 86, 8823–8827.

Swimmer, C., Fenerjian, M.G., Martinez-Cruzado, J.C., Kafatos, F.C. 1990. Evolution of the autosomal chorion cluster in *Drosophila* III. Comparison of the *s*18 gene in evolutionarily distant species and interspecific control of chorion gene amplification. J. Mol. Biol. 215, 225–235.

Tada, S., Li, A., Maiorano, D., Mechali, M., Blow, J.J. 2000. Repression of origin assembly in metaphase depends on inhibition of RLF-B/Cdt1 by geminin. Nature Cell Biol. 3, 107–113.

Theurkauf, W.E., Hawley, R.S. 1992. Meiotic spindle assembly in *Drosophila* females: Behavior of nonexchange chromosomes and the effects of mutations in the *nod* kinesin-like protein. J. Cell Biol. 116, 1167–1180.

Trimarchi, J.M., Lees, J.A. 2002. Sibling rivalry in the E2F family. Nature Reviews Mol. Cell Biol. 3, 11–20.

Tzolovsky, G., Deng, W.M., Schlitt, T., Bownes, M. 1999. The function of the broad-complex during *Drosophila* melanogaster oogenesis. Genetics 153, 1371–1383.

Underwood, E.M., Briot, A.S., Doll, K.Z., Ludwiczak, R.L., Otteson, D.C., Tower, J., Vessey, K.B., Yu, K. 1990. Genetics of 51D-52A, a region containing several maternal-effect genes and two maternal-specific transcripts in Drosophila. Genetics 126, 639–650.

Van Buskirk, C., Hawkins, N.C., Schupbach, T. 2000. Encore is a member of a novel family of proteins and affects multiple processes in *Drosophila* oogenesis. Development 127, 4753–4762.

Van Buskirk, C., Schupbach, T. 1999. Versatility in signalling: multiple responses to EGF receptor activation during *Drosophila* oogenesis. Trends Cell Biol. 9, 1–4.

Wakefield, J.G., Huang, J.-Y., Raff, J.W. 2000. Centrosomes have a role in regulating the destruction of cyclin B in early *Drosophila* embryos. Curr. Biol. 10, 1367–1370.

Weiss, A., Herzig, A., Jacobs, H., Lehner, C.F. 1998. Continuous *Cyclin E* expression inhibits progression through endoreduplication cycles in Drosophila. Curr. Biol. 8, 239–242.

Westphal, C.H. 1997. Cell-ignaling: Atm displays its many talents. Curr. Biol. 7:R789–R792

Whitfield, W., Gonzalez, C., Maldonado-Codina, G., Glover, D. 1990. The A- and B-type cyclins of *Drosophila* are accumulated and destroyed in temporally distinct events that define separable phases of the G2–M transition. EMBO J. 9, 2563–2572.

Whittaker, A.J., Royzman, I., Orr-Weaver, T.L. 2000. *Drosophila* double parked: a conserved, essential replication protein that colocalizes with the origin recognition complex and links DNA replication with mitosis and the down-regulation of S phase transcripts. Genes and Dev. 14, 1765–1776.

Wieschaus, E. 1978. *fs(1)K10*, a germline-dependent female sterile mutation causing abnormal chorion morphology in *Drosophila melanogaster*. Wilhelm Roux's Arch. Dev. Biol. 184, 75–82.

Windle, B., Draper, B.W., Yin, Y.X., O'Gorman, S., Wahl, G.M. 1991. A central role for chromosome breakage in gene amplification, deletion formation, and amplicon integration. Genes and Dev. 5, 160–174.

Wintersberger, E. 2000. Why is there late replication? Chromosoma 109, 300–7.

Wohlschlegel, J.A., Dwyer, B.T., Dhar, S.K., Cvetic, C., Walter, J.C., Dutta, A. 2000. Inhibition of eukaryotic DNA replication by geminin binding to Cdt1. Science 290, 2309–2312.

Yamamoto, R.R., Axton, J.M., Yamamoto, Y., Saunders, R.D., Glover, D.M., Henderson, D.S. 2000. The *Drosophila* mus101 gene, which links DNA repair, replication and condensation of heterochromatin in mitosis, encodes a protein with seven BRCA1 C-terminus domains. Genetics 156, 711–721.

Yu, K.R., Saint, R.B., Sullivan, W. 2000. The Grapes checkpoint coordinates nuclear envelope breakdown and chromosome condensation. Nat. Cell Biol. 2, 609–615.

Zimmet, J., Ravid, K. 2000. Polyploidy: Occurrence in nature, mechanisms, and significance for the megakary-ocyte-platelet system. Exp. Hematol. 28, 3–16.

Zybina, E.V., Zybina, T.G. 1996. Polytene chromosomes in mammalian cells. Int. Rev. of Cytol. 165, 53–119.

Advances in Developmental Biology and Biochemistry, Vol. 12
M. DePamphilis (Editor)

Anterior–posterior patterning in the *Drosophila* embryo

Andrzej Nasiadka, Bruce H. Dietrich and Henry M. Krause

Banting and Best Department of Medical Research, University of Toronto and C.H. Best Institute,
112 College Street, Toronto, Ontario, Canada M5G 1L6

Summary

The cascade of transcription factor interactions that defines the anterior–posterior body plan of the *Drosophila* embryo is one of the most comprehensively investigated transcriptional hierarchies. Its role is to rapidly subdivide the embryo into increasingly smaller metameric units that ultimately comprise the segments of the adult animal. This review follows these transcriptional interactions. We begin with a short review of the factors that control the onset of transcription and then proceed to a comprehensive description of the segmentation gene hierarchy, the genes and proteins that make it up and how they interact to generate a pattern of repeating units, each with its own identity. Although this is meant to be a general review, topics that have not been recently or extensively covered are discussed in some detail. We conclude with a discussion of future prospects.

Contents

1. The onset of zygotic transcription in *Drosophila* embryos

The *Drosophila* segmentation gene hierarchy is unique, even when compared to the segmental hierarchies of other insects. It proceeds amid an array of nuclei that divide within a common cytoplasm. This alleviates the need for complex and time consuming intercellular signaling cascades. Consequently, one class of transcription factors can act directly upon the genes of the next class, optimizing speed and efficiency. The rapid and synchronous nuclear divisions that occur prior to the completion of cellularization (Zalokar and Erk, 1976; Foe and Alberts, 1983; Campos-Ortega and Hartenstein, 1997) produce a monolayer of about 5000–6000 cells in about three hours. This number of cells is ideal for the complex segmental blueprint that is then overlaid. More primitive insects establish their segmental patterns within a predominantly cellular environment (reviewed in Davis and Patel, 1999). Consequently, the process tends to take much longer and likely requires many more genes and gene interactions.

While *Drosophila* nuclei are dividing in their syncytial environment (without plasma membranes), most transcription is initially suppressed. As the last divisions are completed, this block on transcription is lifted and the process of cellularization begins. It is at this time that maternally encoded segmentation gene proteins begin to regulate their downstream targets. The intimate relationship between early nuclear division cycles and the onset of transcription is now becoming clearer. A brief review of these regulatory processes follows (see Fig. 1 for a chronological listing of events).

The first ten nuclear divisions occur in 8–10 min intervals, consisting only of alternating S (DNA-replication) and M (mitosis) phases without intervening G (gap) phases. Mitotic cycles 10–13 are gradually lengthened by the addition and lengthening of post-replicative interphases (Foe and Alberts, 1983; Edgar et al., 1986). It is during cycle 14, which is at least 60 min long, that the blastoderm nuclei become fully enclosed by plasma membranes. At this point, mitotic synchrony is lost, gastrulation begins and high levels of zygotic gene expression are first observed (see Fig. 1). This pivotal period is often referred to as the mid-blastula transition (MBT) (Newport and Kirschner, 1982a, 1982b; Foe and Alberts, 1983; Edgar et al., 1986) or the maternal-to-zygotic transition (MZT) (Edgar and Datar, 1996).

The onset of zygotic transcription has been documented by a variety of methods including in vivo incorporation of radioactive precursors into nuclear RNA (Zalokar, 1976; Anderson and Lengyel, 1979, 1980, 1981; Edgar and Schubiger, 1986), in situ hybridization of nucleic acid probes to embryos (Lamb and Laird, 1976; Ingham et al., 1985;

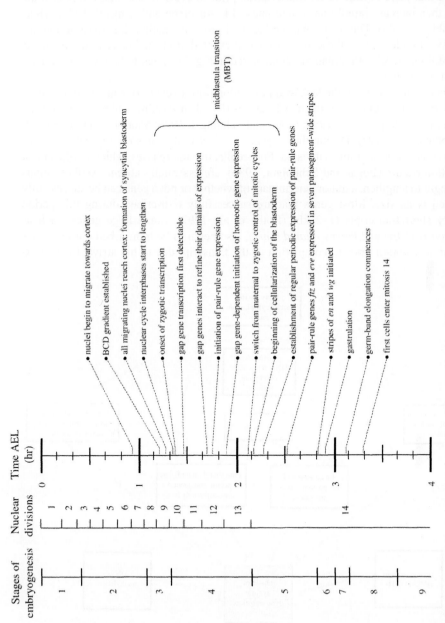

Fig. 1. Timeline of major embryonic and molecular events in the early *Drosophila* embryo. Vertical lines on the left show stages of embryogenesis (Campos-Ortega and Hartenstein, 1997) relative to the number of nuclear divisions completed, and time after egg laying (AEL). The timing of transcriptional events and other related processes are indicated to the right.

Knipple et al., 1985; Weir and Kornberg, 1985; Pritchard and Schubiger, 1996), and electron microscopy of embryo chromatin (McKnight and Miller, 1976, 1979). These studies concluded that there is little or no transcription prior to cycle 10 and that levels of transcription then increase rapidly up to interphase 14 (Anderson and Lengyel, 1979; Edgar and Schubiger, 1986). This is despite the fact that proteins required for transcription are maternally provided. How these proteins are prevented from acting is now becoming clearer. Multiple levels of regulation (summarized in Fig. 2) appear to be involved and are described briefly below.

It has been suggested that the lack of transcription prior to cycle 10 can be explained, at least in part, by the rapid rate of the first 10 cycles (less than 10 min), which leaves no time for transcription (Edgar et al., 1986; Edgar and Schubiger, 1986; Shermoen and O'Farrell, 1991; Rothe et al., 1992). During these cycles, the DNA alternates rapidly between synthesis and condensation prior to division. Post-replicative interphases, which is when most transcription occurs (Edgar and Schubiger, 1986), are essentially absent. As these interphases begin to lengthen, a major determinant of whether or not a gene can be successfully transcribed is its size. Most genes that are successfully transcribed during this period have very short transcripts (1–2 kb) (Rothe et al., 1992). Some large genes, such as *Ultrabithorax* (*Ubx*) and *knirps-related* (*knrl*), also begin to be transcribed during cycles 2–14, but are aborted prior to completion of the transcript due to the arrival of the next

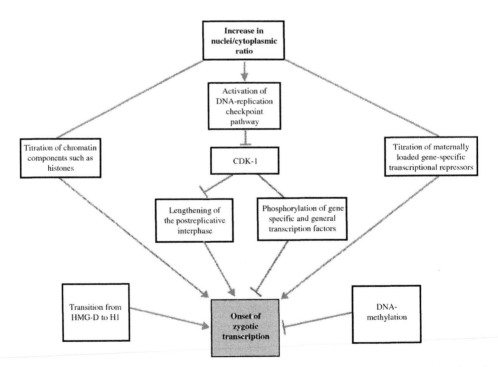

Fig. 2. Factors controlling the onset of zygotic transcription during *Drosophila* embryogenesis. A number of positive (pointed arrows) and negative effects (flat arrows) influence the beginning of transcription. A major determinant is the nuclei-to-cytoplasmic ratio, which affects multiple downstream processes.

mitotic cycle (Shermoen and O'Farrell, 1991; Rothe et al., 1992). Interestingly, *knrl* can rescue mutations of the closely related *kni* gene if the very large *knrl* intron is removed or if mutations in the *Resurrector* (*Res*), *Godzilla* (*God*) or *cyclin B* genes are present (Ruden and Jackle, 1995). *Res, God,* and *cyclin B* mutations appear to act by slowing down the blastoderm stage mitotic cycles allowing sufficient time for the completion of full-length *knrl* transcripts.

A similar conclusion comes from the study of effects of cycloheximide injection, which arrests the replication cycle. If added during or after cycle 10, a wide variety of genes are precociously transcribed (Edgar and Schubiger, 1986). Thus, elongation of post-replicative interphases subsequent to cycle 10 is an important regulator of the initiation and completion of full-length transcripts.

The molecular mechanisms responsible for the gradual lengthening of interphases are now beginning to be unraveled. Central to these mechanisms is the ratio between the number of nuclei and the volume of cytoplasm, and how this ratio affects the activity of the cell cycle regulator CDK1. CDK1 function is required for entry into mitosis (Nurse, 1990; Murray, 1995; Sigrist et al., 1995), and its activity appears to be constitutive through the first 10 cycles of *Drosophila* embryogenesis (Edgar et al., 1994). A currently favored model is that the increasing mass of nuclear materials brought about by nuclear divisions eventually titrates a critical factor; a DNA replication factor for example (Sibon et al., 1997, 1999). This slows down the replication process, leading to the gradual activation of check point factors such as Grapes and MEI-41. These, in turn, are thought to inactivate CDK1 activity by direct phosphorylation (Price et al., 2000) and by inhibiting CDK1-stimulatory factors such as the tyrosine kinase String (Furnari et al., 1997; Sanchez et al., 1997).

Inactivation of CDK1 may affect transcription in two ways, first by lengthening post-replicative interphases so that transcription can occur and second, by preventing CDK1-mediated inhibitory phosphorylation of general and gene-specific transcription factors (Gottesfeld and Forbes, 1997). The latter may also help explain why increasing post-replicative interphases prior to cycle 10 do not result in precocious transcription. Evidence for CDK1-mediated transcription factor phosphorylation has been obtained in other organisms (Leresche et al., 1996; Segil et al., 1996; Akoulitchev and Reinberg, 1998; Long et al., 1998), but has yet to be investigated in *Drosophila*. Interestingly, some general transcription factors such as TBP and RNA polymerase subunit IIc have been reported to be predominantly cytoplasmic prior to cycle 10 (Wang and Lindquist, 1998). Thus, if the activities of these proteins are indeed controlled by CDK1, phosphorylation may be controlling their subcellular localization.

The ratio of nuclei to cytoplasm may also affect transcription via the titration of maternally loaded transcriptional repressors. In support of this possibility is the finding that an increase in the dosage of the maternally provided Tramtrack (TTK) protein delays the onset of expression of the pair-rule gene *fushi tarazu* (*ftz*), whereas a decrease in dosage has the opposite effect (Pritchard and Schubiger, 1996). However, this temporal regulation of transcription by TTK must be somewhat gene-specific as TTK has no effect on expression of the *Kruppel* gene (*Kr*). Thus, if this is a general phenomenon, other maternally loaded repressors, each repressing different sets of genes, would have to exist (Pritchard and Schubiger, 1996). This could account, in part, for differences observed in the timing of transcript initiation for different genes.

Additional mechanisms controlling the temporal pattern of transcription may involve regulation at the level of chromatin. *Drosophila* eggs contain large stores of histones (Davidson, 1986) which are likely to completely coat newly synthesized DNA until a balance between nucleosome levels and DNA is reached. Interestingly, *Drosophila* chromatin assembled immediately following fertilization does not contain the linker histone H1 but instead contains HMG-D, the *Drosophila* homologue of high mobility group proteins of the HMG 1/2 class (Elgin and Hood, 1973; Becker and Wu, 1992; Wagner et al., 1992; Ner and Travers, 1994). HMG-D appears capable of functioning in a manner similar to H1 (Jackson et al., 1979), but yields chromatin that, although less densely packed, is less conducive to transcription than chromatin containing histone H1 (Ner and Travers, 1994). This is consistent with reports that the human HMG-D homologue, HMG1 negatively affects basal and gene-specific transcription factors (Ge and Roeder, 1994; Bustin and Reeves, 1996). At nuclear cycle 10, there is a rapid reduction in HMG-D levels and a complementary accumulation of H1 (Ner and Travers, 1994). The coincidence between this switch in chromatin organizing factors and the onset of transcription strongly suggests that these regulators play a critical role in the control of transcriptional competence.

In many organisms, DNA methylation plays a significant role in transcriptional silencing (Nan et al., 1993; Bird and Wolffe, 1999). *Drosophila*, however, has long been considered to be an exception with little or no methylation having been observed. Recently, however, the presence of 5-methylcytosine has been reported, with levels that are significantly higher in 0–2 h embryos (Hung et al., 1999; Lyko et al., 2000). However, these levels are still relatively low when compared to other organisms, and it has yet to be determined what their impact may be.

In summary, multiple regulatory systems appear to be employed to control the onset of zygotic transcription (see Fig. 2). The effect on transcription of most of these systems is controlled either directly or indirectly by nuclear density. The timing of zygotic transcription is critical. Sufficient time must be provided for nuclear divisions to be completed. Immediately thereafter, genes that control midblastula transition processes must be transcribed. These processes include the degradation of maternal transcripts, reorganization of the cell cycle, cellularization, regulatory events associated with sex-determination, and patterning along the anterior–posterior and dorsal–ventral axes. A description of the transcriptional cascade that controls anterior–posterior patterning follows.

2. Overview of the segmentation gene hierarchy

Transcription of the zygotic genes that control segmentation begins at cycle 10, as mitotic interphases begin to lengthen. Within the next couple of hours, a relatively complete segmental blueprint is established in the form of segmentation gene expression patterns. Segmentally repeating morphological features appear a few hours later, and in less than a day, the segmented features of the larva are complete. This morphology has served as an excellent assay for genetic screens aimed at identifying the genes involved in segmental patterning. Indeed, the depth of our current knowledge in this area can be largely attributed to the scale and success of screens first performed in the early 1980s (Nusslein-Volhard and Wieschaus, 1980; Jurgens et al., 1984; Nusslein-Volhard et al., 1984;

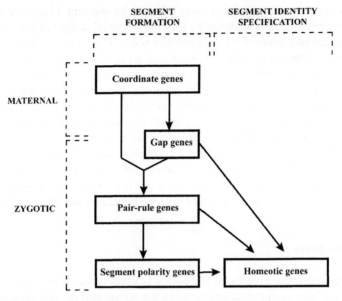

Fig. 3. Schematic representation of the segmentation gene hierarchy. Anterior–posterior patterning of the *Drosophila* embryo is controlled by several classes of genes. At the top of the this hierarchy are the coordinate genes of maternal origin. These control expression of zygotic genes which, in combination with coordinate genes, regulate pair-rule genes. Pair-rule genes, in turn, regulate segment polarity genes. Homeotic genes respond to regulatory cues provided by gap, pair-rule, and segment polarity genes.

Wieschaus et al., 1984). One of the observations made possible by the scale of these screens was that many of the mutant phenotypes had clear similarities, allowing them to be grouped into distinct gene classes (depicted in Fig. 3). Classes that affect anterior–posterior (A–P) segmentation were named coordinate, gap, pair-rule and segment polarity genes. Subsequent genetic and molecular epistasis experiments confirmed that these gene classes function in a hierarchical fashion to control segment number and polarity. A fifth class of genes, discovered earlier and referred to as the homeotic genes, controls the identity of each newly formed segment. These genes are clustered within one of two complexes. Mutations of the bithorax complex (BX-C), which affect posterior segment fates, were first described by Lewis (1978). Subsequent screens for loci that affect anterior segment identity resulted in the identification of a second gene complex, the Antennapedia complex (ANT-C) (Lewis et al., 1980a, 1980b).

The segmentation gene hierarchy (summarized in Fig. 3) begins with the actions of the maternally provided coordinate genes, whose transcription factor products come to be distributed in a graded fashion along the A–P axis. The promoters of the first zygotically expressed genes, the gap genes, respond to different concentrations and combinations of these maternal transcription factors such that they are expressed in single broad stripes within different regions of the maternal protein gradients. Periodic expression patterns are first exhibited by products of the next class of genes, the pair-rule genes. Again, the promoters of these genes respond to different combinations and levels of the transcription factors expressed earlier. The result in this case is exquisitely regular sets of mRNA stripes that repeat with precise periodicity in every other future segment. Finally, segmental periodicity of gene expression is achieved by the combinatorial actions of pair-rule transcription factors on the promoters of segment polarity genes. The segment polarity genes define regional identity within each future segment and act to maintain those cell fates through the rest of development. Unlike the gap and pair-rule genes, only a subset of segment polarity genes encode transcription factors. This is because their functions begin after cells have formed, and hence must include the capacity to send and receive signals between cells.

The gap and pair-rule proteins that establish this segmentally repeating pattern also regulate the homeotic genes. These are expressed within different segmental domains, and their transcription factor products provide each of these domains with unique features and functions. Another class of transcription factors acts subsequently to the initiation of homeotic and segment polarity gene expression patterns to maintain those patterns during the rest of development. These are referred to as the polycomb and trithorax group proteins.

3. Coordinate genes

The anterior–posterior polarity of the oocyte is established during mid-oogenesis as a result of intercellular communication between the oocyte and the surrounding epithelium of somatic follicle cells (Gonzalez-Reyes and St Johnston, 1994). This results in the microtubule-dependent localization of a subset of coordinate gene transcripts within the oocyte. In particular, *bicoid* (*bcd*) mRNA is localized to the anterior end of the oocyte (Berleth et al., 1988) and *oskar* (*osk*) mRNA is localized to the posterior end (Ephrussi et al., 1991; Kim-Ha et al., 1991). These two gene products then act to set up three maternal transcription factor gradients along the A–P axis, Bicoid protein (BCD), maternal Hunchback (HB) and Caudal (CAD) (see Fig. 4). The manner in which each of these protein gradients is established is different.

The anteriorly localized *bcd* transcripts are not translated until after fertilization. Upon translation, the protein diffuses from its site of synthesis, forming an anterior to posterior gradient (Driever and Nusslein-Volhard, 1988; St. Johnston et al., 1989). Although BCD is a homeodomain-containing transcription factor, it is also capable of binding mRNA and regulating translation (Dubnau and Struhl, 1996; Rivera-Pomar et al., 1996). A key target is *caudal* mRNA, which initially is uniformly distributed in the early embryo (Mlodzik et al., 1985). This interaction blocks translation of the *cad* transcript, resulting in a gradient of CAD protein that is lowest in the anterior where BCD is highest (Fig. 4) (Macdonald and Struhl, 1986; Mlodzik and Gehring, 1987). Like BCD, CAD is also a homeodomain-containing transcription factor.

The maternal HB gradient is also shaped by translational repression, but in this case, the repressor is the product of the Oskar-localized transcript *nanos* (*nos*). Transport of NOS from its posterior site of translation produces a gradient, in this case from posterior to anterior (Wang and Lehmann, 1991; Dahanukar and Wharton, 1996). Binding of NOS to uniformly distributed *hb* transcripts prevents translation posteriorly (Murata and Wharton, 1995). This results in decreasing levels of maternal HB in the central to posterior regions of the embryo (Fig. 4). HB is a zinc finger DNA binding protein of the Cys2His2 class (Tautz et al., 1987).

A)

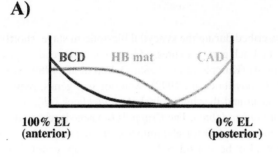

100% EL
(anterior)

0% EL
(posterior)

B)

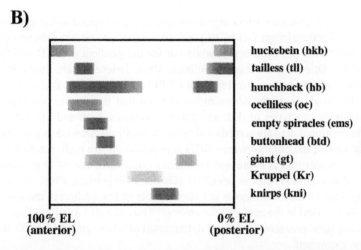

huckebein (hkb)
tailless (tll)
hunchback (hb)
ocelliless (oc)
empty spiracles (ems)
buttonhead (btd)
giant (gt)
Kruppel (Kr)
knirps (kni)

100% EL
(anterior)

0% EL
(posterior)

Fig. 4. Spatial expression of maternal and gap gene proteins. The horizontal axis represents embryo length (EL), with the left end corresponding to the anterior pole of the embryo (100% EL), and the right end to the posterior pole (0% EL). The curves in panel A represent gradients of maternal protein expression established during early embryogenesis. These gradients assign positional values along the anterior–posterior axis of zygotic genes. (BCD stands for Bicoid, HB mat for maternal Hunchback, and CAD for Caudal). The shaded boxes in panel B represent the spatial expression patterns of gap genes at the beginning of blastoderm cellularization (stage 5). These expression patterns are established in response to maternal protein gradients as well as to cross-regulatory interactions among the gap genes.

To summarize, the maternal coordinate gene products combine a multitude of molecular mechanisms to establish three transcription factor gradients (BCD, HB and CAD) along the anterior–posterior axis of the fertilized embryo. The formation of these gradients is facilitated by the syncytial environment of the early embryo. These gradients, in turn, combine to provide the zygotic genome with the positional information required for further patterning of the embryo. Also functioning at this time, but not reviewed here, are the genes that control differentiation of the anterior and posterior termini and genes that control dorsal–ventral patterning. These have been extensively reviewed elsewhere (Perrimon et al., 1995; Anderson, 1998; Govind, 1999; Flores-Saaib and Courey, 2000).

4. Gap genes

Gap genes start to be transcribed during the syncytial blastoderm stage, shortly after the 10th mitotic cycle (see Fig. 1). Each gene is expressed in one or two broad stripes that span several future segments (Fig. 4) (Knipple et al., 1985; Dalton et al., 1989; Mohler et al., 1989; Pignoni et al., 1990; Capovilla et al., 1992). Anterior gap genes, required for the proper development of the head and the thorax, include *hunchback* (*hb*), *ocelliless* (*oc*), *empty spiracles* (*ems*) and *buttonhead* (*btd*). The *Kruppel* (*Kr*) gene is required for thoracic segments and *knirps* (*kni*) and *giant* (*gt*) for abdominal segments (*gt* is also expressed in the anterior and hence can also be considered an anterior gap gene). The majority of proteins encoded by these genes are zinc finger-containing transcription factors (see Table 1).

How are the gap gene promoters organized so that they respond to the graded information of maternal transcription factors to yield discrete stripes within different portions of those gradients? This has been best worked out for the gradient of BCD and the zygotic response of *hb* (Driever and Nusslein-Volhard, 1989; Driever et al., 1989; Struhl et al., 1989). Zygotic *hb* expression overlaps that of BCD and is lost in *bcd* mutant embryos (Tautz, 1988). Analysis of the *hb* promoter showed that it contains two types of BCD binding sites with two types of different general affinities, high and low (Driever et al., 1989). Transcriptional activation mediated by the low-affinity sites takes place in the more anterior portions of the embryo where BCD concentrations are high. Another example of a gap gene promoter that is expressed in anterior portions of the BCD gradient and that makes use of low affinity sites is *ocelliless* (Gao and Finkelstein 1998). In contrast, the high affinity sites of the *hb* promoter are also capable of responding to the lower concentrations of BCD found in the center of the embryo. Thus, the affinity of binding sites within the responding gene promoter is a major determinant of where along the BCD gradient the gene will be expressed.

Another key factor in how genes respond is the number of sites contained within their promoters and how they are spatially distributed (Driever and Nusslein-Volhard, 1989; Driever et al., 1989; Struhl et al., 1989). As the number of BCD binding sites is increased, the strength of the response is increased. However, the affinity of these sites still determines where the posterior limits of the response will be (Driever et al., 1989). Multiple binding sites also lead to sharper stripe borders by increasing the rate of transition between on and off states as BCD concentrations vary. This effect is enhanced further if binding to these sites can occur cooperatively. This produces a sigmoidal transition between the on and off states (Ma et al., 1996; Burz et al., 1998). At the *hb* promoter, cooperativity is achieved via direct contacts between BCD molecules (Yuan et al., 1996; Burz and Hanes, 2001) and through interactions between BCD and CHIP (Torigoi et al., 2000), a protein believed to facilitate enhancer-promoter communication for many genes (Morcillo et al., 1996, 1997; Dorsett, 1999). Consistent with the idea that BCD binds DNA cooperatively is the finding that BCD-dependent gene activation is sensitive to the spacing and orientation of BCD binding sites (Hanes et al., 1994; Burz et al., 1998; Yuan et al., 1999).

Another factor that determines where gap gene promoters can respond within the BCD gradient is how they interact with other transcription factors. Interactions between BCD and these other proteins can affect the DNA binding affinity or activity of either protein. For example, the activation of zygotic *hb* requires a synergistic interaction between BCD and maternal HB (Simpson-Brose et al., 1994). In this case, synergy appears to occur at the level of transcriptional activation rather than cooperative binding (Sauer et al., 1995). Transcriptional synergy may also play an important role in the BCD-dependent activation of *kni* (Burz et al., 1998). *kni* is expressed where BCD concentrations are extremely low and yet, the *kni* promoter does not contain high affinity BCD binding sites. It has been suggested that cooperative interactions with other transcription factors can compensate for this low affinity (Burz et al., 1998).

Further refinement and evolution of gap gene stripe borders is controlled by cross-regulatory interactions among the gap genes (Jackle et al., 1986; Struhl et al., 1992). For example, the posterior *Kr* stripe border is refined by negative regulation mediated by KNI, GT and Tailless (TLL), and the anterior border by zygotic HB, GT and TLL (Jackle et al., 1986; Gaul and Jackle, 1987, 1989, 1991; Hulskamp et al., 1990; Kraut and Levine, 1991; Steingrimsson et al., 1991; Capovilla et al., 1992; Struhl et al., 1992). As with BCD, many of the gap gene proteins may also act in a concentration-dependent manner. For example, different levels of HB negatively regulate different target genes (Hulskamp et al., 1990; Struhl et al., 1992; Schulz and Tautz, 1994; Wu et al., 2001). Low levels of HB are sufficient to repress the posterior *gt* stripe while intermediate levels are required to repress *kni* and high levels to repress *Kr*. Similar concentration-dependent repressive activities have been reported for KNI (Kosman and Small, 1997) and GT (Wu et al., 1998; Hewitt et al., 1999).

Interestingly, some gap proteins can act as both activators and repressors of transcription. This ability varies in a concentration-dependent manner. For example, HB appears to both activate and repress *Kr*; high levels repress *Kr* while low levels activate it (Hulskamp et al., 1990; Struhl et al., 1992; Schulz and Tautz, 1994). These differential effects may be mediated through different binding sites or cofactors. KR, on the other hand, appears capable on its own of switching between repressor and activator functions in a concentration-dependent manner. Using cultured cells, it was shown that KR can activate transcription at low levels and repress transcription at high levels (Sauer and Jackle, 1991). This correlates with the existence of KR as a monomer at low concentrations and as a dimer at high concentrations (Sauer and Jackle, 1993). The opposing actions of these two KR forms may be explained by the apparent ability of KR monomers to interact with TFIIB and KR dimers with TFIIE (Sauer et al., 1996).

KR, KNI and GT have been categorized as "short-range" repressors (Gray et al., 1994, 1995; Arnosti et al., 1996a, 1996b; Cai et al., 1996; Gray and Levine, 1996a, 1996b; Hewitt et al., 1999). This is based on their ability to block the actions of transcriptional activators when nearby in the same enhancer element or when in proximal promoter positions. Most or all of these repressors interact with the maternally expressed corepressor protein, dCtBP (Nibu et al., 1998a, 1998b, 2001; Poortinga et al., 1998; Keller et al., 2000). Recruitment of dCtBP requires the motif PxDLSxK/R/H, a motif that is also found in most other sequence-specific, short-range repressors (Nibu et al., 1998a, 1998b). The mechanism by which CtBP represses transcription is unknown, although interactions with histone deacetylases (Sundqvist et al., 1998; Criqui-Filipe et al., 1999) and Polycomb proteins have been reported (Sewalt et al., 1999).

Table 1

Transcription factors of the segmentation regulatory hierarchy and their cofactors

Gene (abbreviation)	Gene class	Product	Cofactors	References
bicoid (bcd)	maternal	homeodomain protein	Chip	Frigerio et al., 1986; Driever and Nusslein-Volhard, 1988; Berleth et al., 1988; Torigoi et al., 2000
hunchback (hb)	maternal/gap	zinc finger protein	dMi-2	Tautz et al., 1987; Kehle et al., 1998
caudal (cad)	maternal	homeodomain protein		Mlodzik et al., 1985
ocelliless (oc)	gap	homeodomain protein		Cohen and Jurgens, 1990; Finkelstein et al., 1990; Finkelstein and Perrimon, 1990
empty spiracles (ems)	gap	homeodomain protein		Dalton et al., 1989; Walldorf and Gehring, 1992
buttonhead (btd)	gap	zinc finger protein		Cohen and Jurgens, 1990; Wimmer et al., 1993
Kruppel (Kr)	gap	zinc finger protein	dCtBP	Rosenberg et al., 1986; Nibu et al., 1998a
knirps (kni)	gap	zinc finger protein	dCtBP	Nauber et al. 1988; Nibu et al., 1998a; Nibu et al., 1998b
giant (gt)	gap	basic leucine zipper protein	dCtBP	Capovilla et al., 1992; Nibu and Levine, 2001
tailless (tll)	gap	zinc finger protein		Pignoni et al., 1990
huckebein (hkb)	gap	zinc finger protein	Groucho	Weigel et al., 1990; Bronner and Jackle, 1991; Goldstein et al., 1999
hairy (h)	pair-rule	helix-loop-helix protein	Groucho, dCtBP	Holmgren, 1984; Ish-Horowicz et al., 1985; Paroush et al., 1994; Poortinga et al., 1998
even skipped (eve)	pair-rule	homeodomain protein	Groucho, Rpd3	Macdonald et al., 1986; Harding et al., 1986; Mannervik and Levine, 1999; Kobayashi et al., 2001
runt (run)	pair-rule	runt domain protein	Groucho, Brother, Big-brother	Kania et al., 1990; Fujioka et al. 1996; Golling et al., 1996; Aronson et al., 1997
fushi tarazu (ftz)	pair-rule	homeodomain protein	Paired, FTZ-F1	Kuroiwa et al., 1984; Weiner et al., 1984; Copeland et al., 1996; Guichet et al., 1997; Yu et al., 1997

Gene	Class	Protein type	Interacting protein	References
odd skipped (*odd*)	pair-rule	zinc finger protein	FTZ	Coulter et al., 1990
paired (*prd*)	pair-rule	paired domain and homeodomain protein		Frigerio et al., 1986;
sloppy paired (*slp*)	pair-rule	fork head domain protein		Copeland et al., 1996
odd-paired (*opa*)	pair-rule	zinc finger protein		Grossniklaus et al., 1992
engrailed (*en*)	segment polarity	homeodomain protein	Groucho	Benedyk et al., 1994; Fjose et al., 1985; Drees et al., 1987; Jimenez et al., 1997
pangolin (*pan*)	segment polarity	HMG domain protein	Armadillo, Groucho dCBP	Brunner et al., 1997; van de Wetering et al., 1997; Cavallo et al., 1998; Waltzer and Bienz, 1998
armadillo (*arm*)	segment polarity	plakoglobin-homology protein	Pangolin/dTCF	Riggleman et al., 1989; van de Wetering et al., 1997
cubitus interruptus (*ci*)	segment polarity	zinc finger protein	dCBP	Orenic et al., 1990; Eaton and Kornberg, 1990; Akimaru et al., 1997;
gooseberry (*gsb*)	segment polarity	paired domain and homeodomain protein		Baumgartner et al., 1987
labial (*lab*)	homeotic	homeodomain protein	Extradenticle	Mlodzik et al., 1988; Diederich et al., 1989; Popperl et al., 1995; Chan and Mann, 1996
Deformed (*Dfd*)	homeotic	homeodomain protein	Extradenticle	Regulski et al., 1987; Chan et al., 1997; Li et al., 1999
proboscipedia (*pb*)	homeotic	homeodomain protein		Pultz et al., 1988
Sex combs reduced (*Scr*)	homeotic	homeodomain protein	Extradenticle	Kuroiwa et al., 1985; Ryoo and Mann, 1999
Antennapedia (*Antp*)	homeotic	homeodomain protein	Extradenticle	Garber et al., 1983; Ryoo and Mann, 1999
Ultrabithorax (*Ubx*)	homeotic	homeodomain protein	Extradenticle	Bender et al., 1983; Chan et al., 1994b; Chan and Mann, 1996; Ryoo and Mann, 1999
abdominalA (*abdA*)	homeotic	homeodomain protein		Bender et al., 1983; Ryoo and Mann, 1999
AbdominalB (*AbdB*)	homeotic	homeodomain protein	Extradenticle	Bender et al., 1983

5. Pair-rule genes

The pair-rule genes are at the heart of the segmentation gene transcription factor cascade. They are the first set of genes to define segmental regions and to establish polarity within each segment. They also play a pivotal role in establishing unique segmental identities, acting together with gap gene products to control regional expression of specific homeotic genes (see Fig. 3). Unlike the gap genes however, the pair-rule genes are expressed over a prolonged 2–3 h period with patterns of expression and regulatory roles that are very dynamic. Some of these roles and properties have not been reviewed before, and so special attention will be devoted to these issues.

5.1. Pair-rule regulation

The transition from gap to pair-rule gene expression patterns is a remarkable transition in complexity. It requires the conversion of aperiodic gap gene patterns into evenly spaced pair-rule gene stripes (Hafen et al., 1984a; Ingham et al., 1985; Harding et al., 1986; Kilchherr et al., 1986; Gergen and Butler, 1988; Coulter et al., 1990; Grossniklaus et al., 1992). Most of the pair-rule genes begin with seven stripes of expression, each gene with a different segmental phasing (Fig. 5A). Likewise, the mutant phenotypes of these genes are characterized by cuticular deletions and duplications that occur with a double-segment periodicity, each phenotype with a different phasing (Nusslein-Volhard and Wieschaus, 1980).

Pair-rule gene transcription initiates at the end of the syncytial blastoderm stage (see Fig. 5). Early studies suggested that a subset of three pair-rule genes, *hairy (h)*, *even-skipped (eve)* and *runt (run)* are specifically responsible for interpreting the aperiodic gap and coordinate gene cues and converting them into repeating patterns. It was suggested that they do so with the help of large complex promoters that contain multiple stripe-specific enhancers (see Fig. 6). These "primary" pair-rule genes were then thought to regulate the "secondary" pair-rule genes via the regulation of much simpler promoters (Carroll et al., 1986; Carroll and Scott, 1986; Mlodzik et al., 1987; Ingham and Gergen, 1988; Carroll and Vavra, 1989; Hooper et al., 1989). Much of this hypothesis was based on initial studies of the *h* and *eve* promoters (Howard et al., 1988; Goto et al., 1989; Harding et al., 1989; Hooper et al., 1989; Howard and Struhl, 1990; Pankratz et al., 1990; Riddihough and Ish-Horowicz, 1991; Stanojevic et al., 1991; Small et al., 1992; Hartmann et al., 1994).

In the case of the *h* gene, it was first shown that disruptions within different regions of the 5′ promoter, ranging as far upstream as 14 kb, could eliminate the expression of individual stripes (Howard et al., 1988; Hooper et al., 1989). By connecting the corresponding genomic regions to reporter genes, it was then shown that different portions of the very large promoter could indeed drive the expression of individual stripes (Riddihough and Ish-Horowicz, 1991; Langeland et al., 1994; La Rosee et al., 1997). This suggested the existence of multiple enhancer elements, each capable of regulating individual *h* stripes. The *eve* promoter was similarly shown to be composed of stripe-specific enhancers (see Fig. 6) (Small et al., 1991, 1996; Fujioka et al., 1999; Sackerson et al., 1999).

At the molecular level, the mechanisms generating individual pair-rule gene stripes are similar to those that regulate gap gene stripes. In both cases, the responding regulatory elements tend to contain clustered binding sites for positively and negatively acting

A)

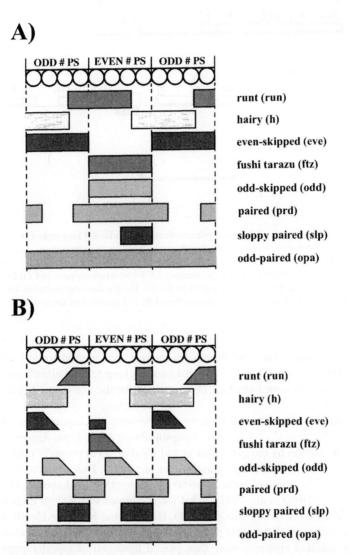

B)

Fig. 5. Spatial expression of pair-rule genes during cellularization of the blastoderm (A) and gastrulation (B). A schematic representation of three consecutive parasegment intervals is shown. Each parasegment is delineated by vertical dashed lines and designated as odd or even numbered at the top of each panel. The row of circles in each panel represents a row of nuclei that pass through three parasegments along the anterior–posterior axis of blastoderm (A) or gastrula stage (B) embryos. Stripes of gene expression are depicted as boxes. During the cellular blastoderm stage (A) most pair-rule genes are expressed in seven stripes with a double-parasegment periodicity. Spatial patterns of different pair-rule genes, although often overlapping, are typically out of phase with one another. Even-numbered parasegments (Even # PS) are marked by expression of *fushi tarazu* (*ftz*), and odd-numbered parasegments (Odd # PS) by expression of *even skipped* (*eve*). During gastrulation (B), pair-rule gene stripes tend to shift and become narrower, with many undergoing a 7 to14 stripe transition (eg. *run, odd,* and *prd*).

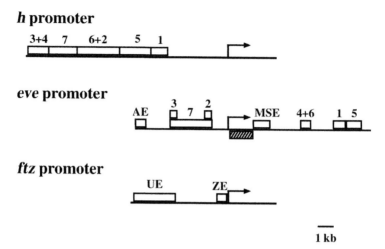

Fig. 6. Structure and organization of three pair-rule promoters. The horizontal lines represent the *hairy* (*h*), *even skipped* (*eve*), and *fushi tarazu* (*ftz*) promoters. Regulatory elements in these promoters, shown as open boxes, are designated either by numbers, which refer to the number of the stripe that they control, or letters. AE stands for autoregulatory element, MSE for minor stripe element, UE for upstream element, and ZE for zebra element. The position of each transcription start site is marked by arrows. The *eve* transcription unit is shown as a hatched box. Note that promoters of *h* and *eve* are larger than that of *ftz* and contain well-defined stripe-specific regulatory elements.

regulators. The proteins that bind these sites interact in a combinatorial fashion, driving expression where combinations are favorable and blocking activity elsewhere. These interactions and principles have been best characterized for the *eve* stripe 2 enhancer (Small et al., 1991, 1992).

Genetic studies were first used to show that *eve* stripe 2 requires positive regulation by *bcd* and *hb* (Small et al., 1991, 1992; Simpson-Brose et al., 1994; Arnosti et al., 1996a) and negative regulation by the gap genes *gt* and *Kr* (Fig. 7) to properly position the stripe borders (Small et al., 1991, 1992; Stanojevic et al., 1991). Deletional analyses narrowed down the minimal response element to a 480 bp fragment located about 1 kb upstream of the *eve* promoter (Small et al., 1992). What then followed was a "tour de force" in terms of promoter analyses. Binding sites for BCD, HB, GT and KR were identified within the 480 bp stripe 2 element (Stanojevic et al., 1991), and each of these was mutated to disrupt binding. The mutated elements were then placed upstream of reporter gene constructs, individually and in a wide assortment of combinations. Many of these mutations had clear-cut effects. For example, disruption of GT binding sites had the same effect as removing the *gt* gene genetically; in both cases the reporter gene stripe expanded anteriorly. Other results were a little more complicated. For example, it appeared that all BCD binding sites needed to be removed to have a significant impact, but instead of diminishing expression as expected, this caused stripe expansion in the posterior direction. It was speculated that this may have been due to the disruption of overlapping elements and the ability of other factors to bind and regulate them. What was clear from these studies, however, was that discrete stripe enhancer elements exist and that they are comprised of clustered binding

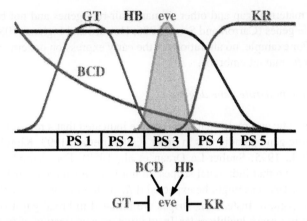

Fig. 7. Activation and positioning of *even skipped* (*eve*) stripe-2. The anterior five-most parasegments (PS) of the embryo are shown as open boxes. The second *eve* stripe is activated in parasegment 3 (PS3) by gradients of BCD and HB. The anterior and posterior borders of *eve* stripe 2 are precisely positioned via repression by GT and KR, respectively. These positive and negative regulatory interactions are summarized underneath.

sites for appropriate regulators that confine expression to defined regions. These enhancers appear to have evolved over time, together with their regulators, such that they now regulate precisely spaced and repeating stripes of expression.

As stated above, the actions of transcription factors on gap and pair-rule gene promoters have many aspects in common. As with the gap gene promoters, multiple activator sites capable of facilitating cooperativity and synergy of action also appear to be important. Also, repressors are thought to act in one of two ways, by competing for overlapping activator binding sites (Small et al., 1991, 1992; Stanojevic et al., 1991), and when not overlapping, by local quenching (Arnosti et al., 1996a).

Other stripes of *eve* as well as those of *h* have been shown to be regulated in much the same way as *eve* stripe 2 (Pankratz et al., 1990; Langeland and Carroll, 1993; Lardelli and Ish-Horowicz, 1993; Langeland et al., 1994; La Rosee et al., 1997; Hader et al., 1998). Unlike the large promoters of the *h* and *eve* genes, however, the *fushi tarazu* (*ftz*) gene was found to be regulated by a less complex type of promoter (Fig. 6). Much of the *ftz* seven stripe pattern is controlled by a relatively small (670 bp) proximal regulatory element referred to as the zebra element (Hiromi et al., 1985; Hiromi and Gehring, 1987). This minimal promoter is capable of initiating the expression of all seven *ftz* stripes, but at relatively low levels. Higher levels of expression are then achieved and maintained by a general enhancer located about 3.5 kb upstream. It was proposed that the proximal zebra element is regulated by the primary pair-rule genes, whereas the upstream enhancer by FTZ itself (Hiromi et al., 1985; Hiromi and Gehring, 1987). In this way, it was postulated that the secondary pair-rule gene promoters could be greatly simplified in terms of size and complexity.

The designation of primary and secondary pair-rule genes was also based on early genetic studies. Pair-rule gene expression analyses in wild-type and pair-rule mutant backgrounds suggested that the initiation of primary pair-rule gene expression is only affected by mutations in maternal, gap and other primary pair-rule genes and not by mutations in the other pair-rule genes (Carroll and Scott, 1986; Howard and Ingham, 1986; Ingham and Gergen, 1988). For example, no alterations in the early expression patterns of *h*, *eve* or *run* were detected in *ftz* mutant embryos.

5.2. Revision of the pair-rule gene dogma

More recent genetic and molecular studies have indicated that a revision in the primary/secondary pair-rule gene paradigm is in order (Gutjahr et al., 1993; Klingler and Gergen, 1993; Yu and Pick, 1995; Saulier-Le Drean et al., 1998). For example, analysis of *ftz* expression has shown that individual stripes do not initiate simultaneously (Krause et al., 1988; Yu and Pick, 1995) as might be expected if *ftz* periodicity were established by a pre-existing pair-rule pattern. Instead, *ftz* stripes are initiated in broad gap-like stripes which are then resolved in a reproducible order. In addition, re-examination of *ftz* transcription in embryos mutant for the primary pair-rule genes revealed that activities of these genes do not appear to affect initiation but rather affect subsequent stages of *ftz* expression (Yu and Pick, 1995). Many of the other pair-rule genes such as *paired* (*prd*) and *odd skipped* (*odd*) have also been shown to exhibit gap-like patterns of early expression (Coulter et al., 1990; Gutjahr et al., 1993; Saulier-Le Drean et al., 1998).

Other results not compatible with the primary/secondary pair-rule gene dogma are the more recent dissections of pair-rule gene promoters. The *run* promoter, for example, shows intermediate levels of complexity. Although it is relatively large, it has not been possible to isolate individual stripe enhancer elements (Klingler et al., 1996). It has been suggested that these response elements must be widely dispersed and therefore impossible to identify by deletional analysis. Ectopic expression studies have also indicated that so-called secondary pair-rule genes may have important roles in the early expression of other pair-rule genes. For example, ectopic expression of ODD shows that it plays an early role in the proper establishment of *eve, h* and *run* expression (Saulier-Le Drean et al., 1998).

Taken together, these results demonstrate that the regulation of all pair-rule genes may depend directly on the activities of maternal and gap loci as well as on complex cross-regulatory interactions among the pair-rule genes themselves.

5.3. Complexity within the pair-rule gene hierarchy

Unlike the more transient expression of the coordinate and gap genes, expression of the pair-rule genes persists for an extended period of time. Transcription of the pair-rule genes begins soon after the gap genes at about 1 h 40 min after fertilization (stage 4; Fig. 1). Their initial gap gene-like expression patterns usually resolve into seven-stripe patterns prior to the completion of cellularization (Fig. 5A). Over the next hour, these stripes become narrower, and in the case of *run, eve, prd, odd* and *sloppy paired* (*slp*), a second set of stripes is generated, either de novo or by the splitting of original stripes. This yields stripes in each of the future segments (Fig. 5B). Portions of all or some of these stripes tend to persist until as late as 5 h after fertilization.

The rapid evolution of pair-rule gene patterns depends predominantly on changing cross-regulatory interactions within the class (diagramed in Fig. 8). Kinetic experiments,

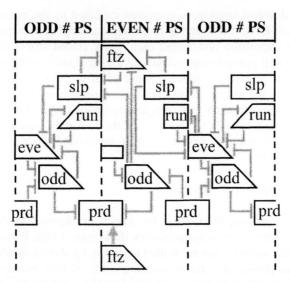

Fig. 8. Gene-regulatory interactions among pair-rule genes at the end of cellular blastoderm and during gastrulation. A schematic representation of three consecutive parasegments separated by vertical dashed lines is shown. Spatial gene expression patterns are depicted as boxes or trapezoids. Sloped sides indicate decreasing levels of expression. Known direct regulatory interactions are depicted as pointed (positive) or flat (negative) arrows. Note that *ftz* is indicated in two positions for the sake of clarity.

in which segmentation gene products were monitored following pulses of ectopic pair-rule gene expression, have shown that successive pair-rule gene interactions can occur every 8–10 min (Nasiadka and Krause, 1999). Hence, the 2–3 h period during which pair-rule genes are expressed allows ample opportunity for a multitude of cross-regulatory interactions. The rapidly evolving expression patterns also result in ever-changing combinations and concentrations of pair-rule proteins expressed within a given region or cell. Since the regulatory output of these proteins is highly dependent upon combinatorial interactions (see below), this means that their functions have the capacity to change with time. Examples of such functional changes have been provided by ectopic expression studies with EVE and ODD (Manoukian and Krause, 1992; Saulier-Le Drean et al., 1998). When expressed prior to cellularization, both proteins activate expression of *ftz*. However, after cellularization, both proteins function as repressors of *ftz*. Charting the full complement of changing pair-rule gene interactions will require considerably more work.

The use of heat shock promoters to provide pulses of pair-rule gene expression also allows precise control of expression levels. This has shown that, like gap proteins, pair-rule proteins can have different outputs at different expression levels. For example, different levels of EVE were shown to have different effects on the pair-rule target genes, *prd, run, ftz,* and *odd* (Manoukian and Krause, 1992). To repress all four, high levels of expression corresponding to those in the center of mature EVE stripes had to be induced. When levels were lowered slightly, *prd* stripes were no longer repressed while the other three were. At lower levels still, such as those found at EVE stripe edges, *prd, ftz* and *run* were unaffected while *odd* remained completely repressed. These differential outputs are comparable to the different target gene responses mediated by the maternal morphogen Bicoid. In the case of EVE, these varying levels and transcriptional outputs occur during different phases of expression and across each of its graded stripes (Manoukian and Krause, 1992). One well-characterized outcome of these properties is the repression of *odd* at the anterior border of each *ftz* stripe. Absence of the ODD repressor allows FTZ to activate one of the key segment polarity genes, *engrailed* in these cells.

In summary, the long duration and complex dynamics of pair-rule gene expression patterns mean that early models of this part of the hierarchy were overly simplified. Further studies will be required before we fully understand all of the pair-rule gene functions and interactions.

5.4. Developmental roles of pair-rule genes

One of the most important functions of the pair-rule proteins is to allocate cells into pre-segmental units referred to as parasegments (Lawrence, 1992). Parasegments are the same size in width as segments but are shifted anteriorly by about a quarter of a segment. They are the first morphologically distinct metameric units to form and also define the borders of homeotic gene expression patterns. The borders of parasegments remain well defined through most of the rest of development such that cells of adjacent parasegments never mix. It is believed that this may be due to differential adhesion properties adopted by cells on either side of the borders. Portions of the parasegment borders will also become the anterior–posterior boundaries of imaginal discs, which later give rise to limbs. Thus, parasegments and their boundaries are important genetic and physical organizers of all segmentally derived patterns. Domains of cell lineage restriction such as these are referred to as compartments (Garcia-Bellido et al., 1973, 1979; Garcia-Bellido, 1975).

Two pair-rule genes, *eve* and *ftz*, play particularly important roles in determining the size, identity and boundaries of parasegments (Martinez-Arias and Lawrence, 1985; Howard and Ingham, 1986; DiNardo and O'Farrell, 1987; Lawrence et al., 1987; Ingham et al., 1988; Lawrence and Johnston, 1989; Kellerman et al., 1990; Hughes and Krause, 2001). During cellularization of the blastoderm, stripes of *eve* and *ftz* fall precisely within parasegmental boundaries, with *eve* expressed in odd-numbered parasegments and *ftz* expressed in even-numbered parasegments (Hafen et al., 1984a; Martinez-Arias and Lawrence, 1985; Carroll and Scott, 1986; Frasch and Levine, 1987; Krause et al., 1988; Karr and Kornberg, 1989; Lawrence and Johnston, 1989). It appears that the relative levels of *eve* and *ftz* expression or activity are the crucial determinants of where parasegmental borders are positioned (Kellerman et al., 1990; Manoukian and Krause, 1992; Hughes and Krause, 2001). Prior to cellularization, *ftz* and *eve* stripes overlap at their borders. When relative expression or activity levels are altered, the positions at which these overlaps are equivalent become shifted along with the ensuing borders (Hughes and Krause, 2001). Interestingly, changes in the rates of programmed cell death can usually compensate for such changes providing that they are small. In these cases, the narrower parasegments exhibit reduced cell death while widened parasegments exhibit increased cell death (Hughes and Krause, 2001). Thus, parasegments appear to be endowed with an ability to sense and modulate their relative sizes.

This ability of parasegments to monitor and correct size does not occur when changes in parasegment width reach or exceed about 20%, as occurs when either *ftz* or *eve* loses or gains significant activity. In these cases, broadened parasegments remain enlarged and narrower intervening parasegments delaminate from the epithelial layer, leaving half of the normal number of parasegments (Hughes and Krause, 2001). Delamination of the reduced parasegments may be driven by differential adhesive properties of cells within the adjacent parasegments and/or by the forces that drive cell reorganization during the process of germband retraction.

A major role for many of the remaining pair-rule proteins is to ensure that *ftz* and *eve* stripes are evenly and precisely positioned, and to then define polarity within each paraseg-ment. The Runt and Hairy proteins, for example, are repressors of *eve* and *ftz* respectively. The phasing of RUN and H stripes helps to define the posterior edges of *eve* and *ftz* stripes, and to then cause a progressive narrowing of their widths (see Figs. 5 and 8). The initial repression of *ftz* and *eve* stripes by H and RUN, respectively, is then augmented and main-tained by the actions of ODD and SLP. This ensures that the late expression of *eve* and *ftz* is limited to the anterior-most cells of each parasegment (Fig. 8). In turn, this also ensures the proper initiation of segment polarity genes at the anterior and posterior edges of each parasegment (see below). Indeed, the different and rapidly changing phasing of each pair-rule protein determines where and when each of the segment polarity genes will be expressed (discussed further below).

5.5. Properties of pair-rule proteins

The pair-rule proteins form a diverse set of DNA binding transcription factors (sum-marized in Table 1). As with the gap proteins, most function predominantly as repres-sors (Fig. 8). These include EVE, H, ODD, SLP and RUN. All except H, however, may also have limited roles as transcriptional activators (Manoukian and Krause, 1992, 1993; Cadigan et al., 1994b; Tsai and Gergen, 1995). The FTZ and PRD proteins, on the other hand, appear to function predominantly as activators, although FTZ is known to repress at least two of its targets (Copeland et al., 1996; Nasiadka and Krause, 1999). Genetic experiments suggest that the eighth pair-rule protein, Odd-paired (OPA), is also likely to function as a transcriptional activator (Ingham et al., 1988; Benedyk et al., 1994).

Recent experiments have provided substantial insight into how the pair-rule repressors regulate gene expression. The *h* protein, for example, is a helix–loop–helix transcription factor (Sasai et al., 1992) that requires the widely utilized co-repressor protein Groucho (GRO) for most of its functions (Paroush et al., 1994; Fisher et al., 1996; Jimenez et al., 1997). This interaction depends on a conserved sequence motif at the Hairy C-terminus, WRPW (Paroush et al., 1994; Fisher et al., 1996). GRO represses both basal and activated transcription (Paroush et al., 1994; Fisher et al., 1996), and has been suggested to function as a cofactor for long-range repressors, repressing target gene expression when as far away as a kilobase from the affected enhancer or promoter (Barolo and Levine, 1997; Zhang and Levine, 1999). GRO also interacts genetically and physically with the *Drosophila* histone deacetylase Rpd3 (Chen et al., 1999a), suggesting that GRO repression involves regulation at the level of chromatin structure.

GRO also interacts with RUN via a conserved, C-terminal VWRPY motif (Aronson et al., 1997). RUN is a member of the Runt domain family of transcriptional regula-tors (Speck and Terryl, 1995), all of which contain a 128 amino acid DNA binding domain referred to as the Runt domain (Ogawa et al., 1993; Bae et al., 1994; Kagoshima et al., 1996). The Runt domain is also required for heterodimerization with one of two structurally and functionally homologous partner proteins, Brother (BRO) or Big-brother (BGB) (Fujioka et al., 1996; Golling et al., 1996; Li and Gergen, 1999; Kaminker et al., 2001). Although BRO and BGB do not bind to DNA on their own, they enhance DNA binding by RUN (Golling et al., 1996). It has been suggested that they may also contribute to transcriptional regulation in other ways (Li and Gergen, 1999; Wheeler et al., 2000).

EVE is a homeodomain protein that functions predominantly as a transcriptional repressor (Biggin and Tjian, 1989; Han et al., 1989; Johnson and Krasnow, 1992; Manoukian and Krause, 1992; Han and Manley, 1993; Fujioka et al., 1995), but as stated earlier, appears to act early (prior to cellularization) as a transcriptional activator (Manoukian and Krause, 1992). Its function as a repressor appears to be attributable, in part, to an interaction with the histone deacetylase RPD3 (Mannervik and Levine, 1999). However, EVE appears to have a number of modular repression domains (Han and Manley, 1993) suggesting the ability to interact with other corepressors. One of these appears to be GRO, since the ability of EVE to repress target genes is significantly reduced when its GRO-interaction motif is removed (Kobayashi et al., 2001). The cofactor(s) that confer the early ability of EVE to activate transcription remain unknown. Interestingly, ODD, which also functions predominantly as a repressor, activates transcription of some of the same target genes as EVE during the same time (Saulier-Le Drean et al., 1998), suggesting the possibility of a common partner.

In terms of the activators FTZ and PRD, more is known about FTZ than PRD. FTZ requires an interaction with the orphan nuclear receptor αFTZ-F1 for the regulation of most of its target genes (Guichet et al., 1997; Yu et al., 1997). The interaction is mediated by a conserved LXXLL nuclear receptor box in the FTZ N-terminus, which contacts the ligand binding domain of FTZ-F1 (Guichet et al., 1997; Schwartz et al., 2001; Suzuki et al., 2001). This interaction with a nuclear receptor protein suggests the possibility of transcriptional activation via the recruitment of histone acetyltransferases or RNA polymerase II-recruiting protein complexes. Interestingly though, FTZ and FTZ-F1 are both required, not only for the activation of FTZ target genes but also for the repression of a subset of target genes (Guichet et al., 1997).

Another protein that has been shown to interact with FTZ is PRD itself (Copeland et al., 1996). Oddly enough, this interaction between the only two pair-rule proteins that generally function as transcriptional activators results in transcriptional repression (Copeland et al., 1996). The mechanism underlying this effect is currently unknown. Presumably, the ability of these proteins to act as both activators and repressors of transcription lies in their ability to interact with different proteins when bound to different target genes or different promoter elements.

6. Segment polarity genes

The last class of genes in the hierarchy required for segmental division (as opposed to segmental identity) is the segment polarity class (see Fig. 3). Mutations in these genes generally result in the deletion of portions of every segment and coincidental mirror-image or tandem duplications of remaining regions (Nusslein-Volhard and Wieschaus, 1980; Perrimon and Mahowald, 1987; Martinez Arias et al., 1988; Perrimon and Smouse, 1989). Since the majority of these genes do not encode transcription factors, we will not delve deeply into their function. Rather, we will focus on their regulation by pair-rule proteins. The majority of such studies deal with the segment polarity genes *engrailed (en)* and *wingless (wg)*, and so we will also focus on the regulation of these two genes. A schematic representation of these interactions is shown in Fig. 9.

Segment polarity genes are expressed in portions of every segment. Transcription of *en* occurs in the anterior-most cells of each parasegment (Fjose et al., 1985; Kornberg et al.,

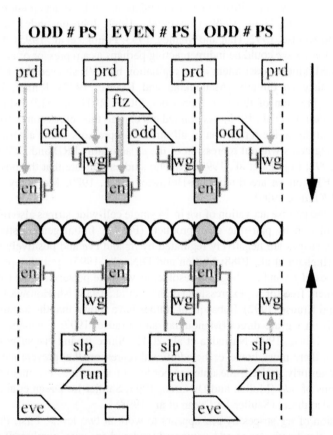

Fig. 9. Initial regulation of *engrailed (en)* and *wingless (wg)*. Expression of *en* and *wg* begins at the end of cellular blastoderm (stage 5) as a result of complex gene-regulatory interactions mediated by pair-rule gene products. A schematic representation of three consecutive parasegments is shown. The circles in the center represent a row of cells along the anterior–posterior axis. These are placed in the center of the figure for clarity with gene interactions diagramed both above and below. Stripes of gene expression are depicted as boxes or trapezoids. Sloped sides indicate decreasing levels of expression. Positive gene-regulatory interactions are marked in green and negative interactions in red. Stripes of *en* and *wg* initiate in single rows on either side of the parasegmental border. (*For a colored version of this figure, see plate section, page 275.*)

1985), while transcription of *wg* is limited to the posterior-most cells of each parasegment (Baker, 1988; van den Heuvel et al., 1989). The interface between *en* and *wg* stripes in adjacent parasegments straddles each parasegment border.

Early genetic studies on *en* expression demonstrated that alternate stripes are regulated by different combinations of pair-rule genes (DiNardo and O'Farrell, 1987; Ingham et al., 1988). In particular, *ftz* and *opa* activities were shown to be required for even-numbered *en* stripes while *prd* and *eve* were required for odd-numbered *en* stripes. This dependence of alternate stripes on two distinct sets of regulatory factors suggested independent cis-acting response elements within the *en* gene promoter.

Although these genetic studies identified genes required for *en* regulation, it was not clear whether the encoded factors play direct or indirect roles in *en* regulation. The *eve* gene, for example, plays a positive role in *en* regulation, but EVE acts predominantly as a repressor (Biggin and Tjian, 1989; Han et al., 1989; Johnson and Krasnow, 1992; Manoukian and Krause, 1992; Han and Manley, 1993; Fujioka et al., 1995). This suggests that *en* regulation by *eve* could be indirect, acting perhaps as a repressor of another repressor gene. Indeed, kinetic assessment of *en* regulation by EVE showed that EVE does not appear to regulate *en* directly (Manoukian and Krause, 1992). Rather, it negatively regulates the expression of three *en* repressors, ODD, RUN and SLP (Manoukian and Krause, 1992; Fujioka et al., 1995). As pointed out earlier, this repression is concentration-dependent, resulting in the repression of all three *en* repressors in the anterior-most cells of all 14 parasegments without repression of the *en* activators PRD and FTZ (Manoukian and Krause, 1992; Fujioka et al., 1995). Kinetic studies indicate that the positive actions of PRD and FTZ on *en* are direct (Ish-Horowicz et al., 1989; Morrissey et al., 1991; Nasiadka and Krause, 1999).

As with *en*, the precise activation of *wg* in 14 single cell-wide stripes also relies on combinatorial regulation by pair-rule gene products (Fig. 9). Initial genetic studies demonstrated that *wg* stripes are negatively regulated by *ftz* and *eve* and positively regulated by *prd* and *opa* (Ingham et al., 1988; Mullen and DiNardo, 1995). Initial expression of *wg* begins as stripes of *ftz* and *eve* begin to resolve, leaving the posterior-most row of cells in each parasegment free for *wg* expression (Ingham et al., 1988; Ish-Horowicz et al., 1989; Manoukian and Krause, 1992). Subsequent studies have shown that the negative effects of FTZ and EVE on *wg* are direct (Manoukian and Krause, 1992; Copeland et al., 1996; Nasiadka and Krause, 1999; Nasiadka et al., 2000). Since *ftz* and *eve* stripes continue to narrow further, there must, however, be additional repressors that prevent *wg* stripes from broadening anteriorly. Indeed, the anterior borders of *wg* stripes are maintained by the negative actions of *odd* (Mullen and DiNardo, 1995; Saulier-Le Drean et al., 1998). This regulation is also direct (Saulier-Le Drean et al., 1998).

The activation of *wg* stripes by *prd* appears to work at two levels. First, PRD has been shown to be a direct activator of odd-numbered (and perhaps even-numbered) stripes of *wg* (Copeland et al., 1996). Second, *prd* acts indirectly by negatively regulating the expression of *odd* (Mullen and DiNardo, 1995). The establishment of *wg* expression also depends on the function of *opa* (Ingham et al., 1988; Benedyk et al., 1994). After its establishment, short-term maintenance of *wg* expression depends on the pair-rule gene *slp* (Cadigan et al., 1994a, 1994b). SLP appears to act directly by activating *wg* expression, as well as indirectly by negatively regulating *ftz*, *eve* and *en* (Grossniklaus et al., 1992; Cadigan et al., 1994a, 1994b).

Long-term maintenance of *wg* and *en* expression is controlled via mutually dependent intercellular pathways and autoregulatory loops (DiNardo et al., 1988; Martinez Arias et al., 1988; Bejsovec and Martinez Arias, 1991; Heemskerk et al., 1991).

In summary, stripes of segment polarity gene expression are controlled in much the same way as those of the pair-rule genes. The actions of repressors appear to predominate (Fig. 9), limiting the actions of the relatively few and more broadly distributed activators to narrow columns of cells. A major difference is that the majority of segment polarity promoters will likely require two sets of enhancers, one for the regulation of odd-numbered stripes and the second for even-numbered stripes. These enhancers are yet to be properly characterized.

7. Homeotic genes

Unlike the segmentation genes described to date, the homeotic genes are required, not to form segments, but to assign each segment with a unique identity. Mutations in these loci result in homeosis, the transformation of one or more segments or structures into the likeness of others (Bateson, 1894). For example, flies that lack *Antennapedia* (*Antp*) have transformations of parasegments 4 and 5 to parasegment 6 identity (Bridges and Morgan, 1923; Lewis, 1978). Conversely, ectopic expression of *Antennapedia* can transform anterior parasegments and structures to parasegment 4 identity (Kaufman et al., 1980; Struhl, 1981b).

The *Drosophila* homeotic genes, collectively referred to as HOM genes, are clustered within two complexes: the *bithorax* and *Antennapedia* complexes (BX-C and ANT-C, respectively). Both complexes are located on the right arm of the third chromosome (Lewis, 1978; Kaufman et al., 1980). The BX-C contains three HOM genes, *Ultrabithorax* (*Ubx*), *abdominal-A* (*abd-A*), and *Abdominal-B* (*Abd-B*). These three genes control the identities of parasegments 5–13 (Lewis, 1978; Bender et al., 1983; Karch et al., 1985; Sanchez-Herrero et al., 1985; Tiong et al., 1985). The homeotic loci in the ANT-C include *labial* (*lab*), *proboscipedia* (*pb*), *Deformed* (*Dfd*), *Sex combs reduced* (*Scr*) and *Antennapedia* (*Antp*) (Lewis et al., 1980a, 1980b; Wakimoto and Kaufman, 1981; Merrill et al., 1987). These genes control regional identities anterior to parasegment 5.

This physical clustering of HOM genes is one of the most fascinating aspects of their function. Homologous HOX gene clusters exist in organisms from worms to man. What is particularly interesting is that there is a precise correlation between the relative position of each homologue in each of these clusters and where those genes are expressed along the anterior–posterior axis of the developing embryo (reviewed in McGinnis and Krumlauf, 1992). Moving from distal to proximal positions along the chromosome (5′ to 3′; see Fig. 10), each gene is expressed in increasingly more anterior domains. *Ubx*, for example, is on the proximal side of *abd-A* on the chromosome, and is more anterior in its pattern of embryonic expression (Akam, 1983; Akam and Martinez Arias, 1985; Harding et al., 1985).

In more primitive insects, there is only one homeotic gene cluster (*Hom-C*), indicating that the ANT-C and BX-C split apart late in insect evolution. The original cluster appears to have formed by gene duplication events, beginning with a single gene and forming

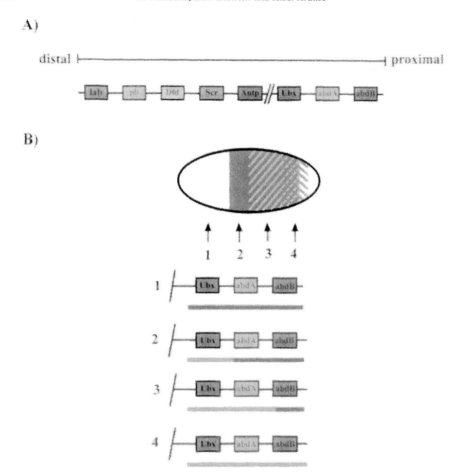

Fig. 10. *Hom-C* genes and regulation by PcG/trxG complexes. (A) The eight genes of the *Hom-C* are displayed with their positions along the chromosome (proximal to distal) indicated. In *Drosophila*, the ANT-C (left) and BX-C (right) complexes are not contiguous (break indicated by //). (B) Postulated control of the BX-C genes by the PcG/trxG. In the anterior region where genes of the BX-C are inactive (1), PcG complexes create a repressive chromatin domain along most of the BX-C (denoted by red line). In more posterior regions, where genes of the BX-C become activated (2 and 3), this repressive complex is replaced by transcriptionally active trxG complexes (green line) such that the boundary between PcG and trxG complexes shifts in a proximal to distal direction. In more primitive insects, this movement of the PcG/trxG boundary may be processive. *lab, labial; pb, proboscipedia; Dfd, Deformed; Scr, Sex combs reduced; Antp, Antennapedia; Ubx, Ultrabithorax; abdA, abdominalA; AbdB, AbdominalB. (For a colored version of this figure, see plate section, page 276.)*

an ancestral complex of about 8 genes. In higher metazoans, this complex has itself been duplicated, twice in mammals for a total of four clusters. The incredible conservation of gene proximity and order implies an important functional context. This is discussed further in the section below on maintenance of *Hom-C* gene expression.

7.1. Initiation of Hom-C gene expression

In *Drosophila*, the initial expression domains of most HOM genes are parasegmental, with some parasegments expressing combinations of two or more genes. (Akam, 1983; Levine et al., 1983; Akam and Martinez Arias, 1985; Beachy et al., 1985; Kuroiwa et al., 1985; Martinez Arias, 1986; Casanova and White, 1987; Chadwick and McGinnis, 1987; Martinez-Arias et al., 1987; Peifer et al., 1987). *Ubx*, for example, is expressed in parasegments 5–12 (Akam, 1983; Akam and Martinez Arias, 1985). The initial expression of HOM genes is mainly controlled by gap and pair-rule genes (Ingham et al., 1986; Ingham and Martinez-Arias, 1986; White and Lehmann, 1986; Harding and Levine, 1988; Jack et al., 1988; Irish et al., 1989). In this respect, the initiation of *Ubx* expression is perhaps the best understood. The expression domain of Kruppel defines a broad central region where *Ubx* expression can occur. Hunchback, expressed in stripes on either side, appears to act as a direct repressor (Irish et al., 1989; Muller and Bienz, 1991; Qian et al., 1991; Zhang et al., 1991). Positive regulation by FTZ (Ingham and Martinez-Arias, 1986; Muller and Bienz, 1992) helps to sharpen the stripe borders and to generate pair-rule periodicity (Muller and Bienz, 1992). Other homeotic genes exhibit similar initial regional peaks of expression, presumably under gap gene control, as well as weaker stripes that repeat with pair-rule periodicity (Ingham and Martinez-Arias, 1986).

The control of HOM gene expression by the gap and pair-rule proteins is transient. After these proteins disappear, further refinement and ongoing regulation is controlled by autoregulation (Bienz and Tremml, 1988; Kuziora and McGinnis, 1988; Chouinard and Kaufman, 1991; Lamka et al., 1992; Tremml and Bienz, 1992) and cross-regulatory interactions (Hafen et al., 1984b; Struhl and White, 1985). Maintenance of these expression patterns also comes under the control of two additional classes of genes—the *Polycomb* group (*PcG*) genes (Lewis, 1978; Struhl, 1981a; Jurgens, 1985; Paro, 1990; Muller and Bienz, 1991) and the *trithorax* group (*trxG*) genes (Ingham, 1985; Kennison and Tamkun, 1988; Tamkun et al., 1992; Breen and Harte, 1993). Proteins encoded by the *trxG* genes maintain expression where transcription has been activated, and proteins of the *PcG* maintain states of transcriptional repression where genes have been repressed. These processes are discussed further below.

7.2. Maintenance of Hom-C gene expression

Mutations in *PcG* genes were first identified by Lewis in 1978 (Lewis, 1978). These cause homeotic transformations of anterior structures into more posterior structures. The majority of *trxG* genes were identified later in screens for *PcG* suppressors (Kennison and Tamkun, 1988). Mutations in *trxG* genes cause homeotic transformations of posterior structures into more anterior structures (reviewed in Francis and Kingston, 2001). The proteins encoded by *PcG* and *trxG* genes form complexes that lock in the repressed (PcG) or activated (*trxG*) states of gene expression. These complexes most likely have opposing effects on nucleosome modifications and packaging (Francis and Kingston, 2001). For example, trxG complexes possess histone acetyltransferase activity (Petruk et al., 2001) and histone deacetylases copurify with PcG complexes (van der Vlag and Otte, 1999; Tie et al., 2001). The effects of these enzyme activities on nucleosome structure and packaging generally lead to major changes in the accessibility of associated regulatory elements to trans-acting factors.

Although recent strides have been made in terms of how PcG and trxG complexes maintain repressed and activated transcriptional states, how they are recruited and activated at specific *Hom-C* loci is still quite poorly understood. Two possibilities are most often discussed. The first involves direct links between HOM gene activators and repressors, and components of the trxG or PcG complexes. The second is an indirect link via epigenetic modifications of bound chromatin by intermediary cofactors recruited by the activators and repressors. This, in turn, would trigger the assembly or activation of PcG or trxG complexes. This epigenetic marking by intermediary cofactors may also work via nucleosome acetylation (Cavalli and Paro, 1999). Corepressors may act to remove existing acetyl groups, while coactivators may increase the number of acetyl groups. This would signal the recruitment or activation of PcG or trxG complexes respectively. Support for this possibility comes from the ability of the *Ubx* repressor Hunchback (HB) to physically associate with dMI-2 (Kehle et al., 1998). dMI-2 has nucleosome remodeling activity in *Drosophila* (Brehm et al., 2000), and is a component of a histone deacetylase complex in *Xenopus* (Wade et al., 1998). EVE, like HB, also appears to repress *Ubx* expression (Manoukian and Krause, unpublished observation) and is also able to interact with a histone deacetylase, in this case RPD3 (Mannervik and Levine, 1999). As RPD3 is also found in PcG complexes (Tie et al., 2001), it is possible that EVE could also play a direct role in PcG complex recruitment.

Hom-C activators may also recruit intermediary histone acetylation modifiers. The best known of these activators is the pair-rule protein FTZ (Ingham and Martinez-Arias, 1986; Muller and Bienz, 1992). As discussed above, the activation of most FTZ targets requires association with the orphan nuclear receptor FTZ-F1 (Guichet et al., 1997; Yu et al., 1997; Schwartz et al., 2001), and most nuclear receptors characterized to date recruit histone acetyltransferases (Chen et al., 1999b; Freedman, 1999; Xu et al., 1999). However, the possibility that FTZ and FTZ-F1 also recruit histone acetyl transferases is yet to be tested.

In the *abx* and *pbx* regulatory elements, located upstream of the *Ubx* promoter, binding sites for the early *Ubx* regulators FTZ, EVE and HB are closely associated with trxG and PcG response elements referred to as TREs and PREs respectively (Muller and Bienz, 1991, 1992; Chan et al., 1994a; Chiang et al., 1995). These response elements are required to maintain activated and repressed transcriptional states over long periods of time (Francis and Kingston, 2001). They tend to be quite large (200–300 bps) with multiple binding sites for PcG or trxG complexes. The proximity between PREs and TREs and the regulatory elements responsible for initial *Hom-C* gene expression may be required for the transition between early gene regulation and the maintenance of these transcriptional states by PcG/trxG complexes.

Other intriguing components of many PREs are short transcribed sequences that contain no open reading frames (Lipshitz et al., 1987). Interestingly, the PcG protein Polycomb contains a chromodomain, a motif that is also found in the dosage compensation proteins MOF and MSL-3 (Gorman et al., 1995; Hilfiker et al., 1997). In the latter two proteins, these chromodomains are required to bind a structural RNA, roX (Akhtar et al., 2000). This RNA molecule is used to target the protein/RNA complexes to the male X-chromosome. Thus, an interesting possibility yet to be explored is that non-coding PRE/TRE transcripts may be involved in the targeting of PcG or trxG complexes to the DNA sequences from which they were transcribed.

The important role played by PcG and trxG complexes in *Hom-C* gene regulation likely explains the physical conservation of homeotic gene clusters in organisms ranging in complexity from worms to humans. In all higher eukaryotes, *Hom-C* genes are organized along the chromosome in the same order as their anterior–posterior patterns of expression (Fig. 10 and McGinnis and Krumlauf, 1992). In vertebrates, this order also correlates with their temporal sequence of activation (Zakany et al., 2001). This is also the case in insects more primitive than *Drosophila* (Pankratz and Jackle, 1993; Tautz and Sommer, 1995; Davis and Patel, 1999). In these insects, segments are not formed simultaneously as in *Drosophila*, but rather one by one, budding off from a posterior proliferative zone. The anterior-most segments are formed first and are pushed forward by the next segment to form. These examples suggest a somewhat different means of coupling the initial regulators of *Hom-C* gene expression and PcG/trxG protein complexes. In these organisms, the colinearity of the complex may allow control by a more processive mechanism. This mechanism, which is currently unknown, may involve the sequential activation of genes, beginning with those that are expressed most anteriorly and ending with those expressed more posteriorly (Fig. 10). As cells of maturing segments move away from the proliferative zone, the transcriptional states of genes in the complex may be locked in by the recruitment or activation of fully functional trxG or PcG complexes. In this way, the dividing line between active and inactive chromatin could move in a seemingly wave-like fashion across the complex.

The idea of a temporal assembly of a functional complex might also apply to the regulation of the HOX gene cluster at 93DE (reviewed in Jagla et al., 2001). The six HOX genes in this cluster are involved in mesodermal patterning and differentiation programs. Although their position within the cluster does not correlate with anterior posterior expression and patterning, they are expressed in a temporal progression with the proximal-most genes expressed first. Although this clustering and sequential activation suggests a similar type of regulation by PcG and trxG proteins, this possibility is yet to be tested. Clearly, there is a great deal of research to be conducted before this correlation between gene clustering and chromatin regulation is understood.

7.3. Properties of Hom-C proteins

All HOM proteins contain highly conserved homeodomains, and consequently, display very similar, often indistinguishable, DNA-binding characteristics in vitro (Desplan et al., 1988; Hoey and Levine, 1988; Kalionis and O'Farrell, 1993; Ekker et al., 1994). In vivo, however, these proteins instruct unique regional identities and structures. How HOM proteins recognize and regulate different target genes has been a major question since the time of their initial molecular characterization (see for example Hayashi and Scott, 1990), and is currently a topic of some debate (see, for example, Biggin and McGinnis, 1997; Mann and Morata, 2000). Although recent studies clearly show that cofactors play a major role in this problem of specificity, how they do so is still being determined. Some results indicate that cofactors play a major role in guiding HOM proteins to different subsets of potential binding sites. Other experiments suggest regulation primarily at the level of transcriptional activity. The major lines of evidence for these "selective binding" and "activity regulation" models are discussed below.

8. Models for homeodomain protein function

8.1. Selective binding model

Perhaps the first lines of evidence in favour of the selective binding model were studies with the yeast homeodomain proteins MATa1 and MATα2. These proteins bind to different sites that have different spacings between subsites. Which sites are bound depends upon which cofactors are present (Goutte and Johnson, 1988; Keleher et al., 1988, 1989; Dranginis, 1990). Another key finding was made with the pair-rule gene *ftz*. The *ftz* gene is located within the ANT-C and contains a highly conserved homeobox, very similar to those of other HOM genes. Interestingly, deletion of the FTZ homeodomain does not destroy FTZ activity (Fitzpatrick et al., 1992; Copeland et al., 1996; Hyduk and Percival-Smith, 1996). In fact, the homeodomain-deleted protein appears able to regulate most key target genes, except when expressed at low levels. Recruitment of FTZ to these target gene promoters is thought to be mediated by interactions with DNA-bound cofactors, the majority of which appear to make sufficient contacts with motifs in FTZ that are outside of the homeodomain. Indeed, the requisite FTZ cofactor, αFTZ-F1, recognizes a motif in the FTZ N-terminus (Guichet et al., 1997; Schwartz et al., 2001; Suzuki et al., 2001), and this interaction is sufficient to recruit FTZ to target gene promoters (Guichet et al., 1997). Most HOM proteins do not show this level of activity if their homeodomains are compromised, due in part to the fact that the homeodomains of these proteins act not only to bind DNA but also to contact key specificity-conferring cofactors (see below). Nevertheless, it is clear that cofactor interactions are important and that many of these have the ability to affect binding site selection.

One of the best examples of a HOM cofactor that regulates binding site selection is the protein Extradenticle (EXD). EXD is a TALE-class homeodomain protein that can bind cooperatively to DNA together with many of the HOM proteins (Chan et al., 1994b; Chang et al., 1995; Lu et al., 1995; Neuteboom et al., 1995; Phelan et al., 1995; Popperl et al., 1995; van Dijk et al., 1995). In vitro binding studies as well as structural studies have revealed how EXD and HOM proteins cooperate in binding to DNA (Chan et al., 1994b; Chang et al., 1995; Knoepfler and Kamps, 1995; Lu et al., 1995; Neuteboom et al., 1995; Phelan et al., 1995; Popperl et al., 1995; van Dijk et al., 1995; Lu and Kamps, 1996; Passner et al., 1999; Piper et al., 1999). The homeodomains of these proteins bind to opposite sides of the DNA double helix and contact one another via a conserved hexapeptide motif located just N-terminal to the HOM protein homeodomain and a pocket in the C-terminal portion of the EXD homeodomain. Since both homeodomains make sequence-specific contacts with the bases and backbone atoms of the DNA substrate, binding affinity and specificity is enhanced.

Given that only a subset of HOM proteins interact with EXD (see Table 1) (van Dijk and Murre, 1994; Chang et al., 1995), and that different heterodimers recognize different sequences, complex formation with EXD results in a redistribution of these proteins to different subsets of potential binding sites (Chan et al., 1994b; Popperl et al., 1995; Chan and Mann, 1996; Chang et al., 1996; Shen et al., 1996). For example, the sequence TGAT*TT*ATGG binds EXD-UBX with much higher affinity than EXD-LAB, while the opposite is true for the sequence TGAT<u>GG</u>ATGG (Chan and Mann, 1996). It has been suggested that some of these differences in heterodimer specificity are due to the uncovering of cryptic DNA-binding specificities built into HOM protein homeodomains, particularly in their N-terminal arms (Mann and Chan, 1996).

Similar effects of EXD on HOM specificity have been reported in vivo. For example, a 37-bp element derived from the *fork head* (*fkh*) promoter, which specifically binds EXD and the homeodomain protein SCR in vitro, requires both protein activities in vivo (Ryoo and Mann, 1999). However, changing two base pairs in the element, converts it from an EXD-SCR binding site to a site that binds different EXD-HOM protein heterodimers in vitro and that can now be regulated by those same protein combinations in vivo (Ryoo and Mann, 1999).

A further line of evidence in support of the selective binding model is that reporter genes composed only of consensus homeodomain binding sites do not appear to respond to HOM proteins in developing embryos (Vincent et al., 1990). Although this suggests that homeodomain binding sites alone are not sufficient for effective binding, it is still possible that HOM proteins bind to these sites, but that they are not otherwise functional.

8.2. Activity regulation model

Support for the activity regulation model comes primarily from in vivo UV-cross linking experiments (Walter et al., 1994; Biggin and McGinnis, 1997; Liang and Biggin, 1998; Carr and Biggin, 1999). These experiments have shown that homeodomain proteins immunoprecipitated together with UV-cross linked DNA are widely distributed along most of the restriction fragments tested. These distributions were much the same as those of the purified proteins bound to DNA in vitro (Walter and Biggin, 1996), suggesting that DNA binding in vivo is also quite promiscuous and not greatly influenced by cofactors. Differential target gene regulation would depend, then, on events that occur subsequent to DNA binding. In particular, the outcome of binding would be wholly dependent upon interactions with proteins bound nearby. Different HOM proteins would interact differently with these proteins, generating outcomes that vary from negligible to strong. The cofactors present would also determine if effects on transcription are positive or negative. In support of this hypothesis, Biggin and colleagues have pointed out that most genes expressed during *Drosophila* embryogenesis exhibit segmentally modulated patterns of expression, consistent with direct regulation by either pair-rule or HOM homeodomain proteins (Liang and Biggin, 1998). However, these patterns could also be attributable to indirect regulatory effects.

Additional support for the activity regulation model comes from studies with the HOM protein Deformed (DFD), which show that it can recognize and occupy sites in vivo in the absence of EXD (Li et al., 1999). However, the bound protein appears to remain in a functionally neutral state. It was suggested that the DFD activation domain is suppressed in a homeodomain-dependent manner, and that this suppression is released upon interactions with EXD (Li et al., 1999). Indeed, when binding sites for EXD and HOM proteins are both present, the ensuing interactions usually result in transcriptional activation. Conversely, HOM protein binding sites lacking adjacent EXD sites tend to repress transcription (Pinsonneault et al., 1997). Thus, EXD appears capable of switching the activities of HOM regulators from repressors to activators (Biggin and McGinnis, 1997; Pinsonneault et al., 1997).

In some cases, it has been shown that HOM proteins can bind the same sites with the same outcomes. For example, repression of *Dll* by UBX and ABD-A appears to be mediated by the same binding sites (Vachon et al., 1992). Similarly, some of the alterations in the SCR-EXD binding site in the *fkh* enhancer that were described above result in a consensus binding site that can be bound similarly by several different EXD-HOM protein dimers in vitro. These are also regulated similarly by the same protein dimers in vivo (Ryoo and Mann, 1999).

Several lines of evidence (including those in the previous section) argue against an extreme version of the activity regulation model. First, the number of molecules available within a cell appears to be orders of magnitude lower than the number of potential HOM protein binding sites (Krause et al., 1988; Nasiadka et al., 2000). Polytene chromosome binding experiments also argue in favour of many fewer binding sites than could potentially be bound (hundreds rather than thousands; Botas and Auwers, 1996; Nasiadka et al., 2000). Consistent with these numbers, a recent microarray experiment performed with the HOM protein Labial showed that only 6% of the genes arrayed changed significantly in their levels of expression in the presence or absence of Labial (Leemans et al., 2001). These responses represented both direct and indirect targets.

A recent study with the FTZ protein argues in favour of both selective binding and activity regulation processes (Nasiadka et al., 2000). This study looked at a set of FTZ target and non-target genes to see if they could be activated by a FTZ protein fused to the powerful transcriptional activation domain of the VP16 protein. If FTZ were widely bound to most potential binding sites, then the presence of the VP16 activation domain should have led to the activation of many new genes that FTZ itself could not regulate. This did not appear to be the case, as the same set of genes that was activated by FTZ was also activated by the fusion protein. Furthermore, similar developmental defects were obtained for both FTZ and FTZVP16, suggesting that other genes not assayed were responding with similar selectivity. It was noted however that the VP16 activation domain made FTZ a much better transcriptional activator of bona fide target genes and also released the protein from some temporal and spatial limitations on its activity. It also changed FTZ from a repressor of some target genes to an activator. Thus, both models of cofactor function appear to have merits.

9. Future problems and prospects

Although the segmentation hierarchy is one of the best understood transcriptional cascades, there remain many gaps in our understanding of the process. For example, while most of the dedicated components of the hierarchy have been identified, many of the ancillary factors have not. Also remaining to be identified are the full spectrum of target genes and the detailed interaction circuitry that occurs between each of these genes and their regulators. Once target genes and their direct regulators have been defined, the functional elements that respond to the regulators, and the nature of the protein–protein interactions that occur on these elements will also have to be unraveled.

The recent completion of the *Drosophila* genome sequence and the development of high-throughput technologies that take full advantage of this information should help tremendously in this endeavor. For example, following the temporal responses of genes on microarrays after thermally activating or inactivating segmentation gene expression may, in a global fashion, identify targets that are likely to be direct. Similarly, sites bound on direct target genes can be identified by the immunoprecipitation of segmentation proteins along with the DNA sequences to which they were cross-linked in vivo. These sequences can be identified by hybridization to DNA microarrays (Ren et al., 2000; Iyer et al., 2001). Comparisons of the bound sequences frequently reveal relevant consensus binding sites that do not always correlate with those bound in vitro.

A related method for identifying sites occupied in vivo involves the use of a DNA modifying enzyme (DAM methylase) fused to the transcription factor of interest (van Steensel and Henikoff, 2000; van Steensel et al., 2001). In this approach, DNA regions bound by the fusion protein are methylated and then identified using methyl-adenine-specific antibodies or restriction enzymes. When used in conjunction with DNA microarrays, this approach identifies potential target genes on a genome-wide scale (van Steensel et al., 2001). The sequencing of additional insect genomes should also help in identifying conserved promoter sequences, thereby implying common regulatory functions. However, testing and dissecting the functions of these elements in vivo will still be time consuming, even if aided by robotics.

Defining all of the physical interactions that occur between segmentation proteins and their cofactors will be an equally important step towards our understanding of the segmentation hierarchy. Several powerful techniques have been developed to analyze protein–protein interactions on a large scale. Among, those is the yeast two-hybrid system (Phizicky and Fields, 1995), recently modified for high throughput analyses (Fromont-Racine et al., 2000; Ito et al., 2000; Uetz et al., 2000; Walhout et al., 2000). Another potential approach is tandem affinity purification (TAP). In this approach, doubly tagged proteins and the cofactors to which they are bound are purified over successive affinity columns under native conditions (Rigaut et al., 1999; Puig et al., 2001). The purified proteins are then identified by mass spectrometry. Although this technique has proven very successful for purifying large multi-component complexes from yeast cells, its usefulness in more complex organisms is yet to be established. Confirmation of such interactions in vivo may also be aided by new technologies such as fluorescence resonance energy transfer (FRET) (Clegg, 1995; Mahajan et al., 1998).

A number of different databases have recently been placed on the World Wide Web that document the rapidly growing lists of segmentation protein and gene interactions (see Table 2). However, these databases are still relatively incomplete and do not yet fully consider the rapid dynamics of segmentation gene interactions or the rigorous distinction between direct and indirect interactions. Ultimately, what one would like to see is a full description of the positive and negative gene interactions displayed both temporally and spatially. If sufficiently detailed and correct, such a database should be able to predict the

A. Nasiadka, B.H. Dietrich and H.M. Krause

Table 2
List of some *Drosophila* bioinformatics resources accessible on the World Wide Web

Site	Description	URL
FlyBase	a database of genetic and molecular data for *Drosophila*	http://flybase.bio.indiana.edu/
The Interactive Fly	an internet guide to *Drosophila* genes and their roles in development	http://sdb.bio.purdue.edu/fly/aimain/1aahome.htm
FlyView	a *Drosophila* image database	http://pbio07.uni-muenster.de/
BDGP	Berkeley *Drosophila* Genome Project	http://www.fruitfly.org/
GeNet	a database on regulatory gene networks operating during *Drosophila* embryogenesis	http://www.csa.ru/Inst/gorb_dep/inbios/lab.htm
FlyNets	a database for molecular interactions in *Drosophila*	http://gifts.univ-mrs.fr/FlyNets/FlyNets_home_page.html
COMPEL	a database on composite regulatory elements affecting gene transcription in eukarytes	http://compel.bionet.nsc.ru/
MEME system	a program for identifying conserved DNA or protein motifs	http://meme.sdsc.edu/meme/website/
Gibbs Motif Sampler	a program for identifying conserved DNA or protein motifs	http://bayesweb.wadsworth.org/gibbs/gibbs.html
Quantgen	a DNA microarray database on gene expression during *Drosophila* development	http://quantgen.med.yale.edu/
CDMC	the Canadian *Drosophila* Microarray Centre	http://www.erin.utoronto.ca/~w3flyma/

spatial and phenotypic outcomes of changes incurred by any segmentation gene product at any developmental time point.

Acknowledgments

We wish to thank members of our laboratory for critical reading of the manuscript. Support for this work was provided by the National Cancer Institute of Canada with funds from the Canadian Cancer Society and by the Canadian Institutes of Health Research.

References

Akam, M.E. 1983. The location of Ultrabithorax transcripts in Drosophila tissue sections. EMBO J. 2, 2075–2084.

Akam, M.E., Martinez Arias, A. 1985. The distribution of *Ultrabithorax* transcripts in *Drosophila* embryos. EMBO J. 4, 1689–1700.

Akhtar, A., Zink, D., Becker, P.B. 2000. Chromodomains are protein-RNA interaction modules. Nature 407, 405–409.

Akimaru, H., Chen, Y., Dai, P., Hou, D.X., Nonaka, M., Smolik, S.M., Armstrong, S., Goodman, R.H., Ishii, S. 1997. Drosophila CBP is a co-activator of cubitus interruptus in hedgehog signalling. Nature 386, 735–738.

Akoulitchev, S., Reinberg, D. 1998. The molecular mechanism of mitotic inhibition of TFIIH is mediated by phosphorylation of CDK7. Genes Dev. 12, 3541–3550.

Anderson, K.V. 1998. Pinning down positional information: dorsal-ventral polarity in the Drosophila embryo. Cell 95, 439–442.

Anderson, K.V., Lengyel, J.A. 1979. Rates of synthesis of major classes of RNA in Drosophila embryos. Dev. Biol. 70, 217–231.

Anderson, K.V., Lengyel, J.A. 1980. Changing rates of histone mRNA synthesis and turnover in Drosophila embryos. Cell 21, 717–727.

Anderson, K.V., Lengyel, J.A. 1981. Changing rates of DNA and RNA synthesis in Drosophila embryos. Dev. Biol. 82, 127–138.

Arnosti, D.N., Barolo, S., Levine, M., Small, S. 1996a. The eve stripe 2 enhancer employs multiple modes of transcriptional synergy. Development 122, 205–214.

Arnosti, D.N., Gray, S., Barolo, S., Zhou, J., Levine, M. 1996b. The gap protein knirps mediates both quenching and direct repression in the Drosophila embryo. EMBO J. 15, 3659–3666.

Aronson, B.D., Fisher, A.L., Blechman, K., Caudy, M., Gergen, J.P. 1997. Groucho-dependent and -independent repression activities of Runt domain proteins. Mol. Cell Biol. 17, 5581–5587.

Bae, S.C., Ogawa, E., Maruyama, M., Oka, H., Satake, M., Shigesada, K., Jenkins, N.A., Gilbert, D.J., Copeland, N.G., Ito, Y. 1994. PEBP2 alpha B/mouse AML1 consists of multiple isoforms that possess differential trans-activation potentials. Mol. Cell Biol. 14, 3242–3252.

Baker, N.E. 1988. Localization of transcripts from the wingless gene in whole Drosophila embryos. Development 103, 289–298.

Barolo, S., Levine, M. 1997. hairy mediates dominant repression in the Drosophila embryo. EMBO J. 16, 2883–2891.

Bateson, W. 1894. *Materials for the Study of Variation with Especial Regard to Discontinuity in the Origin of the Species*, McMillan, London.

Baumgartner, S., Bopp, D., Burri, M., Noll, M. 1987. Structure of two genes at the gooseberry locus related to the paired gene and their spatial expression during Drosophila embryogenesis. Genes Dev. 1, 1247–1267.

Beachy, P.A., Helfand, S.L., Hogness, D.S. 1985. Segmental distribution of bithorax complex proteins during Drosophila development. Nature 313, 545–551.

Becker, P.B., Wu, C. 1992. Cell-free system for assembly of transcriptionally repressed chromatin from Drosophila embryos. Mol. Cell Biol. 12, 2241–2249.

Bejsovec, A., Martinez Arias, A. 1991. Roles of wingless in patterning the larval epidermis of Drosophila. Development 113, 471–485.

Bender, W., Akam, M.E., Karch, F., Beachy, P.A., Peifer, M., Spierer, P., Lewis, E.B. 1983. Molecular genetics of the *bithorax* complex in *Drosophila melanogaster*. Science 221, 23–29.

Benedyk, M.J., Mullen, J.R., DiNardo, S. 1994. odd-paired: a zinc finger pair-rule protein required for the timely activation of engrailed and wingless in Drosophila embryos. Genes Dev. 8, 105–117.

Berleth, T., Burri, M., Thoma, G., Bopp, D., Richstein, S., Frigerio, G., Noll, M., Nusslein-Volhard, C. 1988. The role of localization of bicoid RNA in organizing the anterior pattern of the Drosophila embryo. EMBO J. 7, 1749–1756.

Bienz, M., Tremml, G. 1988. Domain of Ultrabithorax expression in Drosophila visceral mesoderm from autoregulation and exclusion. Nature 333, 576–578.

Biggin, M.D., McGinnis, W. 1997. Regulation of segmentation and segmental identity by Drosophila homeo-proteins: the role of DNA binding in functional activity and specificity. Development 124, 4425–4433.

Biggin, M.D., Tjian, R. 1989. A purified Drosophila homeodomain protein represses transcription in vitro. Cell 58, 433–440.

Bird, A.P., Wolffe, A.P. 1999. Methylation-induced repression—belts, braces, and chromatin. Cell 99, 451–454.

Botas, J., Auwers, L. 1996. Chromosomal binding sites of Ultrabithorax homeotic proteins. Mech. Dev. 56, 129–138.

Breen, T.R., Harte, P.J. 1993. Trithorax regulates multiple homeotic genes in the bithorax and Antennapedia complexes and exerts different tissue-specific, parasegment-specific and promoter-specific effects on each. Development 117, 119–134.

Brehm, A., Langst, G., Kehle, J., Clapier, C.R., Imhof, A., Eberharter, A., Muller, J., Becker, P.B. 2000. dMi-2 and ISWI chromatin remodelling factors have distinct nucleosome binding and mobilization properties. EMBO J. 19, 4332–4341.

Bridges, C., Morgan, T.H. 1923. *The Third-Chromosome Group of Mutant Characters of Drosophila Melanogaster*, Carnegie Institution of Washington Publications, 327, 1–251.

Bronner, G., Jackle, H. 1991. Control and function of terminal gap gene activity in the posterior pole region of the Drosophila embryo. Mech. Dev. 35, 205–211.

Brunner, E., Peter, O., Schweizer, L., Basler, K. 1997. Pangolin encodes a Lef-1 homologue that acts downstream of Armadillo to transduce the Wingless signal in Drosophila. Nature 385, 829–833.

Burz, D.S., Hanes, S.D. 2001. Isolation of mutations that disrupt cooperative DNA binding by the Drosophila bicoid protein. J. Mol. Biol. 305, 219–230.

Burz, D.S., Rivera-Pomar, R., Jackle, H., Hanes, S.D. 1998. Cooperative DNA-binding by Bicoid provides a mechanism for threshold- dependent gene activation in the Drosophila embryo. EMBO J. 17, 5998–6009.

Bustin, M., Reeves, R. 1996. High-mobility-group chromosomal proteins: architectural components that facilitate chromatin function. Prog. Nucleic Acid Res. Mol. Biol. 54, 35–100.

Cadigan, K.M., Grossniklaus, U., Gehring, W.J. 1994a. Functional redundancy: the respective roles of the two sloppy paired genes in Drosophila segmentation. Proc. Natl. Acad. Sci. USA 91, 6324–6328.

Cadigan, K.M., Grossniklaus, U., Gehring, W.J. 1994b. Localized expression of sloppy paired protein maintains the polarity of Drosophila parasegments. Genes Dev. 8, 899–913.

Cai, H.N., Arnosti, D.N., Levine, M. 1996. Long-range repression in the Drosophila embryo. Proc. Natl. Acad. Sci. USA 93, 9309–9314.

Campos-Ortega, J.A., Hartenstein, V. 1997. *The Embryonic Development of Drosophila Melanogaster*, Springer, Berlin; New York, pp. xvii, 405.

Capovilla, M., Eldon, E.D., Pirrotta, V. 1992. The giant gene of Drosophila encodes a b-ZIP DNA-binding protein that regulates the expression of other segmentation gap genes. Development 114, 99–112.

Carr, A., Biggin, M.D. 1999. A comparison of in vivo and in vitro DNA-binding specificities suggests a new model for homeoprotein DNA binding in Drosophila embryos. EMBO J. 18, 1598–1608.

Carroll, S.B., Laymon, R.A., McCutcheon, M.A., Riley, P.D., Scott, M.P. 1986. The localization and regulation of Antennapedia protein expression in Drosophila embryos. Cell 47, 113–122.

Carroll, S.B., Scott, M.P. 1986. Zygotically active genes that affect the spatial expression of the fushi tarazu segmentation gene during early Drosophila embryogenesis. Cell 45, 113–126.

Carroll, S.B., Vavra, S.H. 1989. The zygotic control of Drosophila pair-rule gene expression. II. Spatial repression by gap and pair-rule gene products. Development 107, 673–683.

Casanova, J., White, R.A. 1987. Trans-regulatory functions in the Abdominal-B gene of the bithorax complex. Development 101, 117–122.

Cavalli, G., Paro, R. 1999. Epigenetic inheritance of active chromatin after removal of the main transactivator. Science 286, 955–958.

Cavallo, R.A., Cox, R.T., Moline, M.M., Roose, J., Polevoy, G.A., Clevers, H., Peifer, M., Bejsovec, A. 1998. Drosophila Tcf and Groucho interact to repress Wingless signalling activity. Nature 395, 604–608.

Chadwick, R., McGinnis, W. 1987. Temporal and spatial distribution of transcripts from the *Deformed* gene of *Drosophila*. EMBO J. 6, 779–789.

Chan, C.S., Rastelli, L., Pirrotta, V. 1994a. A Polycomb response element in the Ubx gene that determines an epigenetically inherited state of repression. EMBO J. 13, 2553–2564.

Chan, S.K., Jaffe, L., Capovilla, M., Botas, J., Mann, R.S. 1994b. The DNA binding specificity of Ultrabithorax is modulated by cooperative interactions with extradenticle, another homeoprotein. Cell 78, 603–615.

Chan, S.K., Mann, R.S. 1996. A structural model for a homeotic protein-extradenticle-DNA complex accounts for the choice of HOX protein in the heterodimer. Proc. Natl. Acad. Sci. USA 93, 5223–5228.

Chan, S.K., Ryoo, H.D., Gould, A., Krumlauf, R., Mann, R.S. 1997. Switching the in vivo specificity of a minimal Hox-responsive element. Development 124, 2007–2014.

Chang, C.P., Brocchieri, L., Shen, W.F., Largman, C., Cleary, M.L. 1996. Pbx modulation of Hox homeodomain amino-terminal arms establishes different DNA-binding specificities across the Hox locus. Mol. Cell Biol. 16, 1734–1745.

Chang, C.P., Shen, W.F., Rozenfeld, S., Lawrence, H.J., Largman, C., Cleary, M.L. 1995. Pbx proteins display hexapeptide-dependent cooperative DNA binding with a subset of Hox proteins. Genes Dev. 9, 663–674.

Chen, G., Fernandez, J., Mische, S., Courey, A.J. 1999a. A functional interaction between the histone deacetylase Rpd3 and the corepressor groucho in Drosophila development. Genes Dev. 13, 2218–2230.

Chen, H., Lin, R.J., Xie, W., Wilpitz, D., Evans, R.M. 1999b. Regulation of hormone-induced histone hyperacetylation and gene activation via acetylation of an acetylase. Cell 98, 675–686.

Chiang, A., O'Connor, M.B., Paro, R., Simon, J., Bender, W. 1995. Discrete Polycomb-binding sites in each parasegmental domain of the bithorax complex. Development 121, 1681–1689.

Chouinard, S., Kaufman, T.C. 1991. Control of expression of the homeotic labial (lab) locus of Drosophila melanogaster: evidence for both positive and negative autogenous regulation. Development 113, 1267–1280.

Clegg, R.M. 1995. Fluorescence resonance energy transfer. Curr. Opin. Biotechnol. 6, 103–110.

Cohen, S.M., Jurgens, G. 1990. Mediation of Drosophila head development by gap-like segmentation genes. Nature 346, 482–485.

Copeland, J.W., Nasiadka, A., Dietrich, B.H., Krause, H.M. 1996. Patterning of the Drosophila embryo by a homeodomain-deleted Ftz polypeptide. Nature 379, 162–165.

Coulter, D.E., Swaykus, E.A., Beran-Koehn, M.A., Goldberg, D., Wieschaus, E., Schedl, P. 1990. Molecular analysis of odd-skipped, a zinc finger encoding segmentation gene with a novel pair-rule expression pattern. EMBO J. 9, 3795–3804.

Criqui-Filipe, P., Ducret, C., Maira, S.M., Wasylyk, B. 1999. Net, a negative Ras-switchable TCF, contains a second inhibition domain, the CID, that mediates repression through interactions with CtBP and de-acetylation. EMBO J. 18, 3392–3403.

Dahanukar, A., Wharton, R.P. 1996. The Nanos gradient in Drosophila embryos is generated by translational regulation. Genes Dev. 10, 2610–2620.

Dalton, D., Chadwick, R., McGinnis, W. 1989. Expression and embryonic function of empty spiracles: a Drosophila homeo box gene with two patterning functions on the anterior–posterior axis of the embryo. Genes Dev. 3, 1940–1956.

Davidson, E.H. 1986. Gene activity in early development. Academic Press, Orlando, pp. xiv, 670, [2] folded leaves of plates.

Davis, G.K., Patel, N.H. 1999. The origin and evolution of segmentation. Trends Cell Biol. 9, M68–72.

Desplan, C., Theis, J., O'Farrell, P.H. 1988. The sequence specificity of homeodomain-DNA interaction. Cell 54, 1081–1090.

Diederich, R.J., Merrill, V.K., Pultz, M.A., Kaufman, T.C. 1989. Isolation, structure, and expression of labial, a homeotic gene of the Antennapedia Complex involved in Drosophila head development. Genes Dev. 3, 399–414.

DiNardo, S., O'Farrell, P.H. 1987. Establishment and refinement of segmental pattern in the Drosophila embryo: spatial control of engrailed expression by pair-rule genes. Genes Dev. 1, 1212–1225.

DiNardo, S., Sher, E., Heemskerk-Jongens, J., Kassis, J.A., O'Farrell, P.H. 1988. Two-tiered regulation of spatially patterned engrailed gene expression during Drosophila embryogenesis. Nature 332, 604–609.

Dorsett, D. 1999. Distant liaisons: long-range enhancer-promoter interactions in Drosophila. Curr. Opin. Genet. Dev. 9, 505–514.

Dranginis, A.M. 1990. Binding of yeast a1 and alpha 2 as a heterodimer to the operator DNA of a haploid-specific gene. Nature 347, 682–685.

Drees, B., Ali, Z., Soeller, W.C., Coleman, K.G., Poole, S.J., Kornberg, T. 1987. The transcription unit of the Drosophila engrailed locus: an unusually small portion of a 70,000 bp gene. EMBO J. 6, 2803–2809.

Driever, W., Nusslein-Volhard, C. 1988. A gradient of bicoid protein in Drosophila embryos. Cell 54, 83–93.

Driever, W., Nusslein-Volhard, C. 1989. The bicoid protein is a positive regulator of hunchback transcription in the early Drosophila embryo. Nature 337, 138–143.

Driever, W., Thoma, G., Nusslein-Volhard, C. 1989. Determination of spatial domains of zygotic gene expression in the Drosophila embryo by the affinity of binding sites for the bicoid morphogen. Nature 340, 363–367.

Dubnau, J., Struhl, G. 1996. RNA recognition and translational regulation by a homeodomain protein [see comments] [published erratum appears in Nature 1997 Aug 14;388(6643):697]. Nature 379, 694–699.

Eaton, S., Kornberg, T.B. 1990. Repression of ci-D in posterior compartments of Drosophila by engrailed. Genes Dev. 4, 1068–1077.

Edgar, B.A., Datar, S.A. 1996. Zygotic degradation of two maternal Cdc25 mRNAs terminates Drosophila's early cell cycle program. Genes Dev. 10, 1966–1977.

Edgar, B.A., Kiehle, C.P., Schubiger, G. 1986. Cell cycle control by the nucleo-cytoplasmic ratio in early Drosophila development. Cell 44, 365–372.

Edgar, B.A., Schubiger, G. 1986. Parameters controlling transcriptional activation during early Drosophila development. Cell 44, 871–877.

Edgar, B.A., Sprenger, F., Duronio, R.J., Leopold, P., O'Farrell, P.H. 1994. Distinct molecular mechanism regulate cell cycle timing at successive stages of Drosophila embryogenesis. Genes Dev. 8, 440–452.

Ekker, S.C., Jackson, D.G., von Kessler, D.P., Sun, B.I., Young, K.E., Beachy, P.A. 1994. The degree of variation in DNA sequence recognition among four Drosophila homeotic proteins. EMBO J. 13, 3551–3560.

Elgin, S.C., Hood, L.E. 1973. Chromosomal proteins of Drosophila embryos. Biochemistry 12, 4984–4991.

Ephrussi, A., Dickinson, L.K., Lehmann, R. 1991. Oskar organizes the germ plasm and directs localization of the posterior determinant nanos. Cell 66, 37–50.

Finkelstein, R., Perrimon, N. 1990. The orthodenticle gene is regulated by bicoid and torso and specifies Drosophila head development [see comments]. Nature 346, 485–488.

Finkelstein, R., Smouse, D., Capaci, T.M., Spradling, A.C., Perrimon, N. 1990. The orthodenticle gene encodes a novel homeo domain protein involved in the development of the Drosophila nervous system and ocellar visual structures. Genes Dev. 4, 1516–1527.

Fisher, A.L., Ohsako, S., Caudy, M. 1996. The WRPW motif of the hairy-related basic helix-loop-helix repressor proteins acts as a 4-amino-acid transcription repression and protein- protein interaction domain. Mol. Cell Biol. 16, 2670–2677.

Fitzpatrick, V.D., Percival-Smith, A., Ingles, C.J., Krause, H.M. 1992. Homeodomain-independent activity of the fushi tarazu polypeptide in Drosophila embryos. Nature 356, 610–612.

Fjose, A., McGinnis, W.J., Gehring, W.J. 1985. Isolation of a homoeo box-containing gene from the engrailed region of Drosophila and the spatial distribution of its transcripts. Nature 313, 284–289.

Flores-Saaib, R.D., Courey, A.J. 2000. Regulation of dorso/ventral patterning in the Drosophila embryo by multiple dorsal-interacting proteins. Cell Biochem. Biophys. 33, 1–17.

Foe, V.E., Alberts, B.M. 1983. Studies of nuclear and cytoplasmic behaviour during the five mitotic cycles that precede gastrulation in Drosophila embryogenesis. J. Cell Sci. 61, 31–70.

Francis, N.J., Kingston, R.E. 2001. Mechanisms of transcriptional memory. Nat. Rev. Mol. Cell Biol. 2, 409–421.

Frasch, M., Levine, M. 1987. Complementary patterns of even-skipped and fushi tarazu expression involve their differential regulation by a common set of segmentation genes in Drosophila. Genes Dev. 1, 981–995.

Freedman, L.P. 1999. Increasing the complexity of coactivation in nuclear receptor signaling. Cell 97, 5–8.

Frigerio, G., Burri, M., Bopp, D., Baumgartner, S., Noll, M. 1986. Structure of the segmentation gene paired and the Drosophila PRD gene set as part of a gene network. Cell 47, 735–746.

Fromont-Racine, M., Mayes, A.E., Brunet-Simon, A., Rain, J.C., Colley, A., Dix, I., Decourty, L., Joly, N., Ricard, F., Beggs, J.D., Legrain, P. 2000. Genome-wide protein interaction screens reveal functional networks involving Sm-like proteins. Yeast 17, 95–110.

Fujioka, M., Emi-Sarker, Y., Yusibova, G.L., Goto, T., Jaynes, J.B. 1999. Analysis of an even-skipped rescue transgene reveals both composite and discrete neuronal and early blastoderm enhancers, and multi-stripe positioning by gap gene repressor gradients. Development 126, 2527–2538.

Fujioka, M., Jaynes, J.B., Goto, T. 1995. Early even-skipped stripes act as morphogenetic gradients at the single cell level to establish engrailed expression. Development 121, 4371–4382.

Fujioka, M., Yusibova, G.L., Sackerson, C.M., Tillib, S., Mazo, A., Satake, M., Goto, T. 1996. Runt domain partner proteins enhance DNA binding and transcriptional repression in cultured Drosophila cells. Genes Cells 1, 741–754.

Furnari, B., Rhind, N., Russell, P. 1997. Cdc25 mitotic inducer targeted by chk1 DNA damage checkpoint kinase. Science 277, 1495–1497.

Gao, Q., Finkelstein, R. 1998. Targeting gene expression to the head: the Drosophila orthodenticle gene is a direct target of the Bicoid morphogen. Development 125, 4185–4193.

Garber, R.L., Kuroiwa, A., Gehring, W.J. 1983. Genomic and cDNA clones of the homeotic locus Antennapedia in Drosophila. EMBO J. 2, 2027–2036.

Garcia-Bellido, A. 1975. Genetic control of wing disc development in *Drosophila*. CIBA Foundation Symp. 29, 161–183.

Garcia-Bellido, A., Lawrence, P.A., Morata, G. 1979. Compartments in animal development. Sci. Amer. 241, 102–110.

Garcia-Bellido, A., Ripoll, P., Morata, G. 1973. Developmental compartmentalisation of the wing disk of Drosophila. Nat. New Biol. 245, 251–253.

Gaul, U., Jackle, H. 1987. Pole region-dependent repression of the Drosophila gap gene Kruppel by maternal gene products. Cell 51, 549–555.

Gaul, U., Jackle, H. 1989. Analysis of maternal effect mutant combinations elucidates regulation and function of the overlap of hunchback and Kruppel gene expression in the Drosophila blastoderm embryo. Development 107, 651–662.

Gaul, U., Jackle, H. 1991. Role of gap genes in early *Drosophila* development. Adv. Genet. 27, 239–272.

Ge, H., Roeder, R.G. 1994. The high mobility group protein HMG1 can reversibly inhibit class II gene transcription by interaction with the TATA-binding protein. J. Biol. Chem. 269, 17136–17140.

Gergen, J.P., Butler, B.A. 1988. Isolation of the Drosophila segmentation gene runt and analysis of its expression during embryogenesis. Genes Dev. 2, 1179–1193.

Goldstein, R.E., Jimenez, G., Cook, O., Gur, D., Paroush, Z. 1999. Huckebein repressor activity in Drosophila terminal patterning is mediated by Groucho. Development 126, 3747–3755.

Golling, G., Li, L., Pepling, M., Stebbins, M., Gergen, J.P. 1996. Drosophila homologs of the proto-oncogene product PEBP2/CBF beta regulate the DNA-binding properties of Runt. Mol. Cell Biol. 16, 932–942.

Gonzalez-Reyes, A., St Johnston, D. 1994. Role of oocyte position in establishment of anterior–posterior polarity in Drosophila. Science 266, 639–642.

Gorman, M., Franke, A., Baker, B.S. 1995. Molecular characterization of the male-specific lethal-3 gene and investigations of the regulation of dosage compensation in Drosophila. Development 121, 463–475.

Goto, T., Macdonald, P., Maniatis, T. 1989. Early and late periodic patterns of even skipped expression are controlled by distinct regulatory elements that respond to different spatial cues. Cell 57, 413–422.

Gottesfeld, J.M., Forbes, D.J. 1997. Mitotic repression of the transcriptional machinery. Trends Biochem. Sci. 22, 197–202.

Goutte, C., Johnson, A.D. 1988. a1 protein alters the DNA binding specificity of alpha 2 repressor. Cell 52, 875–882.

Govind, S. 1999. Control of development and immunity by rel transcription factors in Drosophila. Oncogene 18, 6875–6887.

Gray, S., Cai, H., Barolo, S., Levine, M. 1995. Transcriptional repression in the Drosophila embryo. Philos. Trans. R. Soc. Lond. B Biol. Sci. 349, 257–262.

Gray, S., Levine, M. 1996a. Short-range transcriptional repressors mediate both quenching and direct repression within complex loci in Drosophila. Genes Dev. 10, 700–710.

Gray, S., Levine, M. 1996b. Transcriptional repression in development. Curr. Opin. Cell Biol. 8, 358–364.

Gray, S., Szymanski, P., Levine, M. 1994. Short-range repression permits multiple enhancers to function autonomously within a complex promoter. Genes Dev. 8, 1829–1838.

Grossniklaus, U., Pearson, R.K., Gehring, W.J. 1992. The Drosophila sloppy paired locus encodes two proteins involved in segmentation that show homology to mammalian transcription factors. Genes Dev. 6, 1030–1051.

Guichet, A., Copeland, J.W., Erdelyi, M., Hlousek, D., Zavorszky, P., Ho, J., Brown, S., Percival-Smith, A., Krause, H.M., Ephrussi, A. 1997. The nuclear receptor homologue Ftz-F1 and the homeodomain protein Ftz are mutually dependent cofactors. Nature 385, 548–552.

Gutjahr, T., Frei, E., Noll, M. 1993. Complex regulation of early paired expression: initial activation by gap genes and pattern modulation by pair-rule genes. Development 117, 609–623.

Hader, T., La Rosee, A., Ziebold, U., Busch, M., Taubert, H., Jackle, H., Rivera-Pomar, R. 1998. Activation of posterior pair-rule stripe expression in response to maternal caudal and zygotic knirps activities. Mech. Dev. 71, 177–186.

Hafen, E., Kuroiwa, A., Gehring, W.J. 1984a. Spatial distribution of transcripts from the segmentation gene fushi tarazu during Drosophila embryonic development. Cell 37, 833–841.

Hafen, E., Levine, M., Gehring, W.J. 1984b. Regulation of Antennapedia transcript distribution by the bithorax complex in Drosophila. Nature 307, 287–289.

Han, K., Levine, M.S., Manley, J.L. 1989. Synergistic activation and repression of transcription by Drosophila homeobox proteins. Cell 56, 573–583.

Han, K., Manley, J.L. 1993. Transcriptional repression by the Drosophila even-skipped protein: definition of a minimal repression domain. Genes Dev. 7, 491–503.

Hanes, S.D., Riddihough, G., Ish-Horowicz, D., Brent, R. 1994. Specific DNA recognition and intersite spacing are critical for action of the bicoid morphogen. Mol. Cell Biol. 14, 3364–3375.

Harding, K., Hoey, T., Warrior, R., Levine, M. 1989. Autoregulatory and gap gene response elements of the even-skipped promoter of Drosophila. EMBO J. 8, 1205–1212.

Harding, K., Levine, M. 1988. Gap genes define the limits of antennapedia and bithorax gene expression during early development in Drosophila. EMBO J. 7, 205–214.

Harding, K., Rushlow, C., Doyle, H.J., Hoey, T., Levine, M. 1986. Cross-regulatory interactions among pair-rule genes in Drosophila. Science 233, 953–959.

Harding, K., Wedeen, C., McGinnis, W., Levine, M. 1985. Spatially regulated expression of homeotic genes in Drosophila. Science 229, 1236–1242.

Hartmann, C., Taubert, H., Jackle, H., Pankratz, M.J. 1994. A two-step mode of stripe formation in the Drosophila blastoderm requires interactions among primary pair rule genes. Mech. Dev. 45, 3–13.

Hayashi, S., Scott, M.P. 1990. What determines the specificity of action of Drosophila homeodomain proteins? Cell 63, 883–894.

Heemskerk, J., DiNardo, S., Kostriken, R., O'Farrell, P.H. 1991. Multiple modes of engrailed regulation in the progression towards cell fate determination. Nature 352, 404–410.

Hewitt, G.F., Strunk, B.S., Margulies, C., Priputin, T., Wang, X.D., Amey, R., Pabst, B.A., Kosman, D., Reinitz, J., Arnosti, D.N. 1999. Transcriptional repression by the Drosophila giant protein: cis element positioning provides an alternative means of interpreting an effector gradient. Development 126, 1201–1210.

Hilfiker, A., Hilfiker-Kleiner, D., Pannuti, A., Lucchesi, J.C. 1997. mof, a putative acetyl transferase gene related to the Tip60 and MOZ human genes and to the SAS genes of yeast, is required for dosage compensation in Drosophila. EMBO J. 16, 2054–2060.

Hiromi, Y., Gehring, W.J. 1987. Regulation and function of the Drosophila segmentation gene fushi tarazu. Cell 50, 963–974.

Hiromi, Y., Kuroiwa, A., Gehring, W.J. 1985. Control elements of the Drosophila segmentation gene fushi tarazu. Cell 43, 603–613.

Hoey, T., Levine, M. 1988. Divergent homeo box proteins recognize similar DNA sequences in Drosophila. Nature 332, 858–861.

Holmgren, R. 1984. Cloning sequences from the hairy gene of Drosophila. EMBO J. 3, 569–573.

Hooper, K.L., Parkhurst, S.M., Ish-Horowicz, D. 1989. Spatial control of hairy protein expression during embryogenesis. Development 107, 489–504.

Howard, K., Ingham, P. 1986. Regulatory interactions between the segmentation genes fushi tarazu, hairy, and engrailed in the Drosophila blastoderm. Cell 44, 949–957.

Howard, K., Ingham, P., Rushlow, C. 1988. Region-specific alleles of the Drosophila segmentation gene hairy. Genes Dev. 2, 1037–1046.

Howard, K.R., Struhl, G. 1990. Decoding positional information: regulation of the pair-rule gene hairy. Development 110, 1223–1231.

Hughes, S., Krause, H.M. 2001. Establishment and maintenance of parasegmental compartments. Development 128, 1109–1118

Hulskamp, M., Pfeifle, C., Tautz, D. 1990. A morphogenetic gradient of hunchback protein organizes the expression of the gap genes Kruppel and knirps in the early Drosophila embryo. Nature 346, 577–580.

Hung, M.S., Karthikeyan, N., Huang, B., Koo, H.C., Kiger, J., Shen, C.J. 1999. Drosophila proteins related to vertebrate DNA (5-cytosine) methyltransferases. Proc. Natl. Acad. Sci. USA 96, 11940–11945.

Hyduk, D., Percival-Smith, A. 1996. Genetic characterization of the homeodomain-independent activity of the Drosophila fushi tarazu gene product. Genetics 142, 481–492.

Ingham, P., Gergen, J.P. 1988. Interactions between the pair-rule genes *runt, hairy, even-skipped* and *fushi tarazu* and the establishment of periodic pattern in the *Drosophila* embryo. Development (suppl.) 104, 51–60.

Ingham, P.W. 1985. A clonal analysis of the requirement for the trithorax gene in the diversification of segments in Drosophila. J. Embryol. Exp. Morphol. 89, 349–365.

Ingham, P.W., Baker, N.E., Martinez-Arias, A. 1988. Regulation of segment polarity genes in the Drosophila blastoderm by fushi tarazu and even skipped. Nature 331, 73–75.

Ingham, P.W., Howard, K., Ish-Horowicz, D. 1985. Transcription pattern of the *Drosophila* segmentation gene *hairy*. Nature 318, 443–445.

Ingham, P.W., Ish-Horowicz, D., Howard, K.R. 1986. Correlative changes in homeotic and segmentation gene expression in *Kruppel* embryos of *Drosophila*. EMBO J. 5, 1527–1537.

Ingham, P.W., Martinez-Arias, A. 1986. The correct activation of Antennapedia and bithorax complex genes requires the fushi tarazu gene. Nature 324, 592–597.

Irish, V.F., Martinez-Arias, A., Akam, M. 1989. Spatial regulation of the Antennapedia and Ultrabithorax homeotic genes during Drosophila early development. EMBO J. 8, 1527–1537.

Ish-Horowicz, D., Howard, K.R., Pinchin, S.M., Ingham, P.W. 1985. Molecular and genetic analysis of the hairy locus in Drosophila. Cold Spring Harb. Symp. Quant. Biol. 50, 135–144.

Ish-Horowicz, D., Pinchin, S.M., Ingham, P.W., Gyurkovics, H.G. 1989. Autocatalytic ftz activation and metameric instability induced by ectopic ftz expression. Cell 57, 223–232.

Ito, T., Tashiro, K., Muta, S., Ozawa, R., Chiba, T., Nishizawa, M., Yamamoto, K., Kuhara, S., Sakaki, Y. 2000. Toward a protein-protein interaction map of the budding yeast: A comprehensive system to examine two-hybrid interactions in all possible combinations between the yeast proteins. Proc. Natl. Acad. Sci. USA 97, 1143–1147.

Iyer, V.R., Horak, C.E., Scafe, C.S., Botstein, D., Snyder, M., Brown, P.O. 2001. Genomic binding sites of the yeast cell-cycle transcription factors SBF and MBF. Nature 409, 533–538.

Jack, T., Regulski, M., McGinnis, W. 1988. Pair-rule sementation genes regulate the expression of the homoeotic selector gene *Deformed*. Genes Dev. 2, 645–651.

Jackle, H., Tautz, D., Schuh, R., Seifert, E., Lehmann, R. 1986. Cross-regulatory interactions among the gap genes of *Drosophila*. Nature 324, 668–670.

Jackson, J.B., Pollock, J.M., Jr., Rill, R.L. 1979. Chromatin fractionation procedure that yields nucleosomes containing near-stoichiometric amounts of high mobility group nonhistone chromosomal proteins. Biochemistry 18, 3739–3748.

Jagla, K., Bellard, M., Frasch, M. 2001. A cluster of Drosophila homeobox genes involved in mesoderm differentiation programs. Bioessays 23, 125–133.

Jimenez, G., Paroush, Z., Ish-Horowicz, D. 1997. Groucho acts as a corepressor for a subset of negative regulators, including Hairy and Engrailed. Genes Dev. 11, 3072–3082.

Johnson, F.B., Krasnow, M.A. 1992. Differential regulation of transcription preinitiation complex assembly by activator and repressor homeo domain proteins. Genes Dev. 6, 2177–2189.

Jurgens, G. 1985. A group of genes controlling spatial expression of the *bithorax* complex of *Drosophila*. Nature 316, 1533–1535.

Jurgens, G., Wieschaus, E., Nusslein-Volhard, C., Kluding, H. 1984. Mutations affecting the pattern pf the larval cuticle of *Drosophila melanogaster*. II Zygotic loci on the third chromosome. Roux's Archives of Developmental Biology 193, 283–295.

Kagoshima, H., Akamatsu, Y., Ito, Y., Shigesada, K. 1996. Functional dissection of the alpha and beta subunits of transcription factor PEBP2 and the redox susceptibility of its DNA binding activity. J. Biol. Chem. 271, 33074–33082.

Kalionis, B., O'Farrell, P.H. 1993. A universal target sequence is bound in vitro by diverse homeodomains. Mech. Dev. 43, 57–70.

Kaminker, J.S., Singh, R., Lebestky, T., Yan, H., Banerjee, U. 2001. Redundant function of Runt Domain binding partners, Big brother and Brother, during Drosophila development. Development 128, 2639–2648.

Kania, M.A., Bonner, A.S., Duffy, J.B., Gergen, J.P. 1990. The Drosophila segmentation gene runt encodes a novel nuclear regulatory protein that is also expressed in the developing nervous system. Genes Dev. 4, 1701–1713.

Karch, F., Weiffenbach, B., Peifer, M., Bender, W., Duncan, I., Celniker, S., Crosby, M., Lewis, E.B. 1985. The abdominal region of the bithorax complex. Cell 43, 81–96.

Karr, T.L., Kornberg, T.B. 1989. fushi tarazu protein expression in the cellular blastoderm of Drosophila detected using a novel imaging technique. Development 106, 95–103.

Kaufman, T.C., Lewis, R., Wakimoto, B.T. 1980. Cytogenetic analysis of chromosome 3 in *Drosophila melanogaster*: The homeotic gene complex in polytene chromosome 84A-B. Genetics 94, 115–133.

Kehle, J., Beuchle, D., Treuheit, S., Christen, B., Kennison, J.A., Bienz, M., Muller, J. 1998. dMi-2, a hunchback-interacting protein that functions in polycomb repression. Science 282, 1897–1900.

Keleher, C.A., Goutte, C., Johnson, A.D. 1988. The yeast cell-type-specific repressor alpha 2 acts cooperatively with a non-cell-type-specific protein. Cell 53, 927–936.

Keleher, C.A., Passmore, S., Johnson, A.D. 1989. Yeast repressor alpha 2 binds to its operator cooperatively with yeast protein Mcm1. Mol. Cell Biol. 9, 5228–5230.

Keller, S.A., Mao, Y., Struffi, P., Margulies, C., Yurk, C.E., Anderson, A.R., Amey, R.L., Moore, S., Ebels, J.M., Foley, K., Corado, M., Arnosti, D.N. 2000. dCtBP-dependent and -independent repression activities of the Drosophila Knirps protein. Mol. Cell Biol. 20, 7247–7258.

Kellerman, K.A., Mattson, D.M., Duncan, I. 1990. Mutations affecting the stability of the fushi tarazu protein of Drosophila. Genes Dev. 4, 1936–1950.

Kennison, J.A., Tamkun, J.W. 1988. Dosage-dependent modifiers of polycomb and antennapedia mutations in Drosophila. Proc. Natl. Acad. Sci. USA 85, 8136–8140.

Kilchherr, F., Baumgartner, S., Bopp, D., Frei, E., Noll, M. 1986. Isolation of the *paired* gene of *Drosophila* and its spatial expression during early embryogenesis. Nature 321, 493–499.

Kim-Ha, J., Smith, J.L., Macdonald, P.M. 1991. oskar mRNA is localized to the posterior pole of the Drosophila oocyte. Cell 66, 23–35.

Klingler, M., Gergen, J.P. 1993. Regulation of runt transcription by Drosophila segmentation genes. Mech. Dev. 43, 3–19.

Klingler, M., Soong, J., Butler, B., Gergen, J.P. 1996. Disperse versus compact elements for the regulation of runt stripes in Drosophila. Dev. Biol. 177, 73–84.

Knipple, D.C., Seifert, E., Rosenberg, U.B., Preiss, A., Jackle, H. 1985. Spatial and temporal patterns of Kruppel gene expression in early Drosophila embryos. Nature 317, 40–44.

Knoepfler, P.S., Kamps, M.P. 1995. The pentapeptide motif of Hox proteins is required for cooperative DNA binding with Pbx1, physically contacts Pbx1, and enhances DNA binding by Pbx1. Mol. Cell Biol. 15, 5811–5819.

Kobayashi, M., Goldstein, R.E., Fujioka, M., Paroush, Z., Jaynes, J.B. 2001. Groucho augments the repression of multiple Even skipped target genes in establishing parasegment boundaries. Development 128, 1805–1815.

Kornberg, T., Siden, I., O'Farrell, P., Simon, M. 1985. The engrailed locus of Drosophila: in situ localization of transcripts reveals compartment-specific expression. Cell 40, 45–53.

Kosman, D., Small, S. 1997. Concentration-dependent patterning by an ectopic expression domain of the Drosophila gap gene knirps. Development 124, 1343–1354.

Krause, H.M., Klemenz, R., Gehring, W.J. 1988. Expression, modification, and localization of the fushi tarazu protein in Drosophila embryos. Genes Dev. 2, 1021–1036.

Kraut, R., Levine, M. 1991. Mutually repressive interactions between the gap genes *giant* and *Kruppel* define middle body regions of the *Drosophila* embryo. Development 111, 611–622.

Kuroiwa, A., Hafen, E., Gehring, W.J. 1984. Cloning and transcriptional analysis of the segmentation gene fushi tarazu of Drosophila. Cell 37, 825–831.

Kuroiwa, A., Kloter, U., Baumgartner, P., Gehring, W. 1985. Cloning of the homoeotic *Sex combs reduced* gene in *Drosophila* and *in situ* localization of its transcripts. EMBO J. 4.

Kuziora, M.A., McGinnis, W. 1988. Autoregulation of a Drosophila homeotic selector gene. Cell 55, 477–485.

La Rosee, A., Hader, T., Taubert, H., Rivera-Pomar, R., Jackle, H. 1997. Mechanism and Bicoid-dependent control of hairy stripe 7 expression in the posterior region of the Drosophila embryo. EMBO J. 16, 4403–4411.

Lamb, M.M., Laird, C.D. 1976. Increase in nuclear poly(A)-containing RNA at syncytial blastoderm in Drosophila melanogaster embryos. Dev. Biol. 52, 31–42.

Lamka, M.L., Boulet, A.M., Sakonju, S. 1992. Ectopic expression of UBX and ABD-B proteins during Drosophila embryogenesis: competition, not a functional hierarchy, explains phenotypic suppression. Development 116, 841–854.

Langeland, J.A., Attai, S.F., Vorwerk, K., Carroll, S.B. 1994. Positioning adjacent pair-rule stripes in the posterior Drosophila embryo. Development 120, 2945–2955.

Langeland, J.A., Carroll, S.B. 1993. Conservation of regulatory elements controlling hairy pair-rule stripe formation. Development 117, 585–596.

Lardelli, M., Ish-Horowicz, D. 1993. Drosophila hairy pair-rule gene regulates embryonic patterning outside its apparent stripe domains. Development 118, 255–266.

Lawrence, P.A. 1992. *The Making of a Fly: The Genetics of Animal Design*, Blackwell Scientific Publications, Oxford.

Lawrence, P.A., Johnston, P. 1989. Pattern formation in the Drosophila embryo: allocation of cells to parasegments by even-skipped and fushi tarazu. Development 105, 761–767.

Lawrence, P.A., Johnston, P., Macdonald, P., Struhl, G. 1987. Borders of parasegments in Drosophila embryos are delimited by the fushi tarazu and even-skipped genes. Nature 328, 440–442.

Leemans, R., Loop, T., Egger, B., He, H., Kammermeier, L., Hartmann, B., Certa, U., Reichert, H., Hirth, F. 2001. Identification of candidate downstream genes for the homeodomain transcription factor Labial in Drosophila through oligonucleotide-array transcript imaging. Genome. Biol. 2, RESEARCH0015.

Leresche, A., Wolf, V.J., Gottesfeld, J.M. 1996. Repression of RNA polymerase II and III transcription during M phase of the cell cycle. Exp. Cell Res. 229, 282–288.

Levine, M., Hafen, E., Garber, R.L., Gehring, W.J. 1983. Spatial distribution of Antennapedia transcripts during Drosophila development. EMBO J. 2, 2037–2046.

Lewis, E.B. 1978. A gene complex controlling segmentation in Drosophila. Nature 276, 565–570.

Lewis, R.A., Kaufman, T., Denell, R.E., Tallerico, P. 1980a. Genetic analysis of the Antennapedia Gene Complex (ANT-C) and adjacent chromosomal regions of *Drosophila melanogaster*. I. Polytene chromosome segments 84B-D. Genetics 95, 367–381.

Lewis, R.A., Wakimoto, B.T., Denell, R.E., Kaufman, T. 1980b. Genetic analysis of the Antennapedia Gene Complex (ANT-C) and adjacent chromosomal regions of *Drosophila melanogaster*. II. Polytene chromosome segments 84A-84B1,2. Genetics 95, 383–397.

Li, L.H., Gergen, J.P. 1999. Differential interactions between Brother proteins and Runt domain proteins in the Drosophila embryo and eye. Development 126, 3313–3322.

Li, X., Murre, C., McGinnis, W. 1999. Activity regulation of a Hox protein and a role for the homeodomain in inhibiting transcriptional activation. EMBO J. 18, 198–211.

Liang, Z., Biggin, M.D. 1998. Eve and ftz regulate a wide array of genes in blastoderm embryos: the selector homeoproteins directly or indirectly regulate most genes in Drosophila [published erratum appears in Development 1999 Feb;126(5):following table of contents]. Development 125, 4471–4482.

Lipshitz, H.D., Peattie, D.A., Hogness, D.S. 1987. Novel transcripts from the Ultrabithorax domain of the bithorax complex. Genes Dev. 1, 307–322.

Long, J.J., Leresche, A., Kriwacki, R.W., Gottesfeld, J.M. 1998. Repression of TFIIH transcriptional activity and TFIIH-associated cdk7 kinase activity at mitosis. Mol. Cell Biol. 18, 1467–1476.

Lu, Q., Kamps, M.P. 1996. Structural determinants within Pbx1 that mediate cooperative DNA binding with pentapeptide-containing Hox proteins: proposal for a model of a Pbx1-Hox-DNA complex. Mol. Cell Biol. 16, 1632–1640.

Lu, Q., Knoepfler, P.S., Scheele, J., Wright, D.D., Kamps, M.P. 1995. Both Pbx1 and E2A-Pbx1 bind the DNA motif ATCAATCAA cooperatively with the products of multiple murine Hox genes, some of which are themselves oncogenes. Mol. Cell Biol. 15, 3786–3795.

Lyko, F., Ramsahoye, B.H., Jaenisch, R. 2000. DNA methylation in Drosophila melanogaster. Nature 408, 538–540.

Ma, X., Yuan, D., Diepold, K., Scarborough, T., Ma, J. 1996. The Drosophila morphogenetic protein Bicoid binds DNA cooperatively. Development 122, 1195–1206.

Macdonald, P.M., Ingham, P., Struhl, G. 1986. Isolation, structure, and expression of even-skipped: a second pair- rule gene of Drosophila containing a homeo box. Cell 47, 721–734.

Macdonald, P.M., Struhl, G. 1986. A molecular gradient in early Drosophila embryos and its role in specifying the body pattern. Nature 324, 537–545.

Mahajan, N.P., Linder, K., Berry, G., Gordon, G.W., Heim, R., Herman, B. 1998. Bcl-2 and Bax interactions in mitochondria probed with green fluorescent protein and fluorescence resonance energy transfer. Nat. Biotechnol. 16, 547–552.

Mann, R.S., Chan, S.K. 1996. Extra specificity from extradenticle: the partnership between HOX and PBX/EXD homeodomain proteins [published erratum appears in Trends Genet. 1996 Aug;12(8):328]. Trends Genet. 12, 258–262.

Mann, R.S., Morata, G. 2000. The developmental and molecular biology of genes that subdivide the body of Drosophila. Annu. Rev. Cell Dev. Biol. 16, 243–271.

Mannervik, M., Levine, M. 1999. The Rpd3 histone deacetylase is required for segmentation of the Drosophila embryo. Proc. Natl. Acad. Sci. USA 96, 6797–6801.

Manoukian, A.S., Krause, H.M. 1992. Concentration-dependent activities of the even-skipped protein in Drosophila embryos. Genes Dev. 6, 1740–1751.

Manoukian, A.S., Krause, H.M. 1993. Control of segmental asymmetry in Drosophila embryos. Development 118, 785–796.

Martinez Arias, A. 1986. The *Antennapedia* gene is required and expressed in parasegments 4 and 5 of the *Drosophila* embryo. EMBO J. 4, 135–141.

Martinez-Arias, A., Ingham, P.W., Scott, M.P., Akam, M.E. 1987. The spatial and temporal deployment of Dfd and Scr transcripts throughout development of Drosophila. Development 100, 673–683.

Martinez-Arias, A., Lawrence, P.A. 1985. Parasegments and compartments in the Drosophila embryo. Nature 313, 639–642.

Martizez Arias, A., Baker, N.E., Ingham, P.W. 1988. Role of segment polarity genes in the definition and maintenance of cell states in the Drosophila embryo. Development 103, 157–170.

McGinnis, W., Krumlauf, R. 1992. Homeobox genes and axial patterning. Cell 68, 283–302.

McKnight, S.L., Miller, O.L., Jr. 1976. Ultrastructural patterns of RNA synthesis during early embryogenesis of Drosophila melanogaster. Cell 8, 305–319.

McKnight, S.L., Miller, O.L., Jr. 1979. Post-replicative nonribosomal transcription units in D. melanogaster embryos. Cell 17, 551–563.

Merrill, V.K., Turner, F.R., Kaufman, T.C. 1987. A genetic and developmental analysis of mutations in the Deformed locus in Drosophila melanogaster. Dev. Biol. 122, 379–395.

Mlodzik, M., De Montrion, C.M., Hiromi, Y., Krause, H.M., Gehring, W. 1987. The influence on the blastoderm fate map of maternal-effect genes that affect the antero-posterior pattern in *Drosophila*. Genes Dev. 1, 603–614.

Mlodzik, M., Fjose, A., Gehring, W. 1985. Isolation of *caudal*, a *Drosophila* homeobox containing gene with maternal expression, whose transcripts form a concentration gradient at the pre-blastoderm stage. EMBO J. 4, 2961–2969.

Mlodzik, M., Fjose, A., Gehring, W.J. 1988. Molecular structure and spatial expression of a homeobox gene from the labial region of the Antennapedia-complex. EMBO J. 7, 2569–2578.

Mlodzik, M., Gehring, W.J. 1987. Expression of the caudal gene in the germ line of Drosophila: formation of an RNA and protein gradient during early embryogenesis. Cell 48, 465–478.

Mohler, J., Eldon, E.D., Pirrotta, V. 1989. A novel spatial transcription pattern associated with the segmentation gene, *giant*, of Drosophila. EMBO J. 8, 1539–1558.

Morcillo, P., Rosen, C., Baylies, M.K., Dorsett, D. 1997. Chip, a widely expressed chromosomal protein required for segmentation and activity of a remote wing margin enhancer in Drosophila. Genes Dev. 11, 2729–2740.

Morcillo, P., Rosen, C., Dorsett, D. 1996. Genes regulating the remote wing margin enhancer in the Drosophila cut locus. Genetics 144, 1143–1154.

Morrissey, D., Askew, D., Raj, L., Weir, M. 1991. Functional dissection of the paired segmentation gene in Drosophila embryos. Genes Dev. 5, 1684–1696.

Mullen, J.R., DiNardo, S. 1995. Establishing parasegments in Drosophila embryos: roles of the odd- skipped and naked genes. Dev. Biol. 169, 295–308.

Muller, J., Bienz, M. 1991. Long range repression conferring boundaries of Ultrabithorax expression in the Drosophila embryo. EMBO J. 10, 3147–3155.

Muller, J., Bienz, M. 1992. Sharp anterior boundary of homeotic gene expression conferred by the fushi tarazu protein. EMBO J. 11, 3653–3661.

Murata, Y., Wharton, R.P. 1995. Binding of pumilio to maternal hunchback mRNA is required for posterior patterning in Drosophila embryos. Cell 80, 747–756.

Murray, A. 1995. Cyclin ubiquitination: the destructive end of mitosis. Cell 81, 149–152.

Nan, X., Meehan, R.R., Bird, A. 1993. Dissection of the methyl-CpG binding domain from the chromosomal protein MeCP2. Nucleic Acids Res. 21, 4886–4892.

Nasiadka, A., Grill, A., Krause, H.M. 2000. Mechanisms regulating target gene selection by the homeodomain-containing protein Fushi tarazu. Development 127, 2965–2976.

Nasiadka, A., Krause, H.M. 1999. Kinetic analysis of segmentation gene interactions in Drosophila embryos. Development 126, 1515–1526.

Nauber, U., Pankratz, M.J., Kienlin, A., Seifert, E., Klemm, U., Jackle, H. 1988. Abdominal segmentation of the Drosophila embryo requires a hormone receptor-like protein encoded by the gap gene knirps. Nature 336, 489–492.

Ner, S.S., Travers, A.A. 1994. HMG-D, the Drosophila melanogaster homologue of HMG 1 protein, is associated with early embryonic chromatin in the absence of histone H1. EMBO J. 13, 1817–1822.

Neuteboom, S.T., Peltenburg, L.T., van Dijk, M.A., Murre, C. 1995. The hexapeptide LFPWMR in Hoxb-8 is required for cooperative DNA binding with Pbx1 and Pbx2 proteins. Proc. Natl. Acad. Sci. USA 92, 9166–9170.

Newport, J., Kirschner, M. 1982a. A major developmental transition in early Xenopus embryos: I. characterization and timing of cellular changes at the midblastula stage. Cell 30, 675–686.

Newport, J., Kirschner, M. 1982b. A major developmental transition in early Xenopus embryos: II. Control of the onset of transcription. Cell 30, 687–696.

Nibu, Y., Levine, M.S. 2001. CtBP-dependent activities of the short-range Giant repressor in the Drosophila embryo. Proc. Natl. Acad. Sci. USA 98, 6204–6208.

Nibu, Y., Zhang, H., Bajor, E., Barolo, S., Small, S., Levine, M. 1998a. dCtBP mediates transcriptional repression by Knirps, Kruppel and Snail in the Drosophila embryo. EMBO J. 17, 7009–7020.

Nibu, Y., Zhang, H., Levine, M. 1998b. Interaction of short-range repressors with Drosophila CtBP in the embryo. Science 280, 101–104.

Nibu, Y., Zhang, H., Levine, M. 2001. Local action of long-range repressors in the Drosophila embryo. EMBO J. 20, 2246–2253.

Nurse, P. 1990. Universal control mechanism regulating onset of M-phase. Nature 344, 503–508.

Nusslein-Volhard, C., Wieschaus, E. 1980. Mutations affecting segment number and polarity in Drosophila. Nature 287, 795–801.

Nusslein-Volhard, C., Wieschaus, E., Kluding, H. 1984. Mutations affecting the pattern of the larval cuticle of *Drosophila melanogaster.* I Zygotic loci on the second chromosome. Roux's Archives of Developmental Biology 193, 267–282.

Ogawa, E., Maruyama, M., Kagoshima, H., Inuzuka, M., Lu, J., Satake, M., Shigesada, K., Ito, Y. 1993. PEBP2/PEA2 represents a family of transcription factors homologous to the products of the Drosophila runt gene and the human AML1 gene. Proc. Natl. Acad. Sci. USA 90, 6859–6863.

Orenic, T.V., Slusarski, D.C., Kroll, K.L., Holmgren, R.A. 1990. Cloning and characterization of the segment polarity gene cubitus interruptus Dominant of Drosophila. Genes Dev. 4, 1053–1067.

Pankratz, M.J., Jackle, H. 1993. Blatoderm segmentation. In: *The Development of Drosophila Melanogaster* (M. Bate, and A. Martinez-Arias, Eds.). Cold Spring Harbor Laboratory Press, pp. 467–516.

Pankratz, M.J., Seifert, E., Gerwin, N., Billi, B., Nauber, U., Jackle, H. 1990. Gradients of Kruppel and knirps gene products direct pair-rule gene stripe patterning in the posterior region of the Drosophila embryo. Cell 61, 309–317.

Paro, R. 1990. Imprinting a determined state into the chromatin of Drosophila. Trends Genet. 6, 416–421.

Paroush, Z., Finley, R.L., Jr., Kidd, T., Wainwright, S.M., Ingham, P.W., Brent, R., Ish-Horowicz, D. 1994. Groucho is required for Drosophila neurogenesis, segmentation, and sex determination and interacts directly with hairy-related bHLH proteins. Cell 79, 805–815.

Passner, J.M., Ryoo, H.D., Shen, L., Mann, R.S., Aggarwal, A.K. 1999. Structure of a DNA-bound Ultrabithorax-Extradenticle homeodomain complex [see comments]. Nature 397, 714–719.

Peifer, M., Karch, F., Bender, W. 1987. The bithorax complex: Control of segmental identity. Genes Dev. 1, 891–898.

Perrimon, N., Mahowald, A.P. 1987. Multiple functions of segment polarity genes in Drosophila. Dev. Biol. 119, 587–600.

Perrimon, N., Smouse, D. 1989. Multiple functions of a Drosophila homeotic gene, zeste-white 3, during segmentation and neurogenesis. Dev. Biol. 135, 287–305.

Perrimon, N., Lu, X., Hou, X.S., Hsu, J.C., Melnick, M.B., Chou, T.B., Perkins, L.A. 1995. Dissection of the Torso signal transduction pathway in Drosophila. Mol. Reprod. Dev. 42, 515–522.

Petruk, S., Sedkov, Y., Smith, S., Tillib, S., Kraevski, V., Nakamura, T., Canaani, E., Croce, C.M., Mazo, A. 2001. Trithorax and dCBP Acting in a Complex to Maintain Expression of a Homeotic Gene. Science 294, 1331–1334.

Phelan, M.L., Rambaldi, I., Featherstone, M.S. 1995. Cooperative interactions between HOX and PBX proteins mediated by a conserved peptide motif. Mol. Cell Biol. 15, 3989–3997.

Phizicky, E.M., Fields, S. 1995. Protein-protein interactions: methods for detection and analysis. Microbiol. Rev. 59, 94–123.

Pignoni, F., Baldarelli, R.M., Steingrimsson, E., Diaz, R.J., Patapoutian, A., Merriam, J.R., Lengyel, J.A. 1990. The Drosophila gene tailless is expressed at the embryonic termini and is a member of the steroid receptor superfamily. Cell 62, 151–163.

Pinsonneault, J., Florence, B., Vaessin, H., McGinnis, W. 1997. A model for extradenticle function as a switch that changes HOX proteins from repressors to activators. EMBO J. 16, 2032–2042.

Piper, D.E., Batchelor, A.H., Chang, C.P., Cleary, M.L., Wolberger, C. 1999. Structure of a HoxB1-Pbx1 heterodimer bound to DNA: role of the hexapeptide and a fourth homeodomain helix in complex formation. Cell 96, 587–597.

Poortinga, G., Watanabe, M., Parkhurst, S.M. 1998. Drosophila CtBP: a Hairy-interacting protein required for embryonic segmentation and hairy-mediated transcriptional repression. EMBO J. 17, 2067–2078.

Popperl, H., Bienz, M., Studer, M., Chan, S.K., Aparicio, S., Brenner, S., Mann, R.S., Krumlauf, R. 1995. Segmental expression of Hoxb-1 is controlled by a highly conserved autoregulatory loop dependent upon exd/pbx. Cell 81, 1031–1042.

Price, D., Rabinovitch, S., O'Farrell, P.H., Campbell, S.D. 2000. Drosophila wee1 has an essential role in the nuclear divisions of early embryogenesis. Genetics 155, 159–166.

Pritchard, D.K., Schubiger, G. 1996. Activation of transcription in Drosophila embryos is a gradual process mediated by the nucleocytoplasmic ratio. Genes Dev. 10, 1131–1142.

Puig, O., Caspary, F., Rigaut, G., Rutz, B., Bouveret, E., Bragado-Nilsson, E., Wilm, M., Seraphin, B. 2001. The tandem affinity purification (TAP) method: a general procedure of protein complex purification. Methods 24, 218–229.

Pultz, M.A., Diederich, R.J., Cribbs, D.L., Kaufman, T.C. 1988. The proboscipedia locus of the Antennapedia complex: a molecular and genetic analysis. Genes Dev. 2, 901–920.

Qian, S., Capovilla, M., Pirrotta, V. 1991. The bx region enhancer, a distant cis-control element of the Drosophila Ubx gene and its regulation by hunchback and other segmentation genes. EMBO J. 10, 1415–1425.

Regulski, M., McGinnis, N., Chadwick, R., McGinnis, W. 1987. Developmental and molecular analysis of *Deformed*: A homeotic gene controlling *Drosophila* head development. EMBO J. 6, 767–777.

Ren, B., Robert, F., Wyrick, J.J., Aparicio, O., Jennings, E.G., Simon, I., Zeitlinger, J., Schreiber, J., Hannett, N., Kanin, E., Volkert, T.L., Wilson, C.J., Bell, S.P., Young, R.A. 2000. Genome-wide location and function of DNA binding proteins. Science 290, 2306–2309.

Riddihough, G., Ish-Horowicz, D. 1991. Individual stripe regulatory elements in the Drosophila hairy promoter respond to maternal, gap, and pair-rule genes. Genes Dev. 5, 840–854.

Rigaut, G., Shevchenko, A., Rutz, B., Wilm, M., Mann, M., Seraphin, B. 1999. A generic protein purification method for protein complex characterization and proteome exploration. Nat. Biotechnol. 17, 1030–1032.

Riggleman, B., Wieschaus, E., Schedl, P. 1989. Molecular analysis of the armadillo locus: uniformly distributed transcripts and a protein with novel internal repeats are associated with a Drosophila segment polarity gene. Genes Dev. 3, 96–113.

Rivera-Pomar, R., Niessing, D., Schmidt-Ott, U., Gehring, W.J., Jackle, H. 1996. RNA binding and translational suppression by bicoid [see comments]. Nature 379, 746–749.

Rosenberg, U., Schroder, C., Preiss, A., Kienlin, A., Cote, S., Riede, I., Jackle, H. 1986. Structural homology of the product of the *Drosophila Kruppel* gene with *Xenopus* transcription factor IIIA. Nature 319, 336–339.

Rothe, M., Pehl, M., Taubert, H., Jackle, H. 1992. Loss of gene function through rapid mitotic cycles in the Drosophila embryo. Nature 359, 156–159.

Ruden, D.M., Jackle, H. 1995. Mitotic delay dependent survival identifies components of cell cycle control in the Drosophila blastoderm. Development 121, 63–73.

Ryoo, H.D., Mann, R.S. 1999. The control of trunk Hox specificity and activity by Extradenticle. Genes Dev. 13, 1704–1716.

Sackerson, C., Fujioka, M., Goto, T. 1999. The even-skipped locus is contained in a 16-kb chromatin domain. Dev. Biol. 211, 39–52.

Sanchez, Y., Wong, C., Thoma, R.S., Richman, R., Wu, Z., Piwnica-Worms, H., Elledge, S.J. 1997. Conservation of the Chk1 checkpoint pathway in mammals: linkage of DNA damage to Cdk regulation through Cdc25. Science 277, 1497–1501.

Sanchez-Herrero, E., Vernas, I., Marco, R., Morata, G. 1985. Genetic organization of *Drosophila* bithorax complex. Nature 313, 108–113.

Sasai, Y., Kageyama, R., Tagawa, Y., Shigemoto, R., Nakanishi, S. 1992. Two mammalian helix-loop-helix factors structurally related to Drosophila hairy and Enhancer of split. Genes Dev. 6, 2620–2634.

Sauer, F., Hansen, S.K., Tjian, R. 1995. DNA template and activator-coactivator requirements for transcriptional synergism by Drosophila bicoid [see comments]. Science 270, 1825–1828.

Sauer, F., Jackle, H. 1991. Concentration-dependent transcriptional activation or repression by Kruppel from a single binding site. Nature 353, 563–566.

Sauer, F., Jackle, H. 1993. Dimerization and the control of transcription by Kruppel. Nature 364, 454–457.

Sauer, F., Rivera-Pomar, R., Hoch, M., Jackle, H. 1996. Gene regulation in the Drosophila embryo. Philos. Trans. R. Soc. Lond. B Biol. Sci. 351, 579–587.

Saulier-Le Drean, B., Nasiadka, A., Dong, J., Krause, H.M. 1998. Dynamic changes in the functions of Odd-skipped during early Drosophila embryogenesis. Development 125, 4851–4861.

Schulz, C., Tautz, D. 1994. Autonomous concentration-dependent activation and repression of Kruppel by hunchback in the Drosophila embryo. Development 120, 3043–3049.

Schwartz, C.J., Sampson, H.M., Hlousek, D., Percival-Smith, A., Copeland, J.W., Simmonds, A.J., Krause, H.M. 2001. FTZ-Factor1 and Fushi tarazu interact via conserved nuclear receptor and coactivator motifs. EMBO J. 20, 510–519.

Segil, N., Guermah, M., Hoffmann, A., Roeder, R.G., Heintz, N. 1996. Mitotic regulation of TFIID: inhibition of activator-dependent transcription and changes in subcellular localization. Genes Dev. 10, 2389–2400.

Sewalt, R.G., Gunster, M.J., van der Vlag, J., Satijn, D.P., Otte, A.P. 1999. C-Terminal binding protein is a transcriptional repressor that interacts with a specific class of vertebrate Polycomb proteins. Mol. Cell Biol. 19, 777–787.

Shen, W.F., Chang, C.P., Rozenfeld, S., Sauvageau, G., Humphries, R.K., Lu, M., Lawrence, H.J., Cleary, M.L., Largman, C. 1996. Hox homeodomain proteins exhibit selective complex stabilities with Pbx and DNA. Nucleic Acids Res. 24, 898–906.

Shermoen, A.W., O'Farrell, P.H. 1991. Progression of the cell cycle through mitosis leads to abortion of nascent transcripts. Cell 67, 303–310.

Sibon, O.C., Laurencon, A., Hawley, R., Theurkauf, W.E. 1999. The Drosophila ATM homologue Mei-41 has an essential checkpoint function at the midblastula transition. Curr. Biol. 9, 302–312.

Sibon, O.C., Stevenson, V.A., Theurkauf, W.E. 1997. DNA-replication checkpoint control at the Drosophila midblastula transition. Nature 388, 93–97.

Sigrist, S., Jacobs, H., Stratmann, R., Lehner, C.F. 1995. Exit from mitosis is regulated by Drosophila fizzy and the sequential destruction of cyclins A, B and B3. EMBO J. 14, 4827–4838.

Simpson-Brose, M., Treisman, J., Desplan, C. 1994. Synergy between the hunchback and bicoid morphogens is required for anterior patterning in Drosophila. Cell 78, 855–865.

Small, S., Blair, A., Levine, M. 1992. Regulation of even-skipped stripe 2 in the Drosophila embryo. EMBO J. 11, 4047–4057.

Small, S., Blair, A., Levine, M. 1996. Regulation of two pair-rule stripes by a single enhancer in the Drosophila embryo. Dev. Biol. 175, 314–324.

Small, S., Kraut, R., Hoey, T., Warrior, R., Levine, M. 1991. Transcriptional regulation of a pair-rule stripe in Drosophila. Genes Dev. 5, 827–839.

Speck, N.A., Terryl, S. 1995. A new transcription factor family associated with human leukemias. Crit. Rev. Eukaryot. Gene. Expr. 5, 337–364.

St. Johnston, D., Driever, W., Berleth, T., Richstein, S., Nusslein-Volhard, C. 1989. Multiple steps in the localization of bicoid RNA to the anterior pole of the Drosophila oocyte. Development 107, 13–19.

Stanojevic, D., Small, S., Levine, M. 1991. Regulation of a segmentation stripe by overlapping activators and repressors in the Drosophila embryo. Science 254, 1385–1387.

Steingrimsson, E., Pignoni, F., Liaw, G.J., Lengyel, J.A. 1991. Dual role of the Drosophila pattern gene tailless in embryonic termini. Science 254, 418–421.

Struhl, G. 1981a. A gene product required for correct initiation of segmental determination in Drosophila. Nature 293, 36–41.

Struhl, G. 1981b. A homoeotic mutation transforming leg to antenna in Drosophila. Nature 292, 635–638.

Struhl, G., Johnston, P., Lawrence, P.A. 1992. Control of Drosophila body pattern by the hunchback morphogen gradient. Cell 69, 237–249.

Struhl, G., Struhl, K., Macdonald, P.M. 1989. The gradient morphogen bicoid is a concentration-dependent transcriptional activator. Cell 57, 1259–1273.

Struhl, G., White, R.A. 1985. Regulation of the Ultrabithorax gene of Drosophila by other bithorax complex genes. Cell 43, 507–519.

Sundqvist, A., Sollerbrant, K., Svensson, C. 1998. The carboxy-terminal region of adenovirus E1A activates transcription through targeting of a C-terminal binding protein-histone deacetylase complex. FEBS Lett. 429, 183–188.

Suzuki, T., Kawasaki, H., Yu, R.T., Ueda, H., Umesono, K. 2001. Segmentation gene product Fushi tarazu is an LXXLL motif-dependent coactivator for orphan receptor FTZ-F1. Proc. Natl. Acad. Sci. USA 98, 12403–12408.

Tamkun, J.W., Deuring, R., Scott, M.P., Kissinger, M., Pattatucci, A.M., Kaufman, T.C., Kennison, J.A. 1992. brahma: a regulator of Drosophila homeotic genes structurally related to the yeast transcriptional activator SNF2/SWI2. Cell 68, 561–572.

Tautz, D. 1988. Regulation of the Drosophila segmentation gene hunchback by two maternal morphogenetic centres. Nature 332, 281–284.

Tautz, D., Lehmann, R., Schnurch, R., Schuh, E., Seifert, A., Kienlin, K., Jones, K., Jackle, H. 1987. Finger protein of novel structure encoded by *hunchback*, a second member of the gap class of *Drosophila* segmentation genes. Nature 327, 383–389.

Tautz, D., Sommer, R.J. 1995. Evolution of segmentation genes in insects. Trends Genet. 11, 23–27.

Tie, F., Furuyama, T., Prasad-Sinha, J., Jane, E., Harte, P.J. 2001. The Drosophila Polycomb Group proteins ESC and E(Z) are present in a complex containing the histone-binding protein p55 and the histone deacetylase RPD3. Development 128, 275–286.

Tiong, S., Bone, L.M., Whittle, J.R. 1985. Recessive lethal mutations within the bithorax-complex in Drosophila. Mol. Gen. Genet. 200, 335–342.

Torigoi, E., Bennani-Baiti, I.M., Rosen, C., Gonzalez, K., Morcillo, P., Ptashne, M., Dorsett, D. 2000. Chip interacts with diverse homeodomain proteins and potentiates bicoid activity in vivo. Proc. Natl. Acad. Sci. USA 97, 2686–2691.

Tremml, G., Bienz, M. 1992. Induction of labial expression in the Drosophila endoderm: response elements for dpp signalling and for autoregulation. Development 116, 447–456.

Tsai, C., Gergen, P. 1995. Pair-rule expression of the Drosophila fushi tarazu gene: a nuclear receptor response element mediates the opposing regulatory effects of runt and hairy. Development 121, 453–462.

Uetz, P., Giot, L., Cagney, G., Mansfield, T.A., Judson, R.S., Knight, J.R., Lockshon, D., Narayan, V., Srinivasan, M., Pochart, P., Qureshi-Emili, A., Li, Y., Godwin, B., Conover, D., Kalbfleisch, T., Vijayadamodar, G., Yang, M., Johnston, M., Fields, S., Rothberg, J.M. 2000. A comprehensive analysis of protein-protein interactions in Saccharomyces cerevisiae. Nature 403, 623–627.

Vachon, G., Cohen, B., Pfeifle, C., McGuffin, M.E., Botas, J., Cohen, S.M. 1992. Homeotic genes of the Bithorax complex repress limb development in the abdomen of the Drosophila embryo through the target gene Distal-less. Cell 71, 437–450.

van de Wetering, M., Cavallo, R., Dooijes, D., van Beest, M., van Es, J., Loureiro, J., Ypma, A., Hursh, D., Jones, T., Bejsovec, A., Peifer, M., Mortin, M., Clevers, H. 1997. Armadillo coactivates transcription driven by the product of the Drosophila segment polarity gene dTCF. Cell 88, 789–799.

van den Heuvel, M., Nusse, R., Johnston, P., Lawrence, P.A. 1989. Distribution of the wingless gene product in Drosophila embryos: a protein involved in cell-cell communication. Cell 59, 739–749.

van der Vlag, J., Otte, A.P. 1999. Transcriptional repression mediated by the human polycomb-group protein EED involves histone deacetylation. Nat. Genet. 23, 474–478.

van Dijk, M.A., Murre, C. 1994. Extradenticle raises the DNA binding specificity of homeotic selector gene products. Cell 78, 617–624.

van Dijk, M.A., Peltenburg, L.T., Murre, C. 1995. Hox gene products modulate the DNA binding activity of Pbx1 and Pbx2. Mech. Dev. 52, 99–108.

van Steensel, B., Delrow, J., Henikoff, S. 2001. Chromatin profiling using targeted DNA adenine methyltransferase. Nat. Genet. 27, 304–308.

van Steensel, B., Henikoff, S. 2000. Identification of in vivo DNA targets of chromatin proteins using tethered dam methyltransferase. Nat. Biotechnol. 18, 424–428.

Vincent, J.P., Kassis, J.A., O'Farrell, P.H. 1990. A synthetic homeodomain binding site acts as a cell type specific, promoter specific enhancer in Drosophila embryos. EMBO J. 9, 2573–2578.

Wade, P.A., Jones, P.L., Vermaak, D., Wolffe, A.P. 1998. A multiple subunit Mi-2 histone deacetylase from Xenopus laevis cofractionates with an associated Snf2 superfamily ATPase. Curr. Biol. 8, 843–846.

Wagner, C.R., Hamana, K., Elgin, S.C. 1992. A high-mobility-group protein and its cDNAs from Drosophila melanogaster. Mol. Cell Biol. 12, 1915–1923.

Wakimoto, B.T., Kaufman, T.C. 1981. Analysis of larval segmentation in lethal genotypes associated with the antennapedia gene complex in Drosophila melanogaster. Dev. Biol. 81, 51–64.

Walhout, A.J., Sordella, R., Lu, X., Hartley, J.L., Temple, G.F., Brasch, M.A., Thierry-Mieg, N., Vidal, M. 2000. Protein interaction mapping in C. elegans using proteins involved in vulval development. Science 287, 116–122.

Walldorf, U., Gehring, W.J. 1992. Empty spiracles, a gap gene containing a homeobox involved in Drosophila head development. EMBO J. 11, 2247–2259.

Walter, J., Biggin, M.D. 1996. DNA binding specificity of two homeodomain proteins in vitro and in Drosophila embryos. Proc. Natl. Acad. Sci. USA 93, 2680–2685.

Walter, J., Dever, C.A., Biggin, M.D. 1994. Two homeo domain proteins bind with similar specificity to a wide range of DNA sites in Drosophila embryos. Genes Dev. 8, 1678–1692.

Waltzer, L., Bienz, M. 1998. Drosophila CBP represses the transcription factor TCF to antagonize Wingless signalling. Nature 395, 521–525.

Wang, C., Lehmann, R. 1991. Nanos is the localized posterior determinant in Drosophila [published erratum appears in Cell 1992 Mar. 20;68(6):1177]. Cell 66, 637–647.

Wang, Z., Lindquist, S. 1998. Developmentally regulated nuclear transport of transcription factors in Drosophila embryos enable the heat shock response. Development 125, 4841–4850.

Weigel, D., Jurgens, G., Klingler, M., Jackle, H. 1990. Two gap genes mediate maternal terminal pattern information in Drosophila. Science 248, 495–498.

Weiner, A.J., Scott, M.P., Kaufman, T.C. 1984. A molecular analysis of fushi tarazu, a gene in Drosophila melanogaster that encodes a product affecting embryonic segment number and cell fate. Cell 37, 843–851.

Weir, M.P., Kornberg, T. 1985. Patterns of engrailed and fushi tarazu transcripts reveal novel intermediate stages in Drosophila segmentation. Nature 318, 433–439.

Wheeler, J.C., Shigesada, K., Gergen, J.P., Ito, Y. 2000. Mechanisms of transcriptional regulation by Runt domain proteins. Semin. Cell Dev. Biol. 11, 369–375.

White, R.A., Lehmann, R. 1986. A gap gene, hunchback, regulates the spatial expression of Ultrabithorax. Cell 47, 311–321.

Wieschaus, E., Nusslein-Volhard, C., Jurgens, G. 1984. Mutations affecting the pattern of the larval cuticle of *Drosophila melanogaster.* III Zygotic loci on the X- chromosome and the fourth chromosome. Roux's Archives of Developmental Biology 193, 296–307.

Wimmer, E.A., Jackle, H., Pfeifle, C., Cohen, S.M. 1993. A Drosophila homologue of human Sp1 is a head-specific segmentation gene. Nature 366, 690–694.

Wu, X., Vakani, R., Small, S. 1998. Two distinct mechanisms for differential positioning of gene expression borders involving the Drosophila gap protein giant. Development 125, 3765–3774.

Wu, X., Vasisht, V., Kosman, D., Reinitz, J., Small, S. 2001. Thoracic patterning by the Drosophila gap gene hunchback. Dev. Biol. 237, 79–92.

Xu, L., Glass, C.K., Rosenfeld, M.G. 1999. Coactivator and corepressor complexes in nuclear receptor function. Curr. Opin. Genet. Dev. 9, 140–147.

Yu, Y., Li, W., Su, K., Yussa, M., Han, W., Perrimon, N., Pick, L. 1997. The nuclear hormone receptor Ftz-F1 is a cofactor for the Drosophila homeodomain protein Ftz. Nature 385, 552–555.

Yu, Y., Pick, L. 1995. Non-periodic cues generate seven ftz stripes in the Drosophila embryo. Mech. Dev. 50, 163–175.

Yuan, D., Ma, X., Ma, J. 1996. Sequences outside the homeodomain of bicoid are required for protein-protein interaction. J. Biol. Chem. 271, 21660–21665.

Yuan, D., Ma, X., Ma, J. 1999. Recognition of multiple patterns of DNA sites by Drosophila homeodomain protein Bicoid. J. Biochem. (Tokyo) 125, 809–817.

Zakany, J., Kmita, M., Alarcon, P., de la Pompa, J.L., Duboule, D. 2001. Localized and transient transcription of Hox genes suggests a link between patterning and the segmentation clock. Cell 106, 207–217.

Zalokar, M. 1976. Autoradiographic study of protein and RNA formation during early development of Drosophila eggs. Dev. Biol. 49, 425–437.

Zalokar, M., Erk, I. 1976. Division and migration of nuclei during early embryogenesis of *Drosophila melanogaster*. J. Microsc. Biol. Cell 25, 97–106.

Zhang, C.C., Muller, J., Hoch, M., Jackle, H., Bienz, M. 1991. Target sequences for hunchback in a control region conferring Ultrabithorax expression boundaries. Development 113, 1171–1179.

Zhang, H., Levine, M. 1999. Groucho and dCtBP mediate separate pathways of transcriptional repression in the Drosophila embryo. Proc. Natl. Acad. Sci. USA 96, 535–540.

Advances in Developmental Biology and Biochemistry, Vol. 12
M. DePamphilis (Editor)

Embryonic stem cell development in mammals

Colin L. Stewart

Cancer and Developmental Biology Laboratory, Basic Research Division,
National Cancer Institute, Frederick, MD 21702, USA

Summary

Embryonic stem (ES) cells from the mouse embryo are the one pluripotent stem cell that can be routinely maintained in culture. Their principal application has been as a route by which murine genes are mutated and then introduced into the mouse's germline to establish new mutant lines of mice in which the gene's function can be studied in vivo. The recent derivation of human ES lines has stimulated a greater interest in other applications of these cells, namely the study of the molecular basis to cell differentiation and human embryonic development. There is also considerable interest in developing techniques to make ES cells differentiate into specific lineages, to use these derivatives for therapeutic purposes in treating neurodegenerative diseases such as Parkinson's and other diseases such as diabetes. The molecular basis to how murine ES cells regulate their proliferation and differentiation, with regard to the JAK-STAT signaling pathway and POU transcription factor OCT4 are beginning to be understood. These studies will serve as a paradigm to investigating the molecular basis of human embryonic differentiation and to help in characterizing and establishing in culture recently identified stem and progenitor cells found in adult somatic tissues.

Contents

1. Introduction

1.1. Origins and development of pluripotent cell lines from embryos

The understanding of complex biological processes such as the development of higher animals has greatly benefited from genetic analysis. The most successful model systems for animal development have been established in species in which genetic manipulation is relatively easy such as *Drosophila, C. elegans* and most recently the zebrafish *Danio rerio*. For many years genetic analysis of mammalian development has been severely restricted. Relatively low rates of reproduction, long periods of gestation, and limited access to embryos, which develop in the reproductive tract, all complicated analysis of development even in the most commonly used mammal, the mouse *Mus musculus*.

However during the 60s, Leroy Stevens and Barry Pierce made a series of startling observations while studying the origins and development of rare and unusual tumors, called teratocarcinomas, which formed in the gonads of mice (Stevens, 1964; Stevens, 1967b; Stevens, 1968). These tumors were unlike the majority of other cancers, which usually consist of one cell type. Instead the tumors contained many different cell types, such as muscle, epithelium, cartilage and sometimes, particularly in human tumors, tissues such as teeth, and intestine. In addition to having many different cell types, teratocarcinomas are characterized by the presence of embryonal carcinoma (EC) cells (Kleinsmith and Pierce, 1964). In mice, EC cells are morphologically and histologically similar to embryonic cells, in particular the inner cell mass (ICM) of the blastocyst, the embryonic ectoderm or epiblast of the early post implantation embryo, as well as the primordial germ cells (PGCs) of the gonads (Stevens, 1967a). All of these cells are developmentally pluripotent in that they can differentiate into other cell types and tissues, and they all exhibit prominent nucleoli and have a high nuclear to cytoplasmic ratio (Stevens, 1975). When blastocysts, early postimplantation embryos or embryonic gonads were surgically transplanted to sites such as the kidney or testis they formed teratocarcinomas. Furthermore, the clonal transplantation of single EC cells demonstrated that they could give rise to the differentiated cell types found within teratocarcinomas (Kleinsmith and Pierce, 1964). This revealed that EC cells shared many of the developmental and tumor forming capabilities of both early embryonic stages and germ cells.

At the same time, techniques were developed that allowed for cells from one early embryo or blastocyst to be transplanted into another resulting in the formation of a chimera. Such transplantation techniques were instrumental in defining the earliest embryonic lineages of the mouse embryo and showed that cells of the ICM, but not the trophoblast of the blastocyst, contributed to all tissues, including the gametes (Gardner and Johnson, 1972). Trophoblast derivatives were restricted to the placenta. As EC cells showed a strong morphological and developmental similarity to those of the ICM, their developmental potential was assessed by injecting them into blastocysts (Brinster, 1974). When injected or aggregated with preimplantation embryos, EC cells participated in development and formed viable chimeras in which the differentiated derivatives of the injected EC cells contributed to many adult tissues (Papaioannou et al., 1975; Stewart, 1982). Less clear was whether these cells were genetically totipotent, and could contribute to the germ line and form functional gametes. Despite some sporadic reports that EC cells could form functional gametes—initially with cells from ascites tumors maintained in vivo and subsequently with cells maintained for a short term in culture (Mintz and Illmensee, 1975; Stewart and Mintz, 1981)—such claims proved to be elusive and could not be reproduced with many other EC lines (Papaioannou et al., 1978, 1979).

Given the potential of using EC cells either as a ready source to study the differentiation of mammalian embryonic cells in vitro or as a route to introduce novel mutations or mutant genes into the germ line and study their function in vivo, continued efforts were made to derive new EC lines. This potential was realized when the revolutionary discovery was made that cells with similar characteristics could be easily and reproducibly established directly from the ICM of explanted blastocysts (Evans and Kaufman, 1981; Martin, 1981). This advance avoided the tedious process of having to first induce teratocarcinomas, and then explant them in culture to isolate the EC cells. In contrast to EC cells, cell lines derived directly from the ICM readily formed chimeras and most significantly, colonized the germ line producing functional gametes at high and reproducible efficiencies (Bradley et al., 1984). To distinguish them from EC cells, they were called embryonic stem (ES) cells.

1.2. Formation and development of chimeras

Since their discovery, the principal use of ES cells has been as a route of introducing mutated genes into the murine germline, to derive novel lines of mice carrying the mutation and determine the function of the gene in vivo. The success of these lines cannot be overstated with respect to their impact on mammalian genetics. They have provided investigators with a vast array of new lines of mice that are uncovering the functions of genes in development, disease processes and all aspects of mammalian biology. Increasingly sophisticated ways are being devised to regulate the expression of genes or to introduce mutations on a temporal and spatial basis, but these methods will not be covered here, as there are many recent and excellent reviews on the subject (Lewandoski, 2001; Yu and Bradley, 2001).

ES chimeras are produced by either blastocyst injection or aggregation with 4–8 cell stage embryos (Wagner et al., 1985; Stewart, 1993). When aggregated or injected into embryos, ES cells rarely contribute to the trophoblast but do contribute to all other lineages. Whether ES cells can reproducibly give rise to an entire conceptus, as does a transplanted ICM, is less clear. Tetraploid embryos infrequently develop to term (Snow, 1975), however when combined with diploid embryos, the tetraploid derivatives are almost exclusively found in the trophoblast and extraembryonic lineages, with the embryo proper being derived almost entirely from the diploid embryo (in some instances tetraploid derivatives are found in the definitive gut) (James et al., 1995). When ES cells are introduced into tetraploid pre-implantation embryos, many of these embryos develop to term with the newborn being entirely derived from the ES cells (Nagy et al., 1993). Although this method can facilitate the analysis of mutant gene function, the embryos have to be delivered by caesarian and many die shortly after birth. Consequently, such embryos are only produced at low frequencies and only using low passage ES lines. Why ES cells, after few passages in culture cannot fully substitute for the ICM remains unclear. One possibility is that their abnormal development is a consequence of the loss of proper imprinted gene expression during prolonged culture (Dean et al., 1998).

Imprinted genes are a subgroup of mammalian genes in which expression of both parental alleles is not equivalent (for further details see the article by Arney et al in this issue). Depending on the imprinted gene, expression can be either maternal or paternal. If any imbalance arises either by biparental expression or loss of expression from the appropriate parental allele this can result in abnormal and aborted development (Tilghman, 1999). Embryos in which their entire genome is either maternal or paternal fail to develop, and even as chimeras, with normal biparental embryos, their development is severely compromised (Mann et al., 1990). During culture, normal biparental ES cells accumulate changes in imprinted gene expression and it is thought that these alterations contribute to the nonviability of fetuses derived entirely from ES cells. The basis for failure may however be more complex as the majority of these studies have been performed using ES cells from inbred strains. Recent evidence has indicated that ES lines made from F1 hybrid lines of mice are more robust and less prone to forming abnormal embryos, indicating that other inherited factors may be critical (Eggan et al., 2001).

1.3. Embryonic stem cells from other species

With the success of establishing mouse ES cells, many attempts have been made to develop similar stem cells from other mammals. Reports have described the derivation of pluripotent cells with ES morphology from domesticated species such as the sheep, cow and pig (Handyside et al., 1987; Notarianni et al., 1991; First et al., 1994; Chen et al., 1999). Other species including the rat, rabbit, mink and Syrian hamster have also produced such lines (Doetschman et al., 1988; Sukoyan et al., 1992; Graves and Moreadith, 1993; Vassilieva et al., 2000). Although many of these lines differentiate when cultured under the appropriate conditions, chimeras have only been derived from pig, cow and rabbit cells, none of which were chimeric in the germline (Schoonjans et al., 1996; Notarianni et al., 1997; Iwasaki et al., 2000). Much of the motivation for producing ES lines from domesticated animals, which was for their genetic manipulation, has shifted due to the success in cloning by nuclear transplantation of donor nuclei derived from other embryonic or adult cell lines in these species (Wilmut et al., 1997; Lai et al., 2002). Nevertheless, there is still a clear need to produce ES cells from rat embryos, because much of mammalian physiology, such as neuro- and cardiovascular physiology, has been established and continues to be studied using this species, and nuclear transplantation in the rat has been inefficient with no reports of success (Jacob and Kwitek, 2002). Why the derivation of ES cells from other mammalian species has been so difficult is not understood, although genetics may be an important factor. With mice, some inbred strains, such as 129, and to a lesser extent C57Bl6, are easier to use in deriving ES lines than other strains (Brook and Gardner, 1997). Also, species-specific differences in post blastocyst development or the technique of simply explanting embryos in culture, which may be inadequate for establishing ES cells in some species, could contribute to these problems.

However, it is the success in the derivation of ES cells from primate embryos, particularly the human, which has recently attracted much interest. ES cells have been derived from the rhesus and cynomolgus monkeys, common marmoset, and most significantly human blastocysts (Thomson et al., 1995, 1996, 1998; Jacobson et al., 2001; Odorico et al., 2001; Suemori et al., 2001). The cells from all these species have the characteristic high nuclear to cytoplasmic ratio and prominent nucleoli. Primate ES cells differ from those of the mouse in that the colonies they form are flatter, and although they are alkaline phosphatase positive, as are mouse ES lines, they differ in the expression of other markers such as the SSEA antigens and TRA-1 membrane proteins. All stem cell lines depend on mouse embryo fibroblasts to support their proliferation and removal of the feeder layer results in differentiation. Although, primate ES cell renewal is not dependent on Leukemia Inhibitory Factor (LIF), as are mouse ES lines (see below), other, as yet, unidentified factor(s) support their proliferation in the absence of feeder cells (Xu et al., 2001). When injected into mice primate and human ES cells form tumors containing many differentiated cell types, including highly organized structures such as tooth rudiments and intestinal epithelium (Thomson and Marshall, 1998; Amit et al., 2000). Furthermore the primate ES lines can form trophoblast in culture in contrast to mouse stem cell lines. It has not been possible to test their developmental capabilities in chimeras for obvious ethical reasons in humans and for technical and logistical reasons in primates (Thomson and Marshall, 1998).

1.4. Other sources of pluripotential embryonic cells and EG cells

Another embryonic source of pluripotent cells are primordial germ cells (PGCs) that develop into the gametes. In mice PGCs are first detected at E7 in the embryonic ectoderm near the embryonic-extra embryonic boundary. By E8.5-9 the cells have clustered at the base of the allantois at the posterior of the primitive streak and thereafter they mitotically proliferate and migrate to the embryonic gonads. Migration is complete by about E12. In female embryos, once the cells enter the genital ridges, the PGCs initiate meiosis and remain arrested at the first meiotic prophase until puberty. In male embryos, mitotic proliferation of the PGCs in the gonads is arrested until after birth (McLaren, 1992).

When the posterior region and base of the allantois from E8.5-9 embryos were explanted and cultured in medium supplemented with basic Fibroblast Growth Factor (FGF2), Kit ligand (stem cell growth factor), and LIF, compact alkaline phosphate positive colonies that morphologically resembled ES cells were formed. These cells were capable of undergoing extensive differentiation in vitro, and produced teratocarcinomas when injected into mice. They were called EG cells to distinguish their origins from ES cells (Matsui et al., 1992; Resnick et al., 1992).

Proof of the developmental pluripotency of EG lines was demonstrated by their injection into blastocysts and formation of germ line chimeras, which were indistinguishable from those produced by ES cells (Labosky et al., 1994; Stewart et al., 1994). EG lines have also been derived from later stage PGCs that had entered the genital ridge. Such lines are produced at a lower frequency, as these older PGCs are apparently more difficult to establish in culture. Although the later stage EG lines form chimeras, the frequency of chimeras born is low and many exhibit skeletal malformations resembling those found in chimeras produced from androgenetic ES cells or embryos (Mann et al., 1990; Labosky et al., 1994; Tada et al., 1998). One explanation for this difference may be that in the later stage EG lines, imprinted genes have started to undergo "reprogramming" in the PGCs at this stage. Most evidence indicates that for the majority of imprinted genes their future pattern of parental expression is established in the germ cells. Consistent with this observation are the findings that DNA methylation patterns and the expression of some imprinted genes change in the PGCs, once they enter the embryonic gonad (Szabo and Mann, 1995). Also in heterokaryons made between ES cells and T lymphocytes, the resulting hybrid cells resemble ES cells, with T cell specific gene expression being extinguished and ES specific gene expression being induced from the T cell genome. Yet, imprinted gene expression is unchanged in these hybrids. In contrast, heterokaryons made between an EG line of E12.5 origin and T cells, exhibited both altered developmental and imprinted gene expression (Tada et al., 2001). Thus, EG lines isolated from mouse fetal gonads may be able to affect imprinted gene expression, unlike EG lines isolated from earlier stage PGCs, and the effect on imprinted gene expression may contribute to their defective development in chimeras. This area warrants further investigation as the studies on the later stage EG lines have been primarily based on one line, and there are important implications for the potential of human EG lines for therapeutic use.

The derivation of EG cell lines in the mouse was also significant because it revealed an alternative route to isolating pluripotential cell lines from domesticated species, where the isolation of ES cells has been difficult or ineffective. EG lines have been reportedly established from the pig and chicken, although the criteria for their existence has been primarily morphological (Park and Han, 2000; Wheeler and Walters, 2001). The EG lines form chimeras when injected into recipient embryos, although none of the EG lines were clonal in origin and none of the chimeras produced offspring derived from the EG lines. In the zebrafish, cell lines expressing the evolutionarily conserved germ cell marker *vasa* have also been established from early embryos. These zebrafish cells can differentiate in vitro, and form chimeras when injected into recipient embryos, in which a small percentage did transmit the donor cell genome to their offspring. These cells therefore offer a potentially promising alternative route to manipulation of the zebrafish genome, although the cells have a limited lifespan (~3 weeks) in culture restricting their genetic use (Ma et al., 2001).

Human EG (hEG) lines have also been established, providing another source of pluripotential cell lines in addition to ES (hES) cell lines (Shamblott et al., 1998). As they are established from aborted fetuses, the hEG cells may avoid some of the political and ethical issues that make the isolation of human ES cells controversial. However, hEG cells show morphological differences when compared to the human ES cells in that they grow as tight clumps or mounds of cells whereas the hES cells form flatter colonies. As with ES cells hEG cell growth is dependent on a feeder layer of mouse fibroblasts. Unlike hES cells, hEG cells require LIF to sustain their proliferation and are difficult to dissociate into single cells, which has precluded their clonal analysis as stem cells. hEG cells have not yet been reported to form teratocarcinomas (or any tumor) when injected into mice, and they differ in expression of SSEA antigens and TRA-1 markers proteins. Human EG cells spontaneously form embryoid bodies (EBs) at a low frequency (1–5%) and differentiated lines (non clonal) derived from these EBs express markers characteristic of neuronal, muscular, and vascular endothelial lineages (Shamblott et al., 2001).

2. ES cell self-renewal and the requirement for LIF

The establishment of ES and EG lines depends on a feeder layer of fibroblasts, as does the maintenance of their undifferentiated proliferation or self-renewal. This dependence on a feeder layer prompted the search for what factor(s) produced by the fibroblasts maintains the self-renewal of EC/ES cells. Initial efforts showed that medium conditioned by feeder cells sustained EC proliferation in the absence of the feeders and that this activity was due to a secreted polypeptide with a MW of 57 kD (Koopman and Cotton, 1984). Subsequent studies using medium conditioned by other cell lines, including Buffalo Rat Liver and 5637 human bladder carcinoma cells, were also effective at maintaining ES cells in the absence of feeders. These studies led to the identification of a previously identified cytokine, Leukemia Inhibitory Factor (LIF), as the factor essential to sustaining ES proliferation and inhibiting their differentiation (Smith et al., 1988; Williams et al., 1988).

LIF was first characterized by its ability to simultaneously induce the differentiation and inhibit the proliferation of the myeloid leukemia cell line M1. LIF is a secreted protein with a MW of 19 kDa and when produced by cells, it is glycosylated, increasing its MW to 40~60 kDa. It is a member of the interleukin-6 (Il-6) family of cytokines, which include Oncostatin M (OSM), Interleukin-11 (Il-11), Cardiotrophin-1 (CT-1) and Ciliary Neurotrophic Factor (CNTF) (Auernhammer and Melmed, 2000). All share a similar peptide structure, an up-up-down-down four-helix bundle (Robinson et al., 1994). LIF is expressed among a wide variety of eutherian mammals, including humans, and recently cDNAs sharing sequence homology have been identified in marsupials and monotremes (Cui et al., 2001).

LIF deficient fibroblasts are unable to support ES proliferation, and supplementing the culture medium with recombinant LIF, maintains ES self-renewal in the absence of feeder cells (Stewart et al., 1992). The fact that ES cells express the αCTNF receptor and both CT-1 and hOSM bind to the murine LIFrβ explains why CT-1 and hOSM are also effective at maintaining ES cell proliferation (Yoshida et al., 1994; Pennica et al., 1995). Neither IL-6 alone nor Il-11 inhibit ES differentiation as these cytokines both bind to their own specific receptors which are not expressed in ES cells.

LIF binds to cells via the LIF receptor, a heterodimeric complex consisting of the LIF receptor beta (LIFrβ-sometimes also called the alpha receptor) and gp130. LIFrβ and gp130 are both transmembrane receptors with glycosylated molecular weights of ~190 and 130 kDa respectively, and each protein contains a single transmembrane domain (Taga and Kishimoto, 1997). The participation of gp130 in receptor complexes is common to all the receptors of the Il-6 family of cytokines (Heinrich et al., 1998). CNTF also utilizes LIFrβ, but in addition requires the αCNTF receptor, forming a trimeric receptor complex. The cytoplasmic domains of LIFrβ and gp130 show significant homology, and although LIFrβ homodimers can function to inhibit ES differentiation, heterodimers between LIFrβ and gp130 are more effective at restricting differentiation (Starr et al., 1997). Moreover gp130 is itself sufficient, if stimulated with recombinant Il-6 covalently linked to a soluble form of the Il-6 receptor, to sustaining ES cell proliferation (Nichols et al., 1994).

Neither the LIFrβ nor gp130 has intrinsic kinase activity. Instead kinase activity is mediated by the recruitment of JAK kinases, nonreceptor cytoplasmic tyrosine kinases. These kinases associate with the proline rich boxes 1 and 2, regions that are located near the inner cell membrane cytoplasmic domain of the receptors. In mammals at least 4 members of the JAK kinase family (JAKs1-3 and TYK2) have been identified, and JAKs1 and 2, as wells as TYK2 associate with gp130 (Darnell et al., 1994; Imada and Leonard, 2000). The JAKs are relatively large kinases with molecular weights of ~120–140 kDa.

They contain two kinase domains, a functional JH1 domain and a "pseudokinase" domain JH2. Their name Janus kinase originates from the dual kinase faces of the proteins, although a more irreverent term is "just another kinase". Of the 3 JAKs that associate with gp130, gene deletion experiments revealed that only *Jak1* is essential to gp130 signaling in ES cells (Rodig et al., 1998).

The binding of LIF to the LIFrβ enhances LIFrβ's interaction with gp130 resulting in a receptor complex that converts the already associated, but inactive, JAKs to activated tyrosine kinases (Fig. 1). JAK activation occurs by either transphosphorylation or autophosphorylation. With transphosphorylation, the now juxtaposed JAKs cross phosphorylate each other, whereas with autophosphorylation, binding of ligand releases the JAKs from an autoinhibited state resulting in their self-phosphorylation. The activated JAKs phosphorylate at least five tyrosine residues in the cytoplasmic domains of the receptors (Heinrich et al., 1998). Close to boxes 1 and 2 in the receptor is a critical tyrosine (Y757 in gp130 (Saito et al., 1992) and Y769 in the LIFrβ (Tomida et al., 1994) in the mouse) having a YxxV motif (where Y is followed by any two amino acids and valine V). These tyrosines, when phosphorylated, recruit the SHP2 phosphatase and Suppressor of

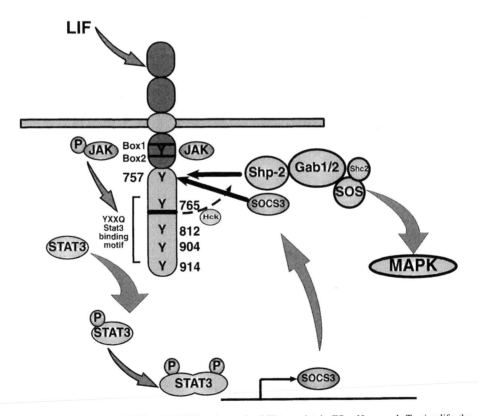

Fig. 1. Activation of the STAT3 and MAPK pathways by LIF to maintain ES self-renewal. To simplify the diagram only part of gp130 is shown as the receptor.

Cytokine Signaling 3 (SOCS3), proteins that have significant roles in regulating intracellular signaling (Nicholson et al., 2000; Schmitz et al., 2000). In addition to Y757, there are four other tyrosines with a YxxQ motif located distally at positions Y765/812/904//914 in mouse gp130 and three similar tyrosines at Y976/996/1023 in the LIFrβ. Phosphorylation of these residues by the JAKs, results in the recruitment of transcription factors, the Signal Transducer and Activators of Transcription (STATs) (Stahl et al., 1995).

The STATs are known as latent transcription factors, which are inactive when they are located in the cytoplasm. In mammals seven members have been identified, STATs1–4, 5a, 5b, 6 and 7. They all have a similar structure with three highly conserved domains, a DNA binding domain, an SH2 domain with an adjacent tyrosine, which when phosphorylated is essential for STAT dimerization, nuclear translocation and activation (Ihle, 2001; Williams, 2000).

In ES cells STAT3 is the predominantly expressed member, and although a minor version has been described (STAT3b) its function is uncertain (Schaefer et al., 1995). STAT1 is also expressed at low levels in ES cells, but is inessential as ES cells lacking a functional STAT1 gene are indistinguishable from normal cells (Niwa et al., 1998). Phosphorylation of the YxxQ tyrosines in the cytoplasmic tails of gp130 and LIFrβ results in the recruitment of STAT3, via its SH2 domain, to the cytoplasmic domains of the receptors bringing the STATs close to the JAKs. The JAKs then phosphorylate STAT3, resulting in dimerization of the STAT3s by reciprocal recognition by the SH2 domains of the phosphorylated tyrosines. The STAT3 dimers translocate to the nucleus, where they bind to specific STAT enhancer binding nuclear elements.

Self-renewal and maintenance of the stem cell phenotype of ES cells is dependent on STAT3 activation as revealed by four different experimental approaches. Gene targeting of STAT3 showed it was not possible to derive ES clones lacking STAT3. ES cells are also sensitive to STAT3 gene dosage, with a reduction in the number of functional alleles affecting the self-renewal of some ES lines (Raz et al., 1999). In contrast over or prolonged expression of STAT3 retards ES cell proliferation. A dominant negative form of STAT3, STAT3F which cannot dimerize, when expressed at sufficient levels in ES cells blocks their self-renewal (Niwa et al., 1998). Mutational analysis of the four distal tyrosine residues in the cytoplasmic domain of gp130 revealed that the two most distal tyrosines were essential to STAT3 activation and ES self-renewal (Niwa et al., 1998; Ernst et al., 1999). And lastly, expression of a STAT3-estrogen receptor fusion protein in ES cells demonstrated that STAT3 alone was sufficient to maintain self-renewal and inhibit differentiation in the absence of LIF (Matsuda et al., 1999). The STAT3 fusion protein could be phosphorylated and activated in ES cells only when estrogen was administered.

The other signaling pathway activated by LIF in ES cells is the Mitogen Activated Protein Kinase pathway (MAPK), a ubiquitous pathway that is stimulated by a multitude of external stimuli (Roovers and Assoian, 2000). In non ES cell types, such as 7-TD1 cells and the pro B cell line Ba/F03, gp130 mediated activation of MAPK is essential to maintaining their proliferation. In the neuronal precursor cell line PC12, Nerve Growth Factor induced MAPK activation promotes PC12 differentiation. Activation of STAT3, by gp130 counteracts the MAPK pathway and retards their differentiation revealing cross-talk between the two signaling pathways (Ihara et al., 1997). In ES cells, MAPK is rapidly induced following LIF stimulation, and mutation analysis revealed that the tyrosine Y757 was essential to activation of this pathway by acting as a docking site for the SHP2 phosphatase (Ernst et al., 1996; Burdon et al., 1999). Recruitment of a phosphatase, which dephosphorylates tyrosines, may counteract the kinase activity of the JAKs and reduce their activation and/or phosphorylation of the STAT binding sites so inhibiting STAT activation. However, it is unclear if such a negative regulatory mechanism operates in ES cells. SHP2 also acts as a docking/recruitment factor for other proteins involved in the activation of the MAPK pathway, including Gabs 1 and 2 and Grb2. Mutation analysis of Y757 revealed that it is not essential to STAT3 phosphorylation in ES cells or to STAT3 activation in vivo (Burdon et al., 1999; Ohtani et al., 2000). Rather STAT3 activity is enhanced in mutants of Y757, and lower levels of phosphorylated STAT3 are required to sustain ES self-renewal due to an increased half-life of the phosphorylated STAT3. Such an effect suggested that MAPK antagonizes the activation of STAT3 in ES cells. Expression of a dominant negative version of SHP2 that blocks the recruitment of endogenous SHP2 and treatment of ES cells with specific MAPK inhibitor, further supported the inhibitory effects of MAPK on STAT3 activity (Burdon et al., 1999). However, a complicating factor is that Y757 is also essential in recruiting SOCS3. SOCS3 serves as a negative feedback regulator, inhibiting STAT3 activation by the JAKs and SOCS3's expression is induced by STAT3 transcription. SOCS3 is expressed in ES cells (Duval et al., 2000) and mutation of Y757 may well prevent SOCS3 binding to gp130, which would prolong STAT3 activation (Burdon et al., 1999; Nicholson et al., 2000). Another factor, the *src* related tyrosine kinase *hck*, which interacts with the cytoplasmic domain of gp130, has also been implicated in regulating LIF responsiveness in ES cells with a hyperactivated form of *hck* reducing ES cell renewal dependence on LIF (Ernst et al., 1994). Clearly, the regulation of STAT3 activation in ES cells is a complex process involving cross-talk between different signaling pathways, with an extensive variety of feedback loops regulating STAT3 levels. Too little STAT3 and ES cells differentiate, too much STAT3 and ES cell proliferation is inhibited (Burdon et al., 1999).

How STAT3 regulates ES cell self-renewal is unclear. It may do so by regulating genes critical to differentiation, to regulating proliferation, or both. STAT3 can have a significant growth promoting effect on cells as constitutive STAT3 expression results in conversion of an immortalized fibroblast cell line to the malignant state (Bromberg et al., 1999). Furthermore, STAT3 regulates genes involved in cell cycle progression and apoptosis such as c-myc, cyclin D1, Bcl2, Bcl-xL, and the cdk inhibitor p21 (Bellido et al., 1998; Kiuchi et al., 1999; Shirogane et al., 1999; Bowman et al., 2000). ES cells however have an unusual cell cycle compared to differentiated cells such as fibroblasts, with a short or non-existent G1 phase, and virtually undetectable cyclin Ds and Cyclin D1/CDK4 activity (Savatier et al., 1994, 1996). Furthermore proliferation of ES cells deficient for all three retinoblastoma (RB) or pocket proteins that regulate G1 to S transition is unaffected, although ES cell differentiation is impaired (Dannenberg et al., 2000). How these differences in cell cycle parameters are established in ES cells and whether STAT3 is required in their regulation remains to be determined.

One of the original goals in identifying which factors regulated ES cell differentiation was to determine whether these same factors regulated embryonic development. LIF transcripts are detected in the trophoblast of the blastocyst, and the ICM expresses the LIFrβ and gp130 (Nichols et al., 1996). Surprisingly embryos deficient for any of these components, including STAT3, either remain viable (LIF), or all develop beyond the blastocyst before dying at later stages in embryogenesis (gp130, LIFrβ, STAT3). LIF deficiency results in female sterility, inasmuch as LIF is required to induce the uterus to become receptive to the implanting blastocyst (Stewart et al., 1992). In contrast mice deficient for gp130 die late in gestation (Yoshida et al., 1996), and LIFrβ deficiency results in perinatal death (Ware et al., 1995). STAT3 deficient embryos die shortly after implantation around E7 (Takeda et al., 1997). These observations reveal that the ICM or epiblast, from which ES cells are established, does not depend on LIF signaling pathways during normal development in vivo. ES cell dependency on LIF would appear to be an adaptation by the epiblast cells to proliferating in vitro. Recent evidence has however suggested that LIF signaling may be essential to the long-term viability of the ICM/epiblast (Nichols et al., 2001). Some 40 mammalian species, including mice, exhibit delayed implantation. Delayed implantation is a reproductive adaptation that allows the mother to arrest embryonic development so as to coincide the birth of the embryo with optimal environmental conditions favoring the newborn's survival. In most species embryonic delay occurs at the blastocyst stage just prior to the onset of implantation. In delayed embryos cell proliferation ceases, the embryos become metabolically quiescent, and depending on the appropriate hormonal milieu, this condition can be maintained for many weeks (Renfree and Shaw, 2000). Viability of the ICM cells in these quiescent blastocysts does however depend on their possessing a functional gp130/LIFrβ gene; delayed embryos lacking functional genes for these components show a significant reduction in ICM cell numbers and lower viability when implantation is renewed compared to wild-type embryos. The LIF signaling pathway may therefore be required to sustain ICM viability in species undergoing implantational delay. LIF dependent ES cell proliferation in vitro may have arisen as a consequence of this cell survival mechanism. This correlation would be consistent with several observations, for example, that STAT3 activates anti-apoptotic factors such as Bcl-2 (Narimatsu et al., 2001). Also primate ES cell lines may not require LIF, for there is no evidence that delayed implantation occurs in these species.

An alternative possibility to this seeming paradox is that more than one signaling pathway may function in maintaining ES self-renewal. Mouse ES cells can be maintained in short term culture for about a week in the complete absence of LIF in conditioned medium prepared from LIF deficient differentiated cells. The factor(s) responsible, called ES Renewal Factor (ESFR) have not yet been identified but is a soluble, trypsin sensitive macromolecule that does not act on ES cells by activating STAT3 (Dani et al., 1998). Whether ESFR or a similar activity acts to maintain hES proliferation remains to be determined. Clearly, more work is required to fully establish what external factors pluripotential ES cells from different species require and which signaling pathways are needed to sustain their self-renewal.

3. Oct4 and other intrinsic factors that maintain the ES cell phenotype

Stem cell pluripotency of ES cells is also dependent on the POU transcription factor OCT4 (or Oct3/4), encoded by the gene *Pou5f1*. The POU (Pit-Oct-Unc) family members are defined by a 150–160 amino acid region common to the mammalian transcription factors Pit-1 and Oct-1, as well as the nematode factor Unc-86. The POUs have a bipartite DNA binding domain consisting of two subdomains (a POU homeobox and an upstream POU domain) connected by a flexible linker. The POU domain and inherent flexibility of the linker allows these factors to interact in several configurations as monomers or as dimers with different coactivators/repressors (Phillips and Luisi, 2000).

Interest in OCT4 was stimulated by the observation that it was expressed in EC cells and that expression declined with EC differentiation (Okamoto et al., 1990). During embryogenesis OCT4 is first detected in the oocyte and then in the 8-cell stage morula. In the mouse blastocyst, OCT4 is localized to the ICM, though in the cow and human blastocysts it is also found in the trophoblast (Hansis et al., 2000; Kirchhof et al., 2000). In early post-implantation embryos, OCT4 expression is restricted to the embryonic ectoderm and subsequently is detectable only in the PGCs (Scholer et al., 1990). It is expressed in both human and mouse ES lines (Pesce et al., 1998; Reubinoff et al., 2000). With its expression being largely confined to these cell types, OCT4 made a good candidate for being important in establishing pluripotency. This hypothesis was experimentally supported by the derivation of mouse embryos lacking OCT4, in which blastocysts did not form a recognizable ICM and consisted only of trophectoderm (Nichols et al., 1998). Additional support for OCT4 having a critical role in regulating pluripotency came from the elegant experimental manipulation of OCT4 levels in murine ES cells. Whereas a 50% reduction in normal levels of OCT4 was sufficient to maintain a stem cell phenotype, complete loss of OCT4 resulted in the ES cells forming trophectoderm-like cells. In contrast, a two-fold increase resulted in the ES cells forming endoderm and mesoderm (Niwa et al., 2000). The levels of OCT4 expression in ES cells were therefore critical in determining their state of cell differentiation (Fig. 2). However OCT4 is necessary but not sufficient to maintain an ES phenotype. ES cells in which STAT3 was reduced, but OCT4 levels were maintained at normal levels, also underwent differentiation revealing that the stem cell phenotype and self-renewal of ES cells requires both STAT3 and OCT4 (Niwa et al., 2000).

The regulation of OCT4 levels and, in turn, which genes OCT4 regulates to maintain ES pluripotency are areas of intense interest. Upstream of the OCT4 transcriptional start site is a region containing two enhancers, the proximal (PE) and distal (DE) (Yeom et al., 1996). The DE enhancer is required for OCT4 expression in ES, ICM and PGC cells, whereas the PE functions in the embryonic ectoderm and EC lines. In addition to the enhancer sites, the proximal promoter is occupied by transcription factors in undifferentiated EC and ES cells, with OCT4 activation in EC cells probably being regulated by the ubiquitous transcription factor Sp1 (Minucci et al., 1996; Pesce et al., 1999). Within the proximal promoter region is a hormone responsive element (HRE) that is recognized by receptors belonging to the steroid-thyroid hormone family, and retinoid and retinoic acid receptors and is also recognized by several members of the orphan nuclear hormone receptor superfamily (Pikarsky et al., 1994; Schoorlemmer et al., 1994; Sylvester and Scholer, 1994). In P19 EC cells the proximal promoter region is occupied by the Undifferentiated Cellular Factor (UCF) complex, which by implication, maintains, with Sp1, OCT4 expression in the EC cells and also contains the orphan nuclear receptor, steroidogenic factor 1 (Sf-1) (Fuhrmann et al., 1999; Barnea and Bergman, 2000). Following retinoic acid (RA) induced differentiation, UCF is replaced by another complex, the Transiently RA-induced factor (TRIF), coinciding with the down regulation of OCT4 expression. The TRIF

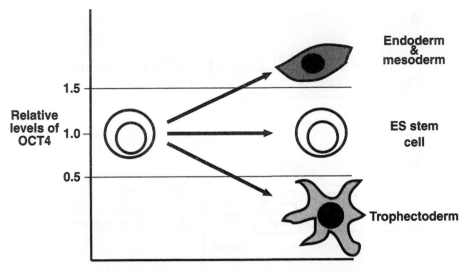

Fig. 2. The levels of OCT4 in ES cells determines their differentiation pathway. Loss of OCT4 results in the formation of cells with trophectoderm characteristics.

complex consists of many proteins, among which is another orphan nuclear receptor, Germ Cell Nuclear Factor (GCNF) (Fuhrmann et al., 2001). OCT4 and GCNF show reciprocal patterns of expression during EC differentiation and overexpression of GCNF inhibits OCT4 expression. In vivo GCNF deficient embryos die due to cardiovascular problems, and the embryos show an expanded expression of OCT4, which is no longer restricted to the PGCs (Fuhrmann et al., 2001). The TRIF complex may itself then be replaced by another complex containing additional nuclear receptors, among which are COUP-TFI and II that may provide a more permanent repression of OCT4 in later stages of differentiation. Overall, the evidence suggests that it is the silencing of OCT4 expression during development that determines its temporal and spatial transcriptional activity (Fig. 3).

Considerable efforts have been expended in determining which genes OCT4 regulates in early development (Du et al., 2001). OCT4 consensus binding sequences have been localized to many different genes, some of which are listed in Table 1. OCT4 can both induce and silence gene expression by itself or by selective association with specific coactivators/repressors. OCT4 represses human chorionic gonadotrophin α and βsubunit genes in human choriocarcinoma by binding directly to the promoter (Liu et al., 1997). It also silences interferon-tau expression in bovine trophoblast, but by a different mechanism. In the latter tissue OCT4 acts by "quenching" Ets-2 trans-activation of the Inf-τ promoter, rather than by directly binding to Inf-τ promoter sequences (Ezashi et al., 2001). Thus, as the trophoblast differentiates OCT4 levels fall, relieving its inhibitory effects on expression of these trophoblast specific genes.

OCT4 is also required to induce and maintain expression of certain genes. Probably the most biologically relevant gene regulated by OCT4 is the Fibroblast Growth Factor 4 (*Fgf4*) that is produced in the ICM and supports, in a paracrine manner, trophectoderm proliferation in the blastocyst (Nichols et al., 1998). The *Fgf4* gene contains an OCT4

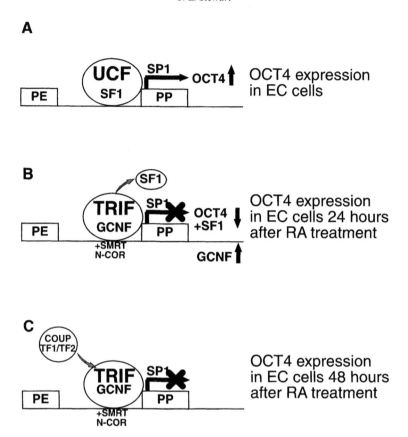

Fig. 3. OCT4 regulation in EC cells after retinoic acid induced differentiation. (A) In untreated EC cells OCT4 is expressed under control of SP1 ubiquitous transcription factor and the UCF complex containing SF-1. (B) Within 24 h of RA treatment UCF is replaced by the TRIF complex containing the GCNF cofactor, as well as the SMRT and N-COR corepressors. OCT4 and SF1 levels decline as GCNF levels increase. (C) By 48 h OCT4 expression is completely suppressed and additional repressor factors COUPsTF1 and 2 are associated with the promoter.

binding site in its 3′ untranslated region and OCT4 is essential for FGF4 expression in ES cells. Adjacent to the *Fgf4* site is another regulatory unit for the high mobility group (HMG) protein, Sox-2 that belongs to the *Sry* family of proteins. OCT4/Sox2 synergy is required for the appropriate expression of FGF4 because absence of either factor results in loss of FGF4 expression (Ambrosetti et al., 1997). A similar dependency between the two factors exists in positively regulating the expression of another transcriptional coactivator *Utf1* that also has an OCT4 binding site in its 3′ UTR (Nishimoto et al., 1999). Not all genes are coregulated by OCT4/Sox2. Osteopontin (*Opn*) expression is induced in EC cells by an OCT4 dimer when cells differentiate into endoderm. However, when OCT4 associates with Sox2, the heterodimeric complex inhibits *Opn* expression by preventing OCT4 dimerization (Botquin et al., 1998) (Table 1).

Table 1
Genes regulated by OCT4 and associated coactivators/repressors

Genes upregulated		Genes suppressed	
Gene	Coactivator	Gene	Corepressor
Fgf4	Sox-2	*hCG-α*	–
Utf-1	Sox-2	*hCG-β*	–
Zfp42/Rex-1	Rox-1	*INF-τ*	Ets-2
Opn	–	*Opn*	Sox-2
Pdgf-α receptor	?	*Cdx-2*	?
Otx-2	?	*Hand-1*	?
Lefty-1	?		
Upp	?		
Terra/226	?		

Thus, maintenance of the ES stem cell phenotype and the changes in gene expression that are initiated following the onset of differentiation are mediated by the levels of OCT4 and its interactions with other transcriptional cofactors such as *Sox-2* and proteins such as HMG 1 and 2 (Butteroni et al., 2000). How and whether STAT3 interacts with OCT4 is not clear despite the fact that ES cells lose their stem cell phenotype when expression of both factors is silenced.

OCT4 may also be acting within the context of more global differences in chromatin organization and transcriptional regulation in ES cells compared to their differentiated derivatives. Histone hypoacetylation, DNA hypermethylation and gene repression are all interlinked (Bird and Wolffe, 1999). ES cells, compared to their differentiated derivatives are remarkably tolerant to the loss of all three DNA methyltransferases, the enzymes responsible for the *de novo* and maintenance methylation of DNA (Okano et al., 1999; Jackson-Grusby et al., 2001). Also ES/EC cells, as well as cells of the pre-implantation embryo, possess a powerful mechanism that silences post-infection retroviral gene expression after infection and integration of the viral genome into the host's DNA (Jahner et al., 1982; Stewart et al., 1982). Whether this silencing evolved solely as a mechanism of genome defense to suppress the potentially deleterious consequences of random retrotransposon insertion into the genome during development is still not clear (Bestor, 2000). Recent evidence has revealed that such a powerful suppressive mechanism with regard to retroviral elements is also found in *Drosophila*, an organism with no significant DNA methylation (Pannell et al., 2000). This observation suggests the existence of an evolutionarily conserved regulatory mechanism with a more fundamental role in regulating gene expression at the start of embryogenesis besides the global function of genome defense.

4. In vitro differentiation of ES cells

One of the original goals in deriving EC/ES cells was to establish a system by which mammalian differentiation could be studied in vitro. Studies with chimeric mice suggested this would be a reasonable objective because, with the exception of trophoblast, ES cells are capable of forming all cell types in embryonic and adult mice. In vitro differentiation has however proven to be more difficult than originally anticipated and has not provided as many insights into mammalian development as initially hoped. ES cell differentiation has been useful in revealing which gene products are essential to the differentiation of certain lineages. For instance ES lines null for the SCL/Tal-1 transcription factor are defective in hematopoiesis (Porcher et al., 1996). Furthermore, null ES lines have also been useful in determining the role of specific genes in later stages of development where the null genotype results in early embryonic lethality. Beta-1 integrin deficient embryos die at the blastocyst stage (Stephens et al., 1995), and ES cells null for β-1 integrin were used to determine the requirement of the gene in cardiomyocyte and keratinocyte differentiation (Fassler et al., 1996; Bagutti et al., 2001). However, interest in directing ES cell differentiation in culture has recently received a considerable stimulus due to the establishment of hES and hEG cells that offer the potential of providing cells that could have therapeutic use in the treatment of diseases such as type 1 diabetes and neurodegenerative diseases such as Parkinson's.

The principal difficulty has been that although ES cells can be maintained as a homogeneous population, making them differentiate into an equally homogeneous population from a particular cell lineage has not been attainable. The protocols used to date result in ES cells consistently differentiating into a heterogeneous mixture of cell types, many of which are poorly characterized. Often it is not known to which embryonic cell the differentiated cells correspond. Differentiation of ES cells is usually induced by removing them from the feeder layer, or source of LIF, and then culturing them in suspension or in hanging drops of media. The ES cells aggregate to form a ball, on the surface of which appears, after 3–4 days, a "rind" of endoderm. These structures are embryoid bodies (EBs), and if maintained the cells continue to proliferate, with the embryoid bodies increasing in size and often forming large fluid filled sacs. These sacs consist of many differentiated cells often with visible blood filled capillaries (Doetschman et al., 1985). If the embryoid bodies are then allowed to attach to the surface of a culture vessel, then additional differentiated cells, including skeletal muscle, beating heart muscle, epithelial and neuronal cells appear from the outgrowths. Such cultures are inevitably highly heterogeneous with regard to cell types formed. Intriguingly, as the embryoid bodies differentiate, they recapitulate the temporal pattern of expression of different genes that occurs during post-implantation embryonic development (Leahy et al., 1999).

Recapitulation of the developmental program is also seen in one of the better characterized differentiation systems, hematopoiesis, in which the order and time in which the various hematopoietic cells appearing during EB differentiation closely parallels, in a highly reproducible and consistent manner, that observed during embryogenesis (Keller et al., 1993). The effectiveness of this system was such that it resulted in the identification and characterization of the long sought hemangioblast, the hematopoietic stem cell that gives rise to vascular endothelial cells, erythroblasts and various other definitive hematopoietic precursors (Choi et al., 1998). These observations on EB differentiation suggest that either the patterns of gene expression and cell differentiation proceed according to some temporal "program" or that appropriate cell-to-cell interactions are established in the EBs that initiate and regulate these changes or combination of both (Table 2).

Many approaches have been employed to drive the differentiation of ES cells preferentially into one lineage. Most of these endeavors involved treating EB cultures with varying

Table 2
Cell types derived from differentiating mouse and human ES cell lines

Cell type	Reference
Adipocyte	(Dani, 1999)
Astrocyte	(Angelov et al., 1998; Brustle et al., 1999)
Cardiomyocyte	(Klug et al., 1996)
Chondrocyte	(Kramer et al., 2000)
Dendritic cell	(Fairchild et al., 2000)
Endothelial	(Yamashita et al., 2000)
Definitive hematopoietic	(Kennedy et al., 1997; Perlingeiro et al., 2001)
Keratinocyte	(Bagutti et al., 1996)
B Lymphocyte	(Cho et al., 1999; Nakano et al., 1994)
T lymphocyte	(Potocnik et al., 1994)
Mast	(Garrington et al., 2000)
Hepatocyte	(Jones et al., 2002)
Neuron (CNS)	(Lee et al., 2000)
Neuron (Dopaminergic)	(Kawasaki et al., 2000; Lee et al., 2000)
Oliogodendrocyte	(Brustle et al., 1999; Liu et al., 2000)
Osteoblast	(Buttery et al., 2001)
Pancreatic islets	(Lumelsky et al., 2001)
Smooth muscle	(Yamashita et al., 2000)
Skeletal muscle	(Myer et al., 1997)
Human cardiomyocyte	(Kehat et al., 2001)
Human neuronal	(Zhang et al., 2001)
Human hematopoietic	(Kaufman et al., 2001)
Embryonic endoderm	(Doetschman et al., 1985; Levinson-Dushnik and Benvenisty, 1997)
Embryonic mesoderm	(Doetschman et al., 1985; Johansson and Wiles, 1995)

times and doses of retinoic acid (RA). The EBs are then allowed to attach in a culture vessel, and the cells that outgrow are treated with different growth factors or combination of factors to enhance the proliferation of the desired cell type. In this manner RA treated EBs, either cultured in the absence of any additional factor or with bFGF were enriched for neurons (Fraichard et al., 1995), whereas, if the EB outgrowths are treated with insulin and triodothyronine then the cultures predominantly form adipocytes (Dani, 1999). Even with these enrichment procedures the resulting cultures consist of many cell types. An alternative has been to isolate the desired cell type from the cultures using lineage specific genes. Selection has been effective at isolating cardiomyocytes and neuronal precursors from EB cultures using ES cell lines expressing a selectable gene under control of either a cardiomyocytes or neural plate specific promoter (Klug et al., 1996; Li et al., 1998).

A promising alternative to inducing ES differentiation, without forming EBs, has been to grow the ES cells as a low-density monolayer. In combination with serum free media (thereby removing exogenous growth factors) a small percentage of ES cells spontaneously form neurospheres, which allowed for the determination of which growth factors may influence some of the earliest stages of mammalian neurogenesis (Tropepe et al., 2001). In another system, ES cells were cocultured as a monolayer with a mesenchymal cell line, resulting in more than 90% of the ES cell forming dopaminergic neurons. The fact that the activity from the mesenchymal cells was soluble suggested the possibility that it could be isolated and used to treat ES cells directly (Kawasaki et al., 2000).

Similarly culturing ES cells on a collagen substrate with specific growth factors, together with specific cell selection, resulted in the efficient isolation of vascular progenitor cells that would form both endothelial and vascular smooth muscle cells capable of vascularization (Yamashita et al., 2000). Increased use will be made of these simplified culture systems that will lead to a clearer understanding as to which factors are required to induce ES cells to form particular lineages. Ultimately, such systems should provide a more uniform source of material for addressing the molecular mechanisms of mammalian differentiation.

Much of the impetus for improving in vitro differentiation procedures has been to develop cells that can be transplanted back into individuals for therapeutic purposes. To date many of these studies have shown that ES derivatives, such as cardiomyocytes, neuronal cells, vascular progenitors and insulin secreting pancreatic B cells (Klug et al., 1996; Kawasaki et al., 2000; Soria et al., 2000; Yamashita et al., 2000) can reintegrate back into the appropriate tissues following transplantation. The one exception has been hematopoietic cells which have not been able to provide long term reconstitution of the hematopoietic system in mice, possibly because the hematopoietic stem cells have to go through additional maturation steps which have not yet been reproduced in culture (Lacaud et al., 2001). Preliminary results have also come from inducing hES cells to differentiate in culture, including those of the hematopoietic and neuronal lineages (Kaufman et al., 2001; Zhang et al., 2001). Furthermore, the human neuronal cells were able to incorporate into neonatal mouse brains following transplantation. The prospect of using hES cells for therapeutic purposes is still in its infancy and many problems will need to be overcome if they are to be effective (Odorico et al., 2001; Smith, 2001).

5. Conclusions

Much needs to be done to understand the mechanisms utilized whereby ES cells differentiate into specific lineages. In murine ES cells the JAK-STAT3 signaling pathway and the OCT4-POU transcription factors are essential to ES cell self-renewal and maintaining the undifferentiated phenotype. Future studies will determine whether these pathways and transcription factors will also be required to sustain the proliferation and phenotype of other stem and progenitor cell types, such as the recently isolated mesenchymal progenitor (Reyes et al., 2001), as well as those found in bone marrow (Lagasse et al., 2000) and adult mouse brain (Clarke et al., 2000). It is particularly intriguing that the JAK-STAT signaling pathway is essential to sustain germ line stem cells in the male *Drosophila* (Kiger et al., 2001; Tulina and Matunis, 2001).

As for inducing differentiation into specific lineages, the observations of Tropepe and colleagues (Tropepe et al., 2001) that a very low percentage of ES cells (0.2%) form neurospheres suggests that differentiation into some lineages may be initially generated by a stochastic mechanism. Subsequently, particular growth factor/cytokine combinations then act to sustain the growth and proliferation of these lineages. Mouse and human ES cells are a resource of enormous importance that is leading to a greater understanding of the molecular basis of cellular differentiation and development of the mammalian embryo. The study of ES cells from both species may ultimately realize the potential of the human cells being of use in the treatment of many debilitating diseases.

Acknowledgments

I thank John Gearhart, Gordon Keller, Michael Wiles, Jeff Mann and Chris Graham for fruitful discussions. Also I am indebted to Leslie Mounkes for her careful and critical reading of the manuscript.

References

Ambrosetti, D.C., Basilico, C., Dailey, L. 1997. Synergistic activation of the fibroblast growth factor 4 enhancer by Sox2 and Oct-3 depends on protein-protein interactions facilitated by a specific spatial arrangement of factor binding sites. Mol. Cell. Biol. 17, 6321–6329.

Amit, M., Carpenter, M.K., Inokuma, M.S., Chiu, C.P., Harris, C.P., Waknitz, M.A., Itskovitz-Eldor, J., Thomson, J.A. 2000. Clonally derived human embryonic stem cell lines maintain pluripotency and proliferative potential for prolonged periods of culture. Dev. Biol. 227, 271–278.

Angelov, D.N., Arnhold, S., Andressen, C., Grabsch, H., Puschmann, M., Hescheler, J., Addicks, K. 1998. Temporospatial relationships between macroglia and microglia during in vitro differentiation of murine stem cells. Dev. Neurosci. 20, 42–51.

Auernhammer, C.J., Melmed, S. 2000. Leukemia-inhibitory factor-neuroimmune modulator of endocrine function. Endocr. Rev. 21, 313–345.

Bagutti, C., Hutter, C., Chiquet-Ehrismann, R., Fassler, R., Watt, F.M. 2001. Dermal fibroblast-derived growth factors restore the ability of beta(1) integrin-deficient embryonal stem cells to differentiate into keratinocytes. Dev. Biol. 231, 321–333.

Bagutti, C., Wobus, A.M., Fassler, R., Watt, F.M. 1996. Differentiation of embryonal stem cells into keratinocytes: comparison of wild-type and beta 1 integrin-deficient cells. Dev. Biol. 179, 184–196.

Barnea, E., Bergman, Y. 2000. Synergy of SF1 and RAR in activation of Oct-3/4 promoter. J. Biol. Chem. 275, 6608–6619.

Bellido, T., O'Brien, C.A., Roberson, P.K., Manolagas, S.C. 1998. Transcriptional activation of the p21(WAF1,CIP1,SDI1) gene by interleukin-6 type cytokines. A prerequisite for their pro-differentiating and anti-apoptotic effects on human osteoblastic cells. J. Biol. Chem. 273, 21137–21144.

Bestor, T.H. 2000. The DNA methyltransferases of mammals. Hum. Mol. Genet. 9, 2395–2402.

Bird, A.P., Wolffe, A.P. 1999. Methylation-induced repression—belts, braces, and chromatin. Cell 99, 451–454.

Botquin, V., Hess, H., Fuhrmann, G., Anastassiadis, C., Gross, M.K., Vriend, G., Scholer, H.R. 1998. New POU dimer configuration mediates antagonistic control of an osteopontin preimplantation enhancer by Oct-4 and Sox-2. Genes Dev. 12, 2073–2090.

Bowman, T., Garcia, R., Turkson, J., Jove, R. 2000. STATs in oncogenesis. Oncogene 19, 2474–2488.

Bradley, A., Evans, M., Kaufman, M.H., Robertson, E. 1984. Formation of germ-line chimaeras from embryo-derived teratocarcinoma cell lines. Nature 309, 255–256.

Brinster, R.L. 1974. The effect of cells transferred into the mouse blastocyst on subsequent development. J. Exp. Med. 140, 1049–1056.

Bromberg, J.F., Wrzeszczynska, M.H., Devgan, G., Zhao, Y., Pestell, R.G., Albanese, C., Darnell, J.E., Jr. 1999. Stat3 as an oncogene. Cell 98, 295–303.

Brook, F.A., Gardner, R.L. 1997. The origin and efficient derivation of embryonic stem cells in the mouse. Proc. Natl. Acad. Sci. USA 94, 5709–5712.

Brustle, O., Jones, K.N., Learish, R.D., Karram, K., Choudhary, K., Wiestler, O.D., Duncan, I.D., McKay, R.D. 1999. Embryonic stem cell-derived glial precursors: a source of myelinating transplants. Science 285, 754–756.

Burdon, T., Stracey, C., Chambers, I., Nichols, J., Smith, A. 1999. Suppression of SHP-2 and ERK signalling promotes self-renewal of mouse embryonic stem cells. Dev. Biol. 210, 30–43.

Butteroni, C., De Felici, M., Scholer, H.R., Pesce, M. 2000. Phage display screening reveals an association between germline-specific transcription factor Oct-4 and multiple cellular proteins. J. Mol. Biol. 304, 529–540.

Buttery, L.D., Bourne, S., Xynos, J.D., Wood, H., Hughes, F.J., Hughes, S.P., Episkopou, V., Polak, J.M. 2001. Differentiation of osteoblasts and in vitro bone formation from murine embryonic stem cells. Tissue Eng. 7, 89–99.

Chen, L.R., Shiue, Y.L., Bertolini, L., Medrano, J.F., BonDurant, R.H., Anderson, G.B. 1999. Establishment of pluripotent cell lines from porcine preimplantation embryos. Theriogenology 52, 195–212.

Cho, S.K., Webber, T.D., Carlyle, J.R., Nakano, T., Lewis, S.M., Zuniga-Pflucker, J.C. 1999. Functional characterization of B lymphocytes generated in vitro from embryonic stem cells. Proc. Natl. Acad. Sci. USA 96, 9797–9802.

Choi, K., Kennedy, M., Kazarov, A., Papadimitriou, J.C., Keller, G. 1998. A common precursor for hematopoietic and endothelial cells. Development 125, 725–732.

Clarke, D.L., Johansson, C.B., Wilbertz, J., Veress, B., Nilsson, E., Karlstrom, H., Lendahl, U., Frisen, J. 2000. Generalized potential of adult neural stem cells. Science 288, 1660–1663.

Cui, S., Hope, R.M., Rathjen, J., Voyle, R.B., Rathjen, P.D. 2001. Structure, sequence and function of a marsupial LIF gene: conservation of IL-6 family cytokines. Cytogenet. Cell. Genet. 92, 271–278.

Dani, C. 1999. Embryonic stem cell-derived adipogenesis. Cells Tissues Organs 165, 173–180.

Dani, C., Chambers, I., Johnstone, S., Robertson, M., Ebrahimi, B., Saito, M., Taga, T., Li, M., Burdon, T., Nichols, J., Smith, A. 1998. Paracrine induction of stem cell renewal by LIF-deficient cells: a new ES cell regulatory pathway. Dev. Biol. 203, 149–162.

Dannenberg, J.H., van Rossum, A., Schuijff, L., te Riele, H. 2000. Ablation of the retinoblastoma gene family deregulates G(1) control causing immortalization and increased cell turnover under growth-restricting conditions. Genes Dev. 14, 3051–3064.

Darnell, J.E., Jr., Kerr, I.M., Stark, G.R. 1994. Jak-STAT pathways and transcriptional activation in response to IFNs and other extracellular signaling proteins. Science 264, 1415–1421.

Dean, W., Bowden, L., Aitchison, A., Klose, J., Moore, T., Meneses, J.J., Reik, W., Feil, R. 1998. Altered imprinted gene methylation and expression in completely ES cell-derived mouse fetuses: association with aberrant phenotypes. Development 125, 2273–2282.

Doetschman, T., Williams, P., Maeda, N. 1988. Establishment of hamster blastocyst-derived embryonic stem (ES) cells. Dev. Biol. 127, 224–227.

Doetschman, T.C., Eistetter, H., Katz, M., Schmidt, W., Kemler, R. 1985. The in vitro development of blastocyst-derived embryonic stem cell lines: formation of visceral yolk sac, blood islands and myocardium. J. Embryol. Exp. Morphol. 87, 27–45.

Du, Z., Cong, H., Yao, Z. 2001. Identification of putative downstream genes of Oct-4 by suppression-subtractive hybridization. Biochem. Biophys. Res. Commun. 282, 701–706.

Duval, D., Reinhardt, B., Kedinger, C., Boeuf, H. 2000. Role of suppressors of cytokine signaling (Socs) in leukemia inhibitory factor (LIF)-dependent embryonic stem cell survival. Faseb. J. 14, 1577–1584.

Eggan, K., Akutsu, H., Loring, J., Jackson-Grusby, L., Klemm, M., Rideout, W.M., III, Yanagimachi, R., Jaenisch, R. 2001. Hybrid vigor, fetal overgrowth, and viability of mice derived by nuclear cloning and tetraploid embryo complementation. Proc. Natl. Acad. Sci. USA 98, 6209–6214.

Ernst, M., Gearing, D.P., Dunn, A.R. 1994. Functional and biochemical association of Hck with the LIF/IL-6 receptor signal transducing subunit gp130 in embryonic stem cells. EMBO J 13, 1574–1584.

Ernst, M., Novak, U., Nicholson, S.E., Layton, J.E., Dunn, A.R. 1999. The carboxyl-terminal domains of gp130–related cytokine receptors are necessary for suppressing embryonic stem cell differentiation. Involvement of STAT3. J. Biol. Chem. 274, 9729–9737.

Ernst, M., Oates, A., Dunn, A.R. 1996. Gp130–mediated signal transduction in embryonic stem cells involves activation of Jak and Ras/mitogen-activated protein kinase pathways. J. Biol. Chem. 271, 30136–30143.

Evans, M.J., Kaufman, M.H. 1981. Establishment in culture of pluripotential cells from mouse embryos. Nature 292, 154–156.

Ezashi, T., Ghosh, D., Roberts, R.M. 2001. Repression of Ets-2-induced transactivation of the tau interferon promoter by Oct-4. Mol. Cell. Biol. 21, 7883–7891.

Fairchild, P.J., Brook, F.A., Gardner, R.L., Graca, L., Strong, V., Tone, Y., Tone, M., Nolan, K.F., Waldmann, H. 2000. Directed differentiation of dendritic cells from mouse embryonic stem cells. Curr. Biol. 10, 1515–1518.

Fassler, R., Rohwedel, J., Maltsev, V., Bloch, W., Lentini, S., Guan, K., Gullberg, D., Hescheler, J., Addicks, K., Wobus, A.M. 1996. Differentiation and integrity of cardiac muscle cells are impaired in the absence of beta 1 integrin. J. Cell. Sci. 109 (Pt 13), 2989–2999.

First, N.L., Sims, M.M., Park, S.P., Kent-First, M.J. 1994. Systems for production of calves from cultured bovine embryonic cells. Reprod. Fertil. Dev. 6, 553–562.

Fraichard, A., Chassande, O., Bilbaut, G., Dehay, C., Savatier, P., Samarut, J. 1995. In vitro differentiation of embryonic stem cells into glial cells and functional neurons. J. Cell. Sci. 108 (Pt 10), 3181–3188.

Fuhrmann, G., Chung, A.C., Jackson, K.J., Hummelke, G., Baniahmad, A., Sutter, J., Sylvester, I., Scholer, H.R., Cooney, A.J. 2001. Mouse germline restriction of Oct4 expression by germ cell nuclear factor. Dev. Cell. 1, 377–387.

Fuhrmann, G., Sylvester, I., Scholer, H.R. 1999. Repression of Oct-4 during embryonic cell differentiation correlates with the appearance of TRIF, a transiently induced DNA-binding factor. Cell. Mol. Biol. (Noisy-le-grand) 45, 717–724.

Gardner, R.L., Johnson, M.H. 1972. An investigation of inner cell mass and trophoblast tissues following their isolation from the mouse blastocyst. J. Embryol. Exp. Morphol. 28, 279–312.

Garrington, T.P., Ishizuka, T., Papst, P.J., Chayama, K., Webb, S., Yujiri, T., Sun, W., Sather, S., Russell, D.M., Gibson, S.B., et al. 2000. MEKK2 gene disruption causes loss of cytokine production in response to IgE and c-Kit ligand stimulation of ES cell-derived mast cells. EMBO J. 19, 5387–5395.

Graves, K.H., Moreadith, R.W. 1993. Derivation and characterization of putative pluripotential embryonic stem cells from preimplantation rabbit embryos. Mol. Reprod. Dev. 36, 424–433.

Handyside, A., Hooper, M.L., Kaufman, M.H., Wilmut, I. 1987. Towards the isolation of embryonal stem cell lines from the sheep. Roux's Arch. Dev. Biol. 196, 185–190.

Hansis, C., Grifo, J.A., Krey, L.C. 2000. Oct-4 expression in inner cell mass and trophectoderm of human blastocysts. Mol. Hum. Reprod. 6, 999–1004.

Heinrich, P.C., Behrmann, I., Muller-Newen, G., Schaper, F., Graeve, L. 1998. Interleukin-6-type cytokine signalling through the gp130/Jak/STAT pathway. Biochem. J. 334 (Pt 2), 297–314.

Ihara, S., Nakajima, K., Fukada, T., Hibi, M., Nagata, S., Hirano, T., Fukui, Y. 1997. Dual control of neurite outgrowth by STAT3 and MAP kinase in PC12 cells stimulated with interleukin-6. EMBO J. 16, 5345–5352.

Ihle, J.N. 2001. The Stat family in cytokine signaling. Curr. Opin. Cell. Biol. 13, 211–217.

Imada, K., Leonard, W.J. 2000. The Jak-STAT pathway. Mol. Immunol. 37, 1–11.

Iwasaki, S., Campbell, K.H., Galli, C., Akiyama, K. 2000. Production of live calves derived from embryonic stem-like cells aggregated with tetraploid embryos. Biol. Reprod. 62, 470–475.

Jackson-Grusby, L., Beard, C., Possemato, R., Tudor, M., Fambrough, D., Csankovszki, G., Dausman, J., Lee, P., Wilson, C., Lander, E., Jaenisch, R. 2001. Loss of genomic methylation causes p53-dependent apoptosis and epigenetic deregulation. Nat. Genet. 27, 31–39.

Jacob, H.J., Kwitek, A.E. 2002. Rat genetics: attaching physiology and pharmacology to the genome. Nat. Rev. Genet. 3, 33–42.

Jacobson, L., Kahan, B., Djamali, A., Thomson, J., Odorico, J.S. 2001. Differentiation of endoderm derivatives, pancreas and intestine, from rhesus embryonic stem cells. Transplant. Proc. 33, 674.

Jahner, D., Stuhlmann, H., Stewart, C.L., Harbers, K., Lohler, J., Simon, I., Jaenisch, R. 1982. De novo methylation and expression of retroviral genomes during mouse embryogenesis. Nature 298, 623–628.

James, R.M., Klerkx, A.H., Keighren, M., Flockhart, J.H., West, J.D. 1995. Restricted distribution of tetraploid cells in mouse tetraploid diploid chimaeras. Dev. Biol. 167, 213–226.

Johansson, B.M., Wiles, M.V. 1995. Evidence for involvement of activin A and bone morphogenetic protein 4 in mammalian mesoderm and hematopoietic development. Mol. Cell. Biol. 15, 141–151.

Jones, E.A., Tosh, D., Wilson, D.I., Lindsay, S., Forrester, L.M. 2002. Hepatic differentiation of murine embryonic stem cells. Exp. Cell. Res. 272, 15–22.

Kaufman, D.S., Hanson, E.T., Lewis, R.L., Auerbach, R., Thomson, J.A. 2001. Hematopoietic colony-forming cells derived from human embryonic stem cells. Proc. Natl. Acad. Sci. USA 98, 10716–10721.

Kawasaki, H., Mizuseki, K., Nishikawa, S., Kaneko, S., Kuwana, Y., Nakanishi, S., Nishikawa, S.I., Sasai, Y. 2000. Induction of midbrain dopaminergic neurons from ES cells by stromal cell-derived inducing activity. Neuron 28, 31–40.

Kehat, I., Kenyagin-Karsenti, D., Snir, M., Segev, H., Amit, M., Gepstein, A., Livne, E., Binah, O., Itskovitz-Eldor, J., Gepstein, L. 2001. Human embryonic stem cells can differentiate into myocytes with structural and functional properties of cardiomyocytes. J. Clin. Invest. 108, 407–414.

Keller, G., Kennedy, M., Papayannopoulou, T., Wiles, M.V. 1993. Hematopoietic commitment during embryonic stem cell differentiation in culture. Mol. Cell. Biol. 13, 473–486.

Kennedy, M., Firpo, M., Choi, K., Wall, C., Robertson, S., Kabrun, N., Keller, G. 1997. A common precursor for primitive erythropoiesis and definitive haematopoiesis. Nature 386, 488–493.

Kiger, A.A., Jones, D.L., Schulz, C., Rogers, M.B., Fuller, M.T. 2001. Stem cell self-renewal specified by JAK-STAT activation in response to a support cell cue. Science 294, 2542–2545.

Kirchhof, N., Carnwath, J.W., Lemme, E., Anastassiadis, K., Scholer, H., Niemann, H. 2000. Expression pattern of Oct-4 in preimplantation embryos of different species. Biol. Reprod. 63, 1698–1705.

T. 1999. STAT3 is required for the gp130–mediated full activation of the c-myc gene. J. Exp. Med. 189, 63–73.

Kleinsmith, L.J., Pierce, G.B. 1964. Multipotentiality of single embryonal carcinoma cells. Cancer Res. 24, 1544–1560.

Klug, M.G., Soonpaa, M.H., Koh, G.Y., Field, L.J. 1996. Genetically selected cardiomyocytes from differentiating embronic stem cells form stable intracardiac grafts. J. Clin. Invest. 98, 216–224.

Koopman, P., Cotton, R.G. 1984. A factor produced by feeder cells which inhibits embryonal carcinoma cell differentiation. Characterization and partial purification. Exp. Cell. Res. 154, 233–242.

Kramer, J., Hegert, C., Guan, K., Wobus, A.M., Muller, P.K., Rohwedel, J. 2000. Embryonic stem cell-derived chondrogenic differentiation in vitro: activation by BMP-2 and BMP-4. Mech. Dev. 92, 193–205.

Labosky, P.A., Barlow, D.P., Hogan, B.L. 1994. Mouse embryonic germ (EG) cell lines: transmission through the germline and differences in the methylation imprint of insulin-like growth factor 2 receptor (Igf2r) gene compared with embryonic stem (ES) cell lines. Development 120, 3197–3204.

Lacaud, G., Robertson, S., Palis, J., Kennedy, M., Keller, G. 2001. Regulation of hemangioblast development. Ann. NY Acad. Sci. 938, 96–107; discussion 108.

Lagasse, E., Connors, H., Al-Dhalimy, M., Reitsma, M., Dohse, M., Osborne, L., Wang, X., Finegold, M., Weissman, I.L., Grompe, M. 2000. Purified hematopoietic stem cells can differentiate into hepatocytes in vivo. Nat. Med. 6, 1229–1234.

Lai, L., Kolber-Simonds, D., Park, K.W., Cheong, H.T., Greenstein, J.L., Im, G.S., Samuel, M., Bonk, A., Rieke, A., Day, B.N., et al. 2002. Production of α-1,3-Galactosyltransferase knockout pigs by nuclear transfer cloning. Science.

Leahy, A., Xiong, J.W., Kuhnert, F., Stuhlmann, H. 1999. Use of developmental marker genes to define temporal and spatial patterns of differentiation during embryoid body formation. J. Exp. Zool. 284, 67–81.

Lee, S.H., Lumelsky, N., Studer, L., Auerbach, J.M., McKay, R.D. 2000. Efficient generation of midbrain and hindbrain neurons from mouse embryonic stem cells. Nat. Biotechnol. 18, 675–679.

Levinson-Dushnik, M., Benvenisty, N. 1997. Involvement of hepatocyte nuclear factor 3 in endoderm differentiation of embryonic stem cells. Mol. Cell. Biol. 17, 3817–3822.

Lewandoski, M. 2001. Conditional control of gene expression in the mouse. Nat. Rev. Genet. 2, 743–755.

Li, M., Pevny, L., Lovell-Badge, R., Smith, A. 1998. Generation of purified neural precursors from embryonic stem cells by lineage selection. Curr. Biol. 8, 971–974.

Liu, L., Leaman, D., Villalta, M., Roberts, R.M. 1997. Silencing of the gene for the alpha-subunit of human chorionic gonadotropin by the embryonic transcription factor Oct-3/4. Mol. Endocrinol. 11, 1651–1658.

Liu, S., Qu, Y., Stewart, T.J., Howard, M.J., Chakrabortty, S., Holekamp, T.F., McDonald, J.W. 2000. Embryonic stem cells differentiate into oligodendrocytes and myelinate in culture and after spinal cord transplantation. Proc. Natl. Acad. Sci. USA 97, 6126–6131.

Lumelsky, N., Blondel, O., Laeng, P., Velasco, I., Ravin, R., McKay, R. 2001. Differentiation of embryonic stem cells to insulin-secreting structures similar to pancreatic islets. Science 292, 1389–1394.

Ma, C., Fan, L., Ganassin, R., Bols, N., Collodi, P. 2001. Production of zebrafish germ-line chimeras from embryo cell cultures. Proc. Natl. Acad. Sci. USA 98, 2461–2466.

Mann, J.R., Gadi, I., Harbison, M.L., Abbondanzo, S.J., Stewart, C.L. 1990. Androgenetic mouse embryonic stem cells are pluripotent and cause skeletal defects in chimeras: implications for genetic imprinting. Cell 62, 251–260.

Martin, G.R. 1981. Isolation of a pluripotent cell line from early mouse embryos cultured in medium conditioned by teratocarcinoma stem cells. Proc. Natl. Acad. Sci. USA 78, 7634–7638.

Matsuda, T., Nakamura, T., Nakao, K., Arai, T., Katsuki, M., Heike, T., Yokota, T. 1999. STAT3 activation is sufficient to maintain an undifferentiated state of mouse embryonic stem cells. EMBO J. 18, 4261–4269.

Matsui, Y., Zsebo, K., Hogan, B.L. 1992. Derivation of pluripotential embryonic stem cells from murine primordial germ cells in culture. Cell. 70, 841–847.

McLaren, A. 1992. Development of primordial germ cells in the mouse. Andrologia 24, 243–247.

Mintz, B., Illmensee, K. 1975. Normal genetically mosaic mice produced from malignant teratocarcinoma cells. Proc. Natl. Acad. Sci. USA 72, 3585–3589.

Minucci, S., Botquin, V., Yeom, Y.I., Dey, A., Sylvester, I., Zand, D.J., Ohbo, K., Ozato, K., Scholer, H.R. 1996. Retinoic acid-mediated down-regulation of Oct3/4 coincides with the loss of promoter occupancy in vivo. EMBO J. 15, 888–899.

Myer, A., Wagner, D.S., Vivian, J.L., Olson, E.N., Klein, W.H. 1997. Wild-type myoblasts rescue the ability of myogenin-null myoblasts to fuse in vivo. Dev. Biol. 185, 127–138.

Nagy, A., Rossant, J., Nagy, R., Abramow-Newerly, W., Roder, J.C. 1993. Derivation of completely cell culture-derived mice from early-passage embryonic stem cells. Proc. Natl. Acad. Sci. USA 90, 8424–8428.

Nakano, T., Kodama, H., Honjo, T. 1994. Generation of lymphohematopoietic cells from embryonic stem cells in culture. Science 265, 1098–1101.

Narimatsu, M., Maeda, H., Itoh, S., Atsumi, T., Ohtani, T., Nishida, K., Itoh, M., Kamimura, D., Park, S.J., Mizuno, K., et al. 2001. Tissue-specific autoregulation of the stat3 gene and its role in interleukin-6-induced survival signals in T cells. Mol. Cell Biol. 21, 6615–6625.

Nichols, J., Chambers, I., Smith, A. 1994. Derivation of germline competent embryonic stem cells with a combination of interleukin-6 and soluble interleukin-6 receptor. Exp. Cell. Res. 215, 237–239.

Nichols, J., Chambers, I., Taga, T., Smith, A. 2001. Physiological rationale for responsiveness of mouse embryonic stem cells to gp130 cytokines. Development 128, 2333–2339.

Nichols, J., Davidson, D., Taga, T., Yoshida, K., Chambers, I., Smith, A. 1996. Complementary tissue-specific expression of LIF and LIF-receptor mRNAs in early mouse embryogenesis. Mech. Dev. 57, 123–131.

Nichols, J., Zevnik, B., Anastassiadis, K., Niwa, H., Klewe-Nebenius, D., Chambers, I., Scholer, H., Smith, A. 1998. Formation of pluripotent stem cells in the mammalian embryo depends on the POU transcription factor Oct4. Cell 95, 379–391.

Nicholson, S.E., De Souza, D., Fabri, L.J., Corbin, J., Willson, T.A., Zhang, J.G., Silva, A., Asimakis, M., Farley, A., Nash, A.D., et al. 2000. Suppressor of cytokine signaling-3 preferentially binds to the SHP-2-binding site on the shared cytokine receptor subunit gp130. Proc. Natl. Acad. Sci. USA 97, 6493–6498.

Nishimoto, M., Fukushima, A., Okuda, A., Muramatsu, M. 1999. The gene for the embryonic stem cell coactivator UTF1 carries a regulatory element which selectively interacts with a complex composed of Oct-3/4 and Sox-2. Mol. Cell. Biol. 19, 5453–5465.

Niwa, H., Burdon, T., Chambers, I., Smith, A. 1998. Self-renewal of pluripotent embryonic stem cells is mediated via activation of STAT3. Genes Dev. 12, 2048–2060.

Niwa, H., Miyazaki, J., Smith, A.G. 2000. Quantitative expression of Oct-3/4 defines differentiation, dedifferentiation or self-renewal of ES cells. Nat. Genet. 24, 372–376.

Notarianni, E., Galli, C., Laurie, S., Moor, R.M., Evans, M.J. 1991. Derivation of pluripotent, embryonic cell lines from the pig and sheep. J. Reprod. Fertil. Suppl. 43, 255–260.

Notarianni, E., Laurie, S., Ng, A., Sathasivam, K. 1997. Incorporation of cultured embryonic cells into transgenic and chimeric, porcine fetuses. Int. J. Dev. Biol. 41, 537–540.

Odorico, J.S., Kaufman, D.S., Thomson, J.A. 2001. Multilineage differentiation from human embryonic stem cell lines. Stem Cells 19, 193–204.

Ohtani, T., Ishihara, K., Atsumi, T., Nishida, K., Kaneko, Y., Miyata, T., Itoh, S., Narimatsu, M., Maeda, H., Fukada, T., et al. 2000. Dissection of signaling cascades through gp130 in vivo: reciprocal roles for STAT3- and SHP2-mediated signals in immune responses. Immunity 12, 95–105.

Okamoto, K., Okazawa, H., Okuda, A., Sakai, M., Muramatsu, M., Hamada, H. 1990. A novel octamer binding transcription factor is differentially expressed in mouse embryonic cells. Cell 60, 461–472.

Okano, M., Bell, D.W., Haber, D.A., Li, E. 1999. DNA methyltransferases Dnmt3a and Dnmt3b are essential for de novo methylation and mammalian development. Cell 99, 247–257.

Pannell, D., Osborne, C.S., Yao, S., Sukonnik, T., Pasceri, P., Karaiskakis, A., Okano, M., Li, E., Lipshitz, H.D., Ellis, J. 2000. Retrovirus vector silencing is de novo methylase independent and marked by a repressive histone code. EMBO J. 19, 5884–5894.

Papaioannou, V.E., Evans, E.P., Gardner, R.L., Graham, C.F. 1979. Growth and differentiation of an embryonal carcinoma cell line (C145b). J. Embryol. Exp. Morphol. 54, 277–295.

Papaioannou, V.E., Gardner, R.L., McBurney, M.W., Babinet, C., Evans, M.J. 1978. Participation of cultured teratocarcinoma cells in mouse embryogenesis. J. Embryol. Exp. Morphol. 44, 93–104.

Papaioannou, V.E., McBurney, M.W., Gardner, R.L., Evans, M.J. 1975. Fate of teratocarcinoma cells injected into early mouse embryos. Nature 258, 70–73.

Park, T.S., Han, J.Y. 2000. Derivation and characterization of pluripotent embryonic germ cells in chicken. Mol. Reprod. Dev. 56, 475–482.

Pennica, D., Shaw, K.J., Swanson, T.A., Moore, M.W., Shelton, D.L., Zioncheck, K.A., Rosenthal, A., Taga, T., Paoni, N.F., Wood, W.I. 1995. Cardiotrophin-1. Biological activities and binding to the leukemia inhibitory factor receptor/gp130 signaling complex. J. Biol. Chem. 270, 10915–10922.

Perlingeiro, R.C., Kyba, M., Daley, G.Q. 2001. Clonal analysis of differentiating embryonic stem cells reveals a hematopoietic progenitor with primitive erythroid and adult lymphoid-myeloid potential. Development 128, 4597–4604.

Pesce, M., Marin Gomez, M., Philipsen, S., Scholer, H.R. 1999. Binding of Sp1 and Sp3 transcription factors to the Oct-4 gene promoter. Cell. Mol. Biol. (Noisy-le-grand) 45, 709–716.

Pesce, M., Wang, X., Wolgemuth, D.J., Scholer, H. 1998. Differential expression of the Oct-4 transcription factor during mouse germ cell differentiation. Mech. Dev. 71, 89–98.

Phillips, K., Luisi, B. 2000. The virtuoso of versatility: POU proteins that flex to fit. J. Mol. Biol. 302, 1023–1039.

Pikarsky, E., Sharir, H., Ben-Shushan, E., Bergman, Y. 1994. Retinoic acid represses Oct-3/4 gene expression through several retinoic acid-responsive elements located in the promoter-enhancer region. Mol. Cell. Biol. 14, 1026–1038.

Porcher, C., Swat, W., Rockwell, K., Fujiwara, Y., Alt, F.W., Orkin, S.H. 1996. The T cell leukemia oncoprotein SCL/tal-1 is essential for development of all hematopoietic lineages. Cell 86, 47–57.

Potocnik, A.J., Nielsen, P.J., Eichmann, K. 1994. In vitro generation of lymphoid precursors from embryonic stem cells. EMBO J. 13, 5274–5283.

Raz, R., Lee, C.K., Cannizzaro, L.A., d'Eustachio, P., Levy, D.E. 1999. Essential role of STAT3 for embryonic stem cell pluripotency. Proc. Natl. Acad. Sci. USA 96, 2846–2851.

Renfree, M.B., Shaw, G. 2000. Diapause. Annu. Rev. Physiol. 62, 353–375.

Resnick, J.L., Bixler, L.S., Cheng, L., Donovan, P.J. 1992. Long-term proliferation of mouse primordial germ cells in culture. Nature 359, 550–551.

Reubinoff, B.E., Pera, M.F., Fong, C.Y., Trounson, A., Bongso, A. 2000. Embryonic stem cell lines from human blastocysts: somatic differentiation in vitro. Nat. Biotechnol. 18, 399–404.

Reyes, M., Lund, T., Lenvik, T., Aguiar, D., Koodie, L., Verfaillie, C.M. 2001. Purification and ex vivo expansion of postnatal human marrow mesodermal progenitor cells. Blood 98, 2615–2625.

Robinson, R.C., Grey, L.M., Staunton, D., Vankelecom, H., Vernallis, A.B., Moreau, J.F., Stuart, D.I., Heath, J.K., Jones, E.Y. 1994. The crystal structure and biological function of leukemia inhibitory factor: implications for receptor binding. Cell 77, 1101–1116.

Rodig, S.J., Meraz, M.A., White, J.M., Lampe, P.A., Riley, J.K., Arthur, C.D., King, K.L., Sheehan, K.C., Yin, L., Pennica, D., et al. 1998. Disruption of the Jak1 gene demonstrates obligatory and nonredundant roles of the Jaks in cytokine-induced biologic responses. Cell 93, 373–383.

Roovers, K., Assoian, R.K. 2000. Integrating the MAP kinase signal into the G1 phase cell cycle machinery. Bioessays 22, 818–826.

Saito, M., Yoshida, K., Hibi, M., Taga, T., Kishimoto, T. 1992. Molecular cloning of a murine IL-6 receptor-associated signal transducer, gp130, and its regulated expression in vivo. J. Immunol. 148, 4066–4071.

Savatier, P., Huang, S., Szekely, L., Wiman, K.G., Samarut, J. 1994. Contrasting patterns of retinoblastoma protein expression in mouse embryonic stem cells and embryonic fibroblasts. Oncogene 9, 809–818.

Savatier, P., Lapillonne, H., van Grunsven, L.A., Rudkin, B.B., Samarut, J. 1996. Withdrawal of differentiation inhibitory activity/leukemia inhibitory factor up-regulates D-type cyclins and cyclin-dependent kinase inhibitors in mouse embryonic stem cells. Oncogene 12, 309–322.

Schaefer, T.S., Sanders, L.K., Nathans, D. 1995. Cooperative transcriptional activity of Jun and Stat3 beta, a short form of Stat3. Proc. Natl. Acad. Sci. USA 92, 9097–9101.

Schmitz, J., Weissenbach, M., Haan, S., Heinrich, P.C., Schaper, F. 2000. SOCS3 exerts its inhibitory function on interleukin-6 signal transduction through the SHP2 recruitment site of gp130. J. Biol. Chem. 275, 12848–12856.

Scholer, H.R., Dressler, G.R., Balling, R., Rohdewohld, H., Gruss, P. 1990. Oct-4: a germline-specific transcription factor mapping to the mouse t-complex. EMBO J. 9, 2185–2195.

Schoonjans, L., Albright, G.M., Li, J.L., Collen, D., Moreadith, R.W. 1996. Pluripotential rabbit embryonic stem (ES) cells are capable of forming overt coat color chimeras following injection into blastocysts. Mol. Reprod. Dev. 45, 439–443.

Schoorlemmer, J., van Puijenbroek, A., van Den Eijnden, M., Jonk, L., Pals, C., Kruijer, W. 1994. Characterization of a negative retinoic acid response element in the murine Oct4 promoter. Mol. Cell. Biol. 14, 1122–1136.

Shamblott, M.J., Axelman, J., Littlefield, J.W., Blumenthal, P.D., Huggins, G.R., Cui, Y., Cheng, L., Gearhart, J.D. 2001. Human embryonic germ cell derivatives express a broad range of developmentally distinct markers and proliferate extensively in vitro. Proc. Natl. Acad. Sci. USA 98, 113–118.

Shamblott, M.J., Axelman, J., Wang, S., Bugg, E.M., Littlefield, J.W., Donovan, P.J., Blumenthal, P.D., Huggins, G.R., Gearhart, J.D. 1998. Derivation of pluripotent stem cells from cultured human primordial germ cells. Proc. Natl. Acad. Sci. USA 95, 13726–13731.

Shirogane, T., Fukada, T., Muller, J.M., Shima, D.T., Hibi, M., Hirano, T. 1999. Synergistic roles for Pim-1 and c-Myc in STAT3-mediated cell cycle progression and antiapoptosis. Immunity 11, 709–719.

Smith, A.G. 2001. Embryo-derived stem cells: of mice and men. Annu. Rev. Cell. Dev. Biol. 17, 435–462.

Smith, A.G., Heath, J.K., Donaldson, D.D., Wong, G.G., Moreau, J., Stahl, M., Rogers, D. 1988. Inhibition of pluripotential embryonic stem cell differentiation by purified polypeptides. Nature 336, 688–690.

Snow, M.H. 1975. Embryonic development of tetraploid mice during the second half of gestation. J. Embryol. Exp. Morphol. 34, 707–721.

Soria, B., Roche, E., Berna, G., Leon-Quinto, T., Reig, J.A., Martin, F. 2000. Insulin-secreting cells derived from embryonic stem cells normalize glycemia in streptozotocin-induced diabetic mice. Diabetes 49, 157–162.

Stahl, N., Farruggella, T.J., Boulton, T.G., Zhong, Z., Darnell, J.E., Jr., Yancopoulos, G.D. 1995. Choice of STATs and other substrates specified by modular tyrosine-based motifs in cytokine receptors. Science 267, 1349–1353.

Starr, R., Novak, U., Willson, T.A., Inglese, M., Murphy, V., Alexander, W.S., Metcalf, D., Nicola, N.A., Hilton, D.J., Ernst, M. 1997. Distinct roles for leukemia inhibitory factor receptor alpha-chain and gp130 in cell type-specific signal transduction. J. Biol. Chem. 272, 19982–19986.

Stephens, L.E., Sutherland, A.E., Klimanskaya, I.V., Andrieux, A., Meneses, J., Pedersen, R.A., Damsky, C.H. 1995. Deletion of beta 1 integrins in mice results in inner cell mass failure and peri-implantation lethality. Genes Dev. 9, 1883–1895.

Stevens, L.C. 1964. Experimental production of testicular teratomas in mice. Proc. Natl. Acad. Sci. USA 52, 654–659.

Stevens, L.C. 1967a. The biology of teratomas. Adv. Morphog. 6, 1–31.

Stevens, L.C. 1967b. Origin of testicular teratomas from primordial germ cells in mice. J. Natl. Cancer Inst. 38, 549–552.

Stevens, L.C. 1968. The development of teratomas from intratesticular grafts of tubal mouse eggs. J. Embryol. Exp. Morphol. 20, 329–341.

Stevens, L.C. 1975. Comparative development of normal and parthenogenetic mouse embryos, early testicular and ovarian teratomas and embryoid bodies. New York, AP.

Stewart, C.L. 1982. Formation of viable chimaeras by aggregation between teratocarcinomas and preimplantation mouse embryos. J. Embryol. Exp. Morphol. 67, 167–179.

Stewart, C.L., Gadi, I., Bhatt, H. 1994. Stem cells from primordial germ cells can reenter the germ line. Dev. Biol. 161, 626–628.

Stewart, C.L. 1993. Production of chimeras between embryonic stem cells and embryos. Meth. Enzymol. 225, 823–855.

Stewart, C.L., Kaspar, P., Brunet, L.J., Bhatt, H., Gadi, I., Kontgen, F., Abbondanzo, S.J. 1992. Blastocyst implantation depends on maternal expression of leukaemia inhibitory factor. Nature 359, 76–79.

Stewart, C.L., Stuhlmann, H., Jahner, D., Jaenisch, R. 1982. De novo methylation, expression, and infectivity of retroviral genomes introduced into embryonal carcinoma cells. Proc. Natl. Acad. Sci. USA 79, 4098–4102.

Stewart, T.A., Mintz, B. 1981. Successive generations of mice produced from an established culture line of euploid teratocarcinoma cells. Proc. Natl. Acad. Sci. USA 78, 6314–6318.

Suemori, H., Tada, T., Torii, R., Hosoi, Y., Kobayashi, K., Imahie, H., Kondo, Y., Iritani, A., Nakatsuji, N. 2001. Establishment of embryonic stem cell lines from cynomolgus monkey blastocysts produced by IVF or ICSI. Dev. Dyn. 222, 273–279.

Sukoyan, M.A., Golubitsa, A.N., Zhelezova, A.I., Shilov, A.G., Vatolin, S.Y., Maximovsky, L.P., Andreeva, L.E., McWhir, J., Pack, S.D., Bayborodin, S.I., et al. 1992. Isolation and cultivation of blastocyst-derived stem cell lines from American mink (Mustela vison). Mol. Reprod. Dev. 33, 418–431.

Sylvester, I., Scholer, H.R. 1994. Regulation of the Oct-4 gene by nuclear receptors. Nucleic Acids Res. 22, 901–911.

Szabo, P.E., Mann, J.R. 1995. Biallelic expression of imprinted genes in the mouse germ line: implications for erasure, establishment, and mechanisms of genomic imprinting. Genes Dev. 9, 1857–1868.

Tada, M., Takahama, Y., Abe, K., Nakatsuji, N., Tada, T. 2001. Nuclear reprogramming of somatic cells by in vitro hybridization with ES cells. Curr. Biol. 11, 1553–1558.

Tada, T., Tada, M., Hilton, K., Barton, S.C., Sado, T., Takagi, N., Surani, M.A. 1998. Epigenotype switching of imprintable loci in embryonic germ cells. Dev. Genes Evol. 207, 551–561.

Taga, T., Kishimoto, T. 1997. Gp130 and the interleukin-6 family of cytokines. Annu. Rev. Immunol. 15, 797–819.

Takeda, K., Noguchi, K., Shi, W., Tanaka, T., Matsumoto, M., Yoshida, N., Kishimoto, T., Akira, S. 1997. Targeted disruption of the mouse Stat3 gene leads to early embryonic lethality. Proc. Natl. Acad. Sci. USA 94, 3801–3804.

Thomson, J.A., Itskovitz-Eldor, J., Shapiro, S.S., Waknitz, M.A., Swiergiel, J.J., Marshall, V.S., Jones, J.M. 1998. Embryonic stem cell lines derived from human blastocysts. Science 282, 1145–1147.

Thomson, J.A., Kalishman, J., Golos, T.G., Durning, M., Harris, C.P., Becker, R.A., Hearn, J.P. 1995. Isolation of a primate embryonic stem cell line. Proc. Natl. Acad. Sci. USA 92, 7844–7848.

Thomson, J.A., Kalishman, J., Golos, T.G., Durning, M., Harris, C.P., Hearn, J.P. 1996. Pluripotent cell lines derived from common marmoset (Callithrix jacchus) blastocysts. Biol. Reprod. 55, 254–259.

Thomson, J.A., Marshall, V.S. 1998. Primate embryonic stem cells. Curr. Top. Dev. Biol. 38, 133–165.

Tilghman, S.M. 1999. The sins of the fathers and mothers: genomic imprinting in mammalian development. Cell 96, 185–193.

Tomida, M., Yamamoto-Yamaguchi, Y., Hozumi, M. 1994. Three different cDNAs encoding mouse D-factor/LIF receptor. J. Biochem. (Tokyo) 115, 557–562.

Tropepe, V., Hitoshi, S., Sirard, C., Mak, T.W., Rossant, J., van der Kooy, D. 2001. Direct neural fate specification from embryonic stem cells: a primitive mammalian neural stem cell stage acquired through a default mechanism. Neuron 30, 65–78.

Tulina, N., Matunis, E. 2001. Control of stem cell self-renewal in Drosophila spermatogenesis by JAK-STAT signaling. Science 294, 2546–2549.

Vassilieva, S., Guan, K., Pich, U., Wobus, A.M. 2000. Establishment of SSEA-1- and Oct-4-expressing rat embryonic stem-like cell lines and effects of cytokines of the IL-6 family on clonal growth. Exp. Cell. Res. 258, 361–373.

Wagner, E.F., Keller, G., Gilboa, E., Ruther, U., Stewart, C.L. 1985. Gene transfer into murine stem cells and mice using retroviral vectors. Paper presented at: 50th Anniversary Symposium, Molecular Biology in Development. Cold Spring Harbor, Cold Spring Harbor Press.

Ware, C.B., Horowitz, M.C., Renshaw, B.R., Hunt, J.S., Liggitt, D., Koblar, S.A., Gliniak, B.C., McKenna, H.J., Papayannopoulou, T., Thoma, B., et al. 1995. Targeted disruption of the low-affinity leukemia inhibitory factor receptor gene causes placental, skeletal, neural and metabolic defects and results in perinatal death. Development 121, 1283–1299.

Wheeler, M.B., Walters, E.M. 2001. Transgenic technology and applications in swine. Theriogenology 56, 1345–1369.

Williams, J.G. 2000. STAT signalling in cell proliferation and in development. Curr. Opin. Genet. Dev. 10, 503–507.

Williams, R.L., Hilton, D.J., Pease, S., Willson, T.A., Stewart, C.L., Gearing, D.P., Wagner, E.F., Metcalf, D., Nicola, N.A., Gough, N.M. 1988. Myeloid leukaemia inhibitory factor maintains the developmental potential of embryonic stem cells. Nature 336, 684–687.

Wilmut, I., Schnieke, A.E., McWhir, J., Kind, A.J., Campbell, K.H. 1997. Viable offspring derived from fetal and adult mammalian cells. Nature 385, 810–813.

Xu, C., Inokuma, M.S., Denham, J., Golds, K., Kundu, P., Gold, J.D., Carpenter, M.K. 2001. Feeder-free growth of undifferentiated human embryonic stem cells. Nat. Biotechnol. 19, 971–974.

Yamashita, J., Itoh, H., Hirashima, M., Ogawa, M., Nishikawa, S., Yurugi, T., Naito, M., Nakao, K. 2000. Flk1-positive cells derived from embryonic stem cells serve as vascular progenitors. Nature 408, 92–96.

Yeom, Y.I., Fuhrmann, G., Ovitt, C.E., Brehm, A., Ohbo, K., Gross, M., Hubner, K., Scholer, H.R. 1996. Germline regulatory element of Oct-4 specific for the totipotent cycle of embryonal cells. Development 122, 881–894.

Yoshida, K., Chambers, I., Nichols, J., Smith, A., Saito, M., Yasukawa, K., Shoyab, M., Taga, T., Kishimoto, T. 1994. Maintenance of the pluripotential phenotype of embryonic stem cells through direct activation of gp130 signalling pathways. Mech. Dev. 45, 163–171.

Yoshida, K., Taga, T., Saito, M., Suematsu, S., Kumanogoh, A., Tanaka, T., Fujiwara, H., Hirata, M., Yamagami, T., Nakahata, T., et al. 1996. Targeted disruption of gp130, a common signal transducer for the interleukin 6 family of cytokines, leads to myocardial and hematological disorders. Proc. Natl. Acad. Sci. USA 93, 407–411.

Yu, Y., Bradley, A. 2001. Engineering chromosomal rearrangements in mice. Nat. Rev. Genet. 2, 780–790.

Zhang, S.C., Wernig, M., Duncan, I.D., Brustle, O., Thomson, J.A. 2001. In vitro differentiation of transplantable neural precursors from human embryonic stem cells. Nat. Biotechnol. 19, 1129–1133.

Advances in Developmental Biology and Biochemistry, Vol. 12
M. DePamphilis (Editor)

Genomic imprinting

Katharine L. Arney, Sylvia Erhardt and M. Azim Surani

*Wellcome Trust/Cancer Research UK Institute of Cancer and Developmental Biology,
University of Cambridge, Tennis Court Road, Cambridge CB2 1QR, UK*

Summary

Since its inception in the 1980s the field of genomic imprinting has grown exponentially, drawing on research ranging from evolutionary and developmental biology to chromatin structure. The study of imprinting in mammals finds parallels in mechanisms of gene regulation in many species, such as yeast, worms, plants and flies. It is this broad range of mechanisms and systems that makes the field of imprinting both fascinating and extremely complex. In this chapter we will address the question of how imprinted domains are organised and regulated, including the role of *cis*-regulatory elements such as differentially methylated regions and enhancers in co-ordinating imprinted gene expression. We then examine how epigenetic modifications of these elements are erased and re-established in a sex-specific manner in the germ line. Following fertilisation these imprints, and therefore the epigenetic asymmetry between the parental genomes, must be maintained through the complex epigenetic changes which occur during preimplantation development. Finally, we discuss the relevance of imprinting for cloning and stem cell technology.

Contents

1. Introduction

1.1. The discovery of imprinting

The classic nuclear transfer experiments carried out in 1980's (McGrath and Solter, 1984; Surani et al., 1984) demonstrated that the maternal and paternal genomes are not equivalent in terms of their contribution to development. Parthenogenetic and androgenetic mouse embryos (containing two female or male haploid genomes respectively) were found to be incapable of directing development to term, implying that some genes required for development are expressed only from the maternal or only from the paternal allele (see Fig. 1). This was confirmed by the subsequent discovery of such monoallelically expressed, or imprinted, genes.

Imprinted regions of the mammalian genome have been identified by means of genetic studies with mice carrying Robertsonian chromosome translocations (Cattanach and Kirk, 1985), resulting in embryos carrying duplications of chromosomal regions inherited solely from one parent (uniparental disomies, or UPDs). Duplication of certain regions results in embryonic death or abnormal phenotypes, implying the presence of imprinted genes, although UPD of the majority of chromosomal regions has no discernible effect. In total, ten such imprinted regions have been identified in mice (Cattanach and Beechey, 1990) although imprinted genes have also been discovered outside the confines of these regions. It is important to note that many imprinted domains identified to date show conservation of synteny between the human and mouse genomes. This indicates that comparative genomic studies and mouse models may provide much useful information about human imprinted genes and their role in disease. Currently over forty imprinted genes have been identified in mice and humans and the list is growing rapidly (http://www.mgu.har.mrc.ac.uk). The discovery that imprinted genes tend to occur in clusters, such as that found on distal mouse chromosome 7, suggests that a small but significant proportion of the mammalian genome may be imprinted, with one group estimating the existence of 100–200 imprinted genes (Hayashizaki et al., 1994). The mammalian X-chromosome provides another example of monoallelic gene expression. In female eutherian mammals, one X-chromosome is always inactivated to compensate for the double dosage of genes compared to males (Heard et al., 1997). Unlike imprinted loci, these genes show no parent-of-origin preference for expression and X-inactivation in the embryo proper is a random process. One exception to this rule occurs in the extraembryonic tissues of the mouse, where the paternal X-chromosome is always inactivated, indicative of some

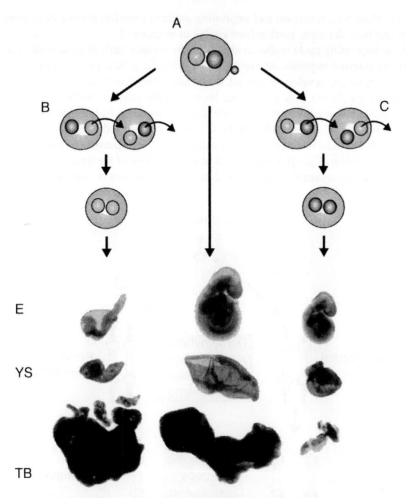

Fig. 1. Both maternal and paternal contributions are required for normal mammalian development. A. A normal fertilised mouse zygote contains both a paternal (blue) and maternal (pink) genome, in the form of pronuclei. The normal embryo (E) at 10.5 days post coitum (dpc), yolk sac (YS) and trophoblast (TB) are shown in the centre panel. B. If pronuclear transfer is used to generate embryos containing two paternal genomes (androgenetic embryos), an abnormally small and retarded embryo is seen at 10.5 dpc. Note the deficient yolk sac and excessive amounts of trophoblast tissue. C. Similarly, two maternal genomes lead to abnormal development (gynogenetic or parthenogenetic embryos). In this case, the embryo appears normal, but small, yet is extremely lacking in trophoblastic tissue at 10.5 dpc. Neither the androgenetic nor parthenogenetic embryos can sustain development much further than this timepoint and are subsequently resorbed by the mother. (*For a coloured version of this figure, see plate section, page 277.*)

basic imprinting mechanism at work (Takagi and Sasaki, 1975). How the co-ordinated epigenetic regulation of these clusters is directed during development raises many important questions. How can two alleles be distinguished when there is no difference in DNA sequence? How are these differences maintained through cell division and reset every generation? How are expression and imprinting patterns established over large domains? And why and how did these mechanisms evolve in mammals?

The basic imprinting cycle is shown in Fig. 2. The somatic cells of a mammal bear both maternal and paternal imprints, interpreted as monoallelic gene expression. In the primordial germ cells of the developing animal, these imprinting marks are completely erased, leading to demethylation in a global and locus-specific manner, biallelic expression of imprinted genes and reactivation of the silent X-chromosome in females. During subsequent gametogenesis, sex-specific imprints are re-established on the chromosomes. Fertilisation then results in a new organism with a correct complement of parental imprints in the next generation. The cycle is repeated in the germ line of the next generation, ensuring that erasure and appropriate resetting of the imprints occurs in every new generation.

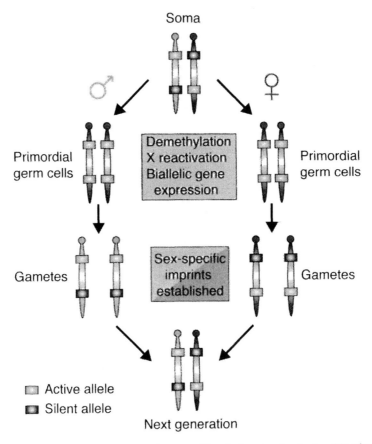

Fig. 2. The imprinting cycle. The somatic cells of a mammal bear both maternal and paternal imprints, interpreted as monoallelic gene expression. In the primordial germ cells of the developing animal, these imprinting marks are completely erased, leading to demethylation in a global and locus-specific manner, biallelic expression of imprinted genes and reactivation of the silent X-chromosome in females. During subsequent gametogenesis, sex-specific imprints are re-established on the chromosomes. Fertilisation results in a new organism with a correct complement of parental imprints in the next generation. (*For a coloured version of this figure, see plate section, page 278.*)

1.2. The evolution of imprinting

A variety of explanations for the evolution and purpose of imprinting, a phenomenon restricted to mammals in the animal kingdom, have been postulated (Hurst, 1997). It is important to note that angiospermic plants also demonstrate genomic imprinting, discussed in greater detail in section 3.4. The progressive trend toward viviparous development and birth of fully-formed young during mammalian evolution may also be related to the evolution of imprinting (John and Surani, 2000). The trophectoderm is the first lineage to differentiate in the blastocyst and contributes to extra-embryonic tissues, and evolutionary development of this lineage was essential to the development of eutherian ("true") mammals. This resulted in the development of relatively large, complex placentae, allowing *in utero* development to proceed to a greater extent than in the monotremes or marsupials. The placenta is the major interface between the foetus and the mother, and therefore of crucial importance for the exchange of nutrients. Important evidence for the role of imprinted genes in placental development is seen when comparing embryos derived from two maternal or two paternal genomes. Androgenetic embryos (comprising two paternal genomes) have only a small amount of extraembryonic tissue, relative to the size of the embryo, while parthenogenetic embryos (two maternal genomes) exhibit placentomegaly (Fundele et al., 1997). It is interesting to note that a relatively large number of imprinted genes are expressed in the placenta- for example, uniparental disomy of chromosome 12 in the mouse results in a complex placental phenotype (Georgiades et al., 2001).

The most persuasive argument to explain the purpose of imprinting is the conflict hypothesis (Haig and Graham, 1991). Based on the uneven parental contributions to the embryo in terms of sustaining development before and after birth, the hypothesis states that genes involved in promoting embryonic growth (at the expense of the mother) will be paternally expressed, while genes responsible for limiting embryonic size (therefore conserving maternal resources) will be maternally expressed. For example, the embryonic growth factor *Igf2* is paternally expressed, while its receptor *Igf2r* (which acts to sequester excess *Igf2*, rather than being involved in *Igf2* signal transduction) is maternally expressed. However, not all paternally expressed genes are growth enhancing, and not all maternally expressed genes are growth suppressors (Hurst and McVean, 1998). For example it is interesting to consider the case of genes for maternal nurturing behaviour, such as *Peg1/Mest* (Lefebvre et al., 1998) and *Peg3* (Li et al., 1999a), which are expressed from the paternally inherited allele. Comparative analysis of imprinted genes in eutherian mammals, marsupials and monotremes has found that the *Igf2* and *Igf2r* genes are correctly imprinted in the opossum, a marsupial which gives birth to under-developed young and carries them in a pouch to maturity (Killian et al., 2000; O'Neill et al., 2000). In contrast, *Igf2r* is not imprinted in the egg-laying monotremes platypus or echidna (Killian et al., 2000). This suggests that imprinting evolved as mammals tended towards *in utero* development. Interestingly, the monotreme *Igf2r* gene cannot bind *Igf2*, therefore not acting to antagonise growth effects. The IGF2R proteins of chicken and amphibians also lack this binding ability, as well as lacking imprinting (Killian et al., 2000). However, the final answer as to why mammals have imprinting while other animals do not is far from clear.

It is also interesting to contemplate the recent hypotheses on the effects of retroviruses and other transposable elements on gene expression and epigenetic regulation, (e.g., Whitelaw and Martin, 2001). Of particular note is the Agouti Variable Yellow (A^{vy}) allele, which manifests as a range of mouse coat colours resulting from a retroviral insertion near the Agouti locus affecting expression of the gene (Morgan et al., 1999). This is due to epigenetic effects and can be re-set in the germ line, meaning that an A^{vy} mouse of a particular coat colour can give rise to mice bearing the full spectrum of possible colours. Notably, this process appears to have some parent-of-origin specific effects: penetrance of the mutant phenotype (yellow coat colour) is much greater when inherited via the mother than from the father. Furthermore, incomplete erasure and resetting of parent-of-origin specific epigenetic marks can be seen in a transgenic mouse line which shows imprinting of the transgene (Kearns et al., 2000). Although there are no transposable or retroviral elements consistently found at imprinted loci that are not found elsewhere in the genome, epigenetic modifications due to transposable elements inserted at imprinted loci may have been differently treated by the parental germ lines and subsequently stabilised during evolution. Such a mechanism might contribute to the complex patterns of expression, organisation and epigenetic modification we see today.

2. Mechanisms of imprinting: paradigms and parallels

Although no two imprinted regions are exactly the same, certain parallels can be drawn between many loci in terms of organisation and regulation. Distinctive epigenetic modifications are found at many imprinted loci, and these probably contribute to the establishment and maintenance of imprints in similar ways. Likewise, it is becoming clear that there are a limited number of mechanisms by which monoallelic expression of imprinted genes is controlled. We will first describe the typical epigenetic modifications found at imprinted loci, then discuss how these might bring about regulated gene expression of these genes.

2.1. DNA methylation during development

Levels of both global and locus-specific methylation do not remain constant during development, but fluctuate widely. At fertilisation, there is a rapid demethylation of the paternal genome, followed by further demethylation of both parental genomes through preimplantation development (see section 4.1). Remethylation of the genome occurs upon implantation of the blastocyst, establishing somatic patterns of methylation. These may subsequently change slightly during cell differentiation, in a tissue- and gene-specific fashion. However, primordial germ cells (PGCs) are hypomethylated, compared to somatic cells, with sex-specific methylation patterns only being established relatively late in gametogenesis. These changes are discussed in more detail in Section 3.

DNA methyltransferases (Dnmts), the enzymes which methylate DNA, are of key importance in establishing and maintaining both global and gene-specific patterns of methylation. Mutation of the putative maintenance methyltransferase *Dnmt1* results in genome-wide demethylation, leading to biallelic expression of some imprinted genes (Li et al., 1992) and at least transient inactivation of all X chromosomes in somatic cells (Beard et al., 1995). Embryonic stem (ES) cells lacking *Dnmt1* are viable and proliferate normally in an undifferentiated state. However, during differentiation these cells undergo apoptosis possibly due to activation of a p53-dependent pathway (Jackson-Grusby et al., 2001). Reintroduction of *Dnmt1* cDNA restores general genome-wide methylation, but not the methylation patterns associated with imprinted loci. Appropriate methylation imprints are only established in these cells after transmission through the germ line. This indicates that events that take place exclusively during gametogenesis are essential for imprinting (Tucker et al., 1996).

When retroviral DNA is integrated into *Dnmt1* deficient cells it becomes partially methylated, indicating the existence of other methyltransferase enzymes. The more recently discovered *de novo* methyltransferases *Dnmt3a* and *Dnmt3b* are highly expressed in undifferentiated ES cells. After differentiation, and in adult somatic tissues, only low levels of these enzymes are detectable (Okano et al., 1999). Mice lacking both *Dnmt3a* and *3b* exhibit a similar early lethal phenotype as that observed in *Dnmt1* deficient animals (Okano et al., 1999). However, ES cells and early embryos lacking both these genes are unable to methylate newly integrated retroviral DNA, pointing to a role for *Dnmt3a* and *3b* in *de novo* methylation. But although *Dnmt3a* and *Dnmt3b* are essential for *de novo* methylation in early embryos, the differential methylation pattern of *H19* was not affected in ES cells deficient in both *Dnmt3a* and *3b*. Again this suggests that germ line passage is needed for the initiation of some essential epigenetic modifications, and invokes the need for other methyltransferase enzymes. A newly identified zinc finger protein called Dnmt3-like (Dnmt3L) shares significant similarity to both ATRX and the Dnmt3 family. Dnmt3L expression is only detected in ovary, testis and thymus (the only other tissue besides gametes where recombination occurs naturally), making it a very promising candidate for germ line specific function and a possible role in imprinting (Aapola et al., 2000, 2001).

2.2. DNA methylation at imprinted loci

A number of distinct epigenetic traits have been described at imprinted loci, the most obvious of which is DNA methylation. Allele-specific DNA methylation (Li et al., 1993) has been described at imprinted loci. Indeed the presence of differential methylation is a key indicator of imprinted status. Generally, the silenced parental allele is more methylated than the expressed allele, although this is not the case for some imprinted loci, for example the *Igf2* and *Igf2r* genes (Feil et al., 1994; Sasaki et al., 1992; Stoger et al., 1993). It is noteworthy that a relatively high percentage of the imprinted genes discovered to date have methylation on the maternal allele, regardless of the allele which is expressed (Reik and Walter, 2001). Interestingly, regions of differential methylation (DMRs) at CpG dinucleotides commonly occur in or near short tandem repeats (Neumann et al., 1995) which are purported to have an important role in imprinting (Stoger et al., 1993). Some differentially methylated regions at imprinted domains constitute Imprinting Control Regions (ICRs), such as the Differentially Methylated Domain (DMD) lying 5' of *H19*, which controls imprinted expression of both *H19* and *Igf2*. Somewhat confusingly, DNA methylation functions in two opposite ways at this locus, being required not only to mediate silencing of the methylated paternal *H19* allele but also for activation of the paternal *Igf2* allele (discussed further in Section 2.6).

Currently, the mechanisms which direct differential methylation to imprinted loci are unknown. There is growing evidence that DNA methylation can be targeted by specific

DNA sequence elements (such as B1 repeat elements, Yates et al., 1999), although it is not clear whether such a mechanism is at work at imprinted loci. There is some speculation that the type of direct repeats often found adjacent to differentially methylated regions at imprinted loci are responsible for attracting methylation although a direct link is yet to be demonstrated (Neumann et al., 1995). To date, the only clearly defined signal responsible for directing DNA methylation at an imprinted gene is the DNS (*de novo* methylation signal) found as part of the bipartite imprinting box at the *Igf2r* locus (Birger et al., 1999). It is debatable as to what extent DNA methylation is a primary imprint, although the data from mice lacking the maintenance methyltransferase *Dnmt1* suggests that it is certainly important for correct expression of some imprinted genes (Li et al., 1993).

2.3. Other characteristics of imprinted loci

Parent-of-origin specific differences in bulk chromatin conformation, in terms of regions of hypersensitivity to the enzyme DNase I and nucleosome positioning (Hark and Tilghman, 1998; Kanduri et al., 2000a; Khosla et al., 1999; Koide et al., 1994; Szabo et al., 1998) have also been detected at imprinted loci, and a sex-specific DNA binding protein was also recently described (Birger et al., 1999). Fascinatingly, imprinted regions which are important for embryonic viability also show a distortion of Mendelian transmission ratios in meiosis (Naumova et al., 2001). Sex-specific differences in meiotic recombination frequency are yet another characteristic of imprinted loci (Paldi et al., 1995; Robinson and Lalande, 1995), as are differences in timing of DNA replication between maternal and paternal alleles (Bickmore and Carothers, 1995; Gunaratne et al., 1995; Kitsberg et al., 1993). It is difficult to establish whether such replication timing differences are a cause of mono-allelic gene expression, or merely an effect of differential chromatin organisation at these regions. It is interesting to note that generally within eukaryotic nuclei heterochromatic regions are late-replicating (Taddei et al., 1999), although it remains to be demonstrated whether late-replication plays a role in actually establishing a heterochromatic state. Certainly a number of mechanisms are utilised to ensure the maintenance of heterochromatin through replication. There is also a peculiar contradiction, in that although it is principally the maternal alleles of imprinted loci which are methylated, the maternal allele replicates earlier in S-phase than the paternal. It is therefore more likely that these replication timing differences are established in the germ line and reinforced during the first S-phase in the zygote (Simon et al., 1999), when the entire maternal pronucleus enters S-phase prior to the paternal pronucleus (Adenot et al., 1997; Ferreira and Carmo-Fonseca, 1997). More generally, it is still unclear whether any of these allele-specific epigenetic marks, with the possible exception of DNA methylation, are of primary importance in mediating imprinted gene expression or are the secondary results of an underlying, more fundamental, mark.

2.4. Cis-regulatory elements at imprinted domains

Monoallelic expression of imprinted genes in the correct tissues and developmental stage requires the presence of a variety of key *cis*-regulatory elements. These include elements responsible for directing allele-specific epigenetic modifications such as methylation.

Other important regulatory elements include enhancers for tissue-specific expression and elements responsible for directing those enhancers to the appropriate promoter. The discovery that imprinted genes are commonly located in large genomic clusters, and show some level of co-ordinate regulation, suggests that *cis*-regulatory elements at imprinted loci are capable of directing their effects over large distances. In some ways this regulation is similar to that seen at complex genetic loci such as the beta-globin cluster, in which gene expression and epigenetic status is controlled over a great distance by the action of a Locus Control Region (LCR; reviewed in Li et al., 1999b). In other organisms such as *Drosophila* long range regulatory elements have been discovered at a number of loci (Dorsett, 1999).

Unfortunately, it is difficult to identify *cis*-acting sequences at imprinted loci by straight-forward DNA sequence analysis, although comparing mouse and human genomic sequence has revealed novel regulatory elements (Drewell et al., 2002; Ishihara et al., 2000). To date, most specific sequences have been identified experimentally, using trans-genic and pronuclear injection approaches. For example, "imprinting boxes" at both the *Igf2r* and PWS/AS loci have been discovered in this way. Imprinting boxes are described as regions which contain allele-specific recognition sequences and are subject to parent-of-origin specific modifications during gametogenesis (Barlow, 1993). In the mouse, the most successful technique for identifying regulatory elements is transgenesis. This has included the use of large transgenes derived from Yeast and Bacterial Artificial Chromosomes (YACs and BACs) to define regions capable of imprinting ectopically in a stable fashion (for example Ainscough et al., 1997; John et al., 2001). Deletion of regions from these large transgenes can narrow down the search for important elements, as demon-strated by the delineation of a skeletal muscle-specific repressor element by deletion from an *H19-Igf2* YAC transgene (Ainscough et al., 2000).

In humans, careful genomic analysis of patients with defects in imprinting has revealed regulatory elements capable of affecting imprinted gene expression over long distances (for example Buiting et al., 1995). Such studies in humans and mice have revealed the existence of Imprinting Centres (ICs)- regulatory regions capable of directing imprinted gene expression over a large domain. ICs appear to act in disparate ways at different loci, and a variety of regulatory functions are harboured within what might be termed an IC. The ICs at a variety of imprinted loci have been characterised- the best under-stood are those located within mouse distal chromosome 7 (human chromosome 11p15.5) at *H19/Igf2* and the BWS locus, and at the *SNRPN* (PWS/AS) locus on human chromo-some 15q11-13/mouse central chromosome 7. The combined human and mouse genetic studies have suggested that deletions at ICs can be transmitted through the germ line of one sex without effect, yet result in an imprinted disease phenotype upon inheri-tance from the opposite sex. This supports the notion that certain sequences within the IC are explicitly required to reset the imprint during gametogenesis (for example Buiting et al., 1995).

Recent work on the PWS/AS locus in humans and mice suggests that ICs may be bipar-tite, requiring two separate but interacting elements for correct function (Fig. 3), (Shemer et al., 2000). Genetic mapping analysis of PWS and AS patients led to the delineation of a 4.3 kb short region of overlap (SRO) at the *SNRPN* promoter (Buiting et al., 1995). Mutation of this region on the paternal allele leads to establishment of a maternal

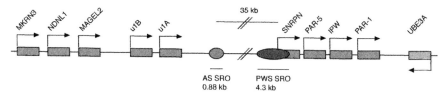

Fig. 3. The human Prader-Willi/Angelman syndrome (PWS/AS) locus on chromosome 15q11-13. The majority of the genes are paternally expressed, with only *UBE3A* transcribed from the maternal allele. The bipartite imprinting centre at the locus comprises the short regions of overlap (SRO) identified in PWS and AS patients. Deletion of 4.3 kb of the promoter and exon 1 of *SNRPN* (PWS SRO, larger ellipse) prevents the establishment of the paternal imprint in the male germ line, leading to silencing of genes on the paternal allele. Deletion of the AS SRO (smaller ellipse), 35 kb upstream of the *SNRPN* promoter, alleviates repression of genes on the maternal allele. Mouse mini-transgenes containing a small region of the mouse PWS SRO and the human AS SRO can demonstrate correct imprinting (Shemer *et al.*, 2000).

methylation pattern and subsequent lack of expression of paternal genes, resulting in PWS. Further analysis defined an 0.88 kb SRO located 35 kb upstream of the *SNRPN* promoter (Buiting et al., 1999; Ohta et al., 1999). When mutated on the maternal allele repression of genes on the maternal chromosome is alleviated, resulting in AS. Composite transgenes carrying the human AS-SRO and the mouse minimal *SNRPN* promoter are capable of imprinting correctly as judged by parent-of-origin specific methylation patterns and reporter expression, and also demonstrate correct imprint switching and replication timing (Shemer et al., 2000). The discovery of discrete elements which can mediate epigenetic events raises the interesting question of whether these elements can act interchangeably between imprinted loci.

2.5. Templates for imprinting?

The observation that the majority of methylation imprints at imprinted loci are maternal in origin, regardless of which allele is expressed, has led to the proposal of several models as to how imprinting might be regulated by allele-specific methylation and *cis*-regulatory elements (Reik et al., 2001). These may be seen as templates, upon which the mechanism of imprinting at a number of loci may be based. These models do not explain how differential methylation and other phenomena are recruited to imprinted regions, but attempt to describe how epigenetic and *cis*-regulatory information is interpreted.

Firstly, genes may be directly silenced on the methylated maternal allele via the recruitment of silencing factors (Fig. 4A). Differential methylation near the promoter of a gene may be sufficient to silence that gene. Alternatively, in clusters of imprinted genes, methylation at just one imprinting centre may be sufficient to direct repression over the entire region, thus exerting a long-range effect (Fig. 4B). It is interesting to speculate on the evolution of these clusters and long-range regulatory elements. Given that the IC's can control gene expression over large distances, it may have been necessary for genes to remain together through evolution, in order to maintain their imprinted status. This relatively simple model of imprinting is thought to apply to clusters such as the paternally expressed Prader-Willi/Angelman Syndrome (PWS/AS) region on human chromosome 15q11-13

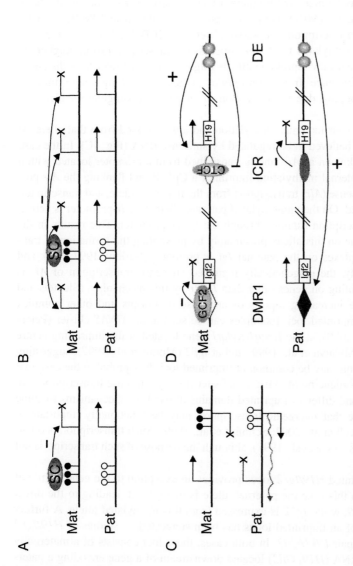

Fig. 4. Templates for imprinting. A. Methylation (filled lollipops) on the maternal allele (mat) recruits silencing complexes (SC) to exert localised gene repression (-). B. Maternal methylation may recruit silencing complexes (SC) which can exert long-range repressive effects over several genes in a cluster (-). C. Maternal methylation acts to silence a downstream intronic promoter for an antisense transcript (wiggly line). On the paternal allele (pat), this transcript is unimpeded and in turn leads to the silencing of the paternal sense promoter. D. The mechanism of imprinting at *H19/Igf2*. On the maternal allele, the Imprinting Control Region (ICR) is unmethylated (empty ellipse) allowing the protein CTCF to bind and form a boundary/insulator. This prevents the downstream enhancers (circles) from interacting with *Igf2*, instead directing them to act on *H19* and activate maternal transcription (+). In addition, the unmethylated Differentially Methylated Region 1 (DMR1, empty diamond) at the *Igf2* promoter binds silencing factors such as GCF2 to enhance repression from the maternal allele (-). On the paternal allele, the ICR is methylated (filled ellipse). This prevents the binding of CTCF, abrogating boundary function, and also recruits silencing factors to repress transcription from the paternal *H19* allele. In the absence of the boundary the downstream enhancers are free to interact with *Igf2* (+). Although regions of the *Igf2* gene are methylated, for example DMR1 (filled diamond), this does not interfere with transcription from the paternal allele.

(mouse central chromosome 7). Microdeletions in humans of a relatively small region near the *SnurpN* promoter lead to misregulation of genes located throughout the approximately 2 megabase domain (Buiting et al., 1995). This suggests that one key centre is responsible for mediating the silencing of genes on the maternal chromosome. Repression exerted by methylated regions may be mediated by methyl-CpG binding factors such as the family of methyl binding domain (MBD)-containing proteins (Ballestar and Wolffe, 2001). These might include the transcriptional repressors MeCP2, MBD2 or MBD3, which are all capable of binding to methylated DNA and suppressing transcription from neighbouring promoters. It is not clear how repressive effects might be directed over long distances by such mechanisms. Evidence from *Drosophila* suggests that long range interactions are mediated by looping due to interaction of dispersed *cis*-regulatory elements (Pirrotta, 1998).

Secondly, genes which bear maternal methylation and are expressed from that allele, as exemplified by *Igf2r*, are believed to be regulated by antisense RNA (Fig. 4C). In the case of the *Igf2r* this RNA is known as *Air*, and is transcribed from a promoter located within intron 2 of *Igf2r*. The maternal methylation imprint at a CpG island flanking the *Air* promoter acts to prevent antisense (*Air*) transcription from the maternal allele, and transcription of *Igf2r* occurs unhindered. On the unmethylated paternal allele the *Air* promoter is active, leading to antisense transcription running through the *Igf2r* promoter. This results in the silencing of the *Igf2r* gene on this allele, presumably by preventing the formation of transcriptional activator complexes on the paternal *Igf2r* promoter (Barlow, 1997; Reik and Walter, 1998). Importantly, the epigenetically regulated antisense transcription of *Air* is sufficient to confer imprinting on genes other than *Igf2r* by this mechanism (Sleutels and Barlow, 2001). There are increasing reports of antisense transcripts and other complex transcripts identified at imprinted loci. Examples can be seen at the *GNAS* cluster (Peters et al., 1999; Wroe et al., 2000), at the *Kcnq1/Kvlqt1* gene located at the imprinting centre within the BWS region (Mitsuya et al., 1999) and at *Igf2* (Moore et al., 1997), suggesting that this mode of regulation may be common at imprinted loci. It is probably the case that these two typical models- silencing of clusters and silencing by antisense transcription- are not mutually exclusive, and different imprinted domains utilise both mechanisms to some extent. It is also possible that intergenic transcription may be functionally important at imprinted loci (e.g., Drewell et al., 2002; Schmidt et al., 1999). Such transcription is documented at the globin locus (Ashe et al., 1997), although the purpose of such transcripts is not fully clear.

The reciprocally imprinted *H19/Igf2* locus provides an exception to the general rule of maternal methylation. In this case the paternal allele is methylated, leading to the direct silencing of paternal *H19*, while *Igf2* is expressed from this methylated allele. A further template for regulation of an imprinted locus has been suggested by studies of *H19/Igf2* and the analogous gene pair *Dlk1/Gtl2*. In both cases, these loci consists of a maternally expressed untranslated RNA (*H19*, *Gtl2*) located downstream of a gene encoding a paternally transcribed embryonic growth factor (*Igf2*, *Dlk1*; Schmidt et al., 2000; Takada et al., 2000). Both loci have regions of differential methylation, with the paternal allele being methylated. However, although it is attractive to speculate that the mechanism of imprinting is identical for these similar gene pairs there are actually significant differences between the regions, discussed below.

2.6. H19 and Igf2

H19 and *Igf2* are perhaps the best studied of all imprinted genes, and the mechanism of their imprinting is gradually becoming apparent- the basic mechanism is shown in Fig. 4D. The two genes are known to share the same enhancers, located downstream of *H19*, for expression in endodermal and some mesodermal tissues (Ishihara et al., 2000; Leighton et al., 1995). On the unmethylated maternal allele, the ICR located at *H19* acts as a boundary/insulator element, mediated by the zinc finger protein CTCF (Bell and Felsenfeld, 2000; Hark et al., 2000; Holmgren et al., 2001; Kanduri et al., 2000b). This boundary function blocks the access to *Igf2* by the downstream enhancers, meaning they are targeted to the *H19* promoter. On the paternal allele, the ICR becomes methylated, preventing CTCF binding and subsequent boundary formation. This leaves the downstream enhancers free to interact with *Igf2*. The methylated ICR also acts as a tissue-specific silencer, repressing *H19* expression in principally endodermal tissues (Drewell et al., 2000). Additionally it should be noted that the *Igf2* promoter is not differentially methylated, unlike that of *H19*, and remains potentially active on both parental alleles (Sasaki et al., 1992). A mesoderm-specific silencer element was recently identified within DMR1 of *Igf2*. The protein GCF2 is thought to mediate the silencing activity of this regulatory elements. It is not clear how this relates to the boundary model as presumably the boundary is blocking all downstream enhancers from acting on *Igf2*. The recent discovery of novel mesodermal enhancers located between *H19* and *Igf2* sheds some light on the important role of this silencer and also suggests that imprinting at the *H19/Igf2* locus may not be as clear-cut as currently imagined (Drewell et al., 2002).

It was recently published that similar CTCF consensus binding sites had been identified upstream of the *Gtl2* promoter, indicating the exciting possibility that these two loci may be regulated in a similar way (Wylie et al., 2000). However, closer analysis reveals that these CTCF sites are in fact located within the first intron of the gene (Ferguson-Smith and Surani, 2001), making it unclear as to how these might mediate imprinting in a similar fashion to those at *H19*. Although the CTCF-mediated boundary model may be exclusive to *H19/Igf2*, it is highly likely that certain mechanisms employed at this locus will be utilised elsewhere. These might include epigenetically regulated silencer elements, for example as seen at the *ZAC/HYMAI* locus on human chromosome 6q24/mouse chromosome 10 a differentially methylated CpG island acts as a silencer for adjacent genes (Arima et al., 2001). It is also likely that the regulatory motif of restricting and controlling access of enhancers to promoters by means of boundaries/insulators is also reiterated at other imprinted loci.

As with all generalisations there are exceptions. The paternally transcribed Neuronatin, expressed primarily in the developing brain and limbs, appears to be a solitary imprinted gene. In addition, *Neuronatin* lies within an intron of a larger, biallelically expressed gene (John et al., 2001). This discovery is contrary to the current dogma suggesting that all imprinted genes are located within clusters, and raises the question that there may be other isolated imprinted genes scattered throughout the genome. Given how little we know about the fundamental differences between the parental genomes, it may also be possible that many "normal" genes are in fact imprinted to some extent. The fact that overt phenotypes are not observed for disomies of all chromosomes is not necessarily incompatible with this hypothesis. Such an idea is supported by the finding that some genes located within complex clusters, such as the BWS locus on mouse distal chromosome 7, show relaxed patterns of imprinting, with mono-allelic expression in only a subset of tissues at restricted points during development (Paulsen et al., 2000). The imprinted *U2afrs-1* gene also shares some similarity with *Neuronatin*, as it is also a single imprinted gene located within the intron of a biallelically expressed gene (Nabetani et al., 1997).

3. Epigenetic modifications and the germ line

What is clear is that the imprint, whatever it may be, is placed on the chromosomes in the germ line according to the sex of the parent. In the case of females, imprints are established during the growth phase of oocyte development (Bao et al., 2000; Obata et al., 1998), while in males this occurs during spermatogenesis (Ueda et al., 2000). Imprints must also be completely erased in the developing germ line of the offspring, in order to be reset appropriately according to the sex of the embryo (Fig. 2). In addition, the imprints must be interpreted and maintained throughout development, particularly during the complex epigenetic changes encountered during preimplantation development.

To understand imprinting we have to understand the key events of establishment and erasure of epigenetic marks occurring in the developing mammalian germ line. The primordial germ cell (PGC) lineage arises from the proximal epiblast of the mouse embryo at around 7.2 days post coitum (dpc). The PGCs then migrate into the embryo and colonise the developing genital ridges alongside somatic gonadal support cells. The only time and place where imprints can be erased is in PGCs. But the erasure of imprints is only one of a number of features unique to germ cells. They are also the only cells that undergo genetic recombination during meiosis. Perhaps the most fundamental unique property of the germ line is immortality, in the form of its capacity to form a whole new organism, including the germ line of the next generation.

3.1. Origins of the germ line

In most of these aspects, mouse PGCs are very similar to PGCs of many other organisms for example their migration activity and even some proteins important for PGC development are conserved (reviewed in Wylie, 1999). For example the vasa and nanos proteins are essential in the germ lines of *Drosophila*, Zebrafish, *Xenopus* and mouse, although the time point of expression varies (Tanaka et al., 2000; Toyooka et al., 2000). However, unlike most other organisms, there is no evidence for maternally deposited germ plasm inherited by germ cells in mammals (Zernicka-Goetz, 1998). Heterotropic transplantation experiments showed that distal cells of the mouse embryo are able to give rise to germ cells if they are grafted into proximal epiblast tissue at 6.5 dpc. In contrast, proximal epiblast cells transplanted distally will not give rise to germ cells but instead take on a distal cell fate (Tam and Zhou, 1996). These elegant experiments show that by 6.5 dpc murine PGCs are not yet determined and suggests that the fate of PGCs precursors are induced by signals from surrounding cells. Little is known about the signals required for initial specification of the germ line in the mouse. Members of the bone morphogenetic protein (BMP) family are expressed in the extraembryonic ectoderm and are necessary for the formation of PGCs (Lawson et al., 1999; Ying et al., 2000) and may also be sufficient for initial specification of germ cell fate (Yoshimizu et al., 2001). Even less is known about the signals influencing early PGCs to maintain/regain a pluripotent state, as characterised by the expression of transcription factors such as Oct4 (Pesce and Scholer, 2000).

3.2. Erasure of imprints occurs only in the germ line

It is likely that the early founder population of murine PGCs, specified from 7.2 dpc onwards, inherits parental imprints as found in somatic cells. The DNA of these early PGCs is highly methylated and shows somatic imprinting patterns, reflecting the shared ancestry of PGCs with the surrounding embryonic cells (Brandeis et al., 1993; Kafri et al., 1992; Monk et al., 1987; Tada et al., 1998; Tam and Zhou, 1996). However, during the development and differentiation of the germ cells it is crucial to erase these existing marks, in order to ensure that germ cells pass on the correct signals of their own sex.

The initial step of erasure of the methylation marks occurs in PGCs at about 11.5/ 12.5 dpc, after the majority of PGCs have reached the genital ridge (Fig. 5; Hajkova et al., 2002). These modifications include genome-wide demethylation (Kafri et al., 1992; Monk et al., 1987) erasure of allele-specific methylation at imprinted genes (Szabo and Mann, 1995; Tada et al., 1997) and the re-activation of the inactive X chromosome in females (Brockdorff and Duthie, 1998). The nature of the signal which initiates genomic erasure in PGCs is also a mystery. One could imagine that the surrounding environment of the developing gonad may interact with the PGCs, leading to epigenetic changes. The erasure takes place only shortly before mitotic or meiotic arrest in the male or female gametes, respectively, processes which may also be initiated by signals from the surrounding gonadal cells (Matsui, 1998; Nakatsuji and Chuma, 2001). This careful timing of demethylation may be important for preventing chromosome structural abnormalities due to the error-prone replication of hypomethylated DNA (Chen et al., 1998).

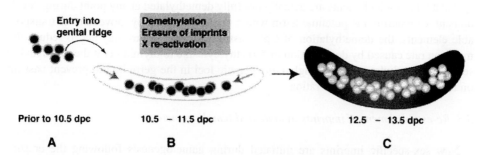

Fig. 5. Demethylation of PGCs in vivo. A. Migrating PGCs have a somatic methylation pattern (indicated by red nuclei) at the time they colonise the sexually bipotential developing genital ridges (10.5–11.5 days post coitum). B. Reprogramming of the PGC genome may occur in response to signals from the surrounding gonadal cells (blue arrows). This reprogramming is seen as demethylation of the genome, erasure of parental imprints and reactivation of the silent X chromosome in female cells. C. By 12.5 dpc the demethylation of single copy loci is complete (indicated by pink nuclei). Note that the gonads are sexually dimorphic at this point. (*For a coloured version of this figure, see plate section, page 279.*)

There are two useful cell culture systems for studying this genomic erasure and reprogramming. ES (Embryonic Stem) cells are derived from the inner cell mass of mouse blastocysts, while EG (Embryonic Germ) cells are derived from developing PGCs. Both cell types can contribute to all body tissues, including the germ line, in chimaera experiments (Rossant and Spence, 1998). This represents a gain of function in the case of EG cells, as uncultured PGCs are unable to do this. EG and ES cells also express Oct4, a transcription factor which is known to be a marker of pluripotency (Pesce and Scholer, 2001). In addition, all X chromosomes are re-activated in female ES/EG cells and extensive hypomethylation of the genome including biallelic expression of some imprinted genes can be detected. There is a strong *trans*-modification activity present in both ES cells and EG cells (derived from 12.5 dpc embryos), as demonstrated by cell fusion experiments in which EG cells or ES cells were fused to thymic lymphocytes. This resulted in extensive reprogramming of the somatic nucleus, imparting all the typical features of pluripotent cells mentioned above including the potency to contribute to chimeras (Tada et al., 1997, 2001). However, the dominant ability to erase genomic imprints is exclusive to EG cells-fusion experiments with ES cells do not result in erasure of methylation imprints. This very important finding demonstrates that indeed erasure of imprints occurs only in the germ line.

Although demethylation and genomic erasure in the germ line is essential for re-establishment of imprints, it is also a highly risky procedure for the organism. Retrotransposable elements become activated by demethylation, which could lead to their mobilisation and subsequent disruption of the genome (Whitelaw and Martin, 2001). More methylation is found at transposable elements than genes, which is unsurprising if one considers that 40–50% of the genome is made up of these elements, while only 2% is coding sequence. Obviously methylation at these elements is important for maintaining genomic stability, so how is this compatible with the need to erase and reset imprints? This may be achieved by utilising different mechanism of demethylation, remethylation and protection for imprinted loci and retroelements. The methylation status of retroelements in the germ line and during preimplantation embryogenesis has not been fully investigated, and one could speculate that such elements are actually not fully demethylated at any point during development. Countering the potential harm which might be caused by movement of transposable elements, the demethylation of CpGs associated with genes may act to reduce the mutation rate caused by deamination of 5-methylcytosine to thymidine. It therefore makes sense to demethylate at least single-copy genetic loci in the germ line to prevent passing on mutations to the next generation.

3.3. Re-establishment of imprints in male and female gametes

New sex-specific imprints are initiated during gametogenesis following the erasure step. In the male germ line this takes place at the prospermatogonial stage from 16.5 dpc onwards (Brandeis et al., 1993; Kafri et al., 1992). At around this time the male germ cells re-enter the mitotic cell cycle before entering meiosis. Although parental methylation imprints are erased at this time, some data suggest that the two alleles of *H19* become remethylated asynchronously, with the paternally inherited allele being remethylated earlier. This is indicative of some remaining epigenetic marks distinct from DNA methylation

(Davis et al., 2000). In the female germ line maternal imprints are apparently introduced after the quiescent oocyte resumes growth after birth. Little is known about this event, principally because the timing of oocyte maturation can vary substantially (Obata et al., 1998). The direction of the correct parental imprints to the appropriate genomic loci requires both germ line-specific factors and *cis*-regulatory elements to target these factors. Numerous deletion studies have highlighted the importance of certain *cis*-elements in erasing and resetting germ line imprints. Although to date no specific protein factors have been found restricted to the germ line of one sex or the other, one can imagine a complex network of factors which are not necessarily germ line specific but expressed in a certain combination at a certain time that is exclusive to the germ line. *Dnmt3L*, mentioned in Section 2.1, is also a good candidate for a protein involved in re-establishment of imprints in the germ line.

3.4. Alternative mechanisms for epigenetic regulation

Although DNA methylation is vital for mammalian development as demonstrated in Dnmt target mutations it cannot be the only existing mechanism by which organisms fulfil epigenetic gene regulation. In the examples of the preimplantation embryo and the germ line discussed here, it is clear that loci are demethylated at many key stages. In addition, although most imprinted genes have differentially methylated regions there are exceptions (for instance Mash2, Tanaka et al., 1999). Methylation is mostly utilised for gene silencing and it is known from studying other organisms that alternative mechanisms of silencing are found specifically in the germ line. Importantly, many eukaryotes do not methylate their DNA at all and still exhibit epigenetic gene regulation. Therefore it might be interesting to consider DNA methylation as one mark which interacts with other conserved epigenetic regulatory networks that operate in other organisms.

Protein modifications are both reversible and versatile, making this another suitable mechanism for epigenetic regulation in the germ line. One of the best-studied proteins in this context are histones- the proteins responsible for packaging DNA into higher order structure. The connection between histone modification and DNA methylation (and therefore imprinting) is growing rapidly (reviewed by Zhang and Reinberg, 2001). Another conserved mechanism of epigenetic modulation is found in the form of the Polycomb group proteins, first characterised in *Drosophila* (reviewed by Ringrose and Paro, 2001). The Polycomb group (PcG) of proteins is a highly conserved protein family, involved in organising higher order chromatin structure. A connection between these proteins and imprinting was recently shown for the mammalian PcG member eed ("embryonic ectoderm development"; Wang et al., 2001). The murine eed is a highly conserved member of the family and was first described as *Extra sex combs* (*esc*) in *Drosophila* (Struhl, 1983). Homologues are present in many other model organisms. Mice with a targeted mutation of *eed* die at around 8.5 dpc of gestation displaying disrupted anterioposterior patterning of the primitive streak (Schumacher et al., 1996). Recently, a role for *eed* in imprinting of the paternal X-chromosome in extraembryonic tissues was discovered. Using a paternally inherited X-linked GFP transgene, which is normally silenced in the extraembryonic tissues of female embryos, Wang et al. could show that this inactivation is lost in *eed* null embryos. The normally silenced paternally inherited X-chromosome was reactivated in the extraembryonic tissues, pointing to an important role for *eed* in maintaining imprinted gene silencing in at least this case (Wang et al., 2001). A connection to random X inactivation or genomic imprinting in the embryo remains to be examined.

Eed is always found in a complex with Enhancer of zeste (E(z), Ezh in mammals), another polycomb group protein. Significantly, these are the two most highly conserved members of the PcG family, and are present in a wide variety of organisms. The eed/Ezh complex can be detected earlier in embryonic development than other PcG complexes in a number of model organisms, suggesting a fundamental role in development (Jones et al., 1998; Sewalt et al., 1998; Shao et al., 1999; van Lohuizen et al., 1998). Similar to *eed*, targeted mutation of the mouse *Ezh* homologue leads to early lethality in embryos and a failure to maintain pluripotency in null ES cells (O'Carroll et al., 2001). Interestingly, the Ezh protein also has a SET domain, the catalytic domain responsible for methylating histones, although no histone methyltransferase activity has been reported for Ezh to date. In mice *eed* also interacts with a histone deacetylase (van der Vlag and Otte, 1999), suggesting there may be a complex regulatory network involving PcG proteins and higher order chromatin structure. Evidence of a further connection between other epigenetic mechanisms and PcG proteins has come from studies of *Drosophila* dMi-2, another member of the SWI2/SNF2 family of ATPases. dMi2 binds preferentially to methylated DNA and is a component of a histone deacetylase complex. This binds to the transcriptional repressor hunchback and interacts genetically with Polycomb group proteins in flies, suggesting PcG repression works in conjunction with histone deacetylation (Kehle et al., 1998). In vitro experiments on *Xenopus* extracts and mammalian cultured cells also show that the vertebrate Mi-2 homologue connects DNA methylation, chromatin remodelling and histone deacetylation in vertebrates (Wade et al., 1999). PcG proteins and related factors may therefore be of crucial importance in the germ line and the preimplantation embryo, where epigenetic regulation is not achieved by DNA methylation, although the roles of such proteins in the mouse remain to be determined.

Further indications of the importance of PcG proteins in the germ line come from studies of other organisms. In *C. elegans*, for example, these proteins are key mediators of transcriptional repression in the germ line (Seydoux and Strome, 1999). Mes-2 and Mes-6 represent the entire family of PcG proteins in *C. elegans* and are the homologues of *Ezh* and *Eed*, respectively. *Mes-2* and *Mes-6* are essential for normal proliferation and viability of the germ line: mutations cause a "grandchildless" phenotype. In addition they are important for silencing transgene arrays in the germ line (Holdeman et al., 1998; Kelly and Fire, 1998; Korf et al., 1998). Similarly, the PcG homologues of the flowering plant *Arabidopsis* play a role in germ cell formation and epigenetic imprinting. Homologues of *ezh* and *eed* are found here in the form of the *Medea* (*Mea*) and *Fie* genes, respectively (Grossniklaus et al., 1998; Ohad et al., 1999). The presence of a wild-type *Mea* allele in the genome of the female gametophyte (egg and central cell which gives rise to the endosperm after fertilisation) is essential for the survival of the embryo, regardless of the presence of a wild type paternal allele (Grossniklaus et al., 1998). This suggests that maternal *Mea* products are essential before zygotic expression starts and are stored in the female gametophyte, or that the *Mea* locus is imprinted, with only the maternal copy expressed. In fact both are true. Mea protein is provided maternally in the early embryo (Grossniklaus et al., 1998) and *Mea* is also expressed zygotically only from the maternal allele in the endosperm (Kinoshita et al., 1999) and young seeds (Vielle-Calzada et al., 1999). Monoallelic expression was also observed for *fie* in early seed development (Luo et al., 1999). Genetic approaches with *Methyltransferase1* (*Met1*) mutants showed a requirement for DNA methylation for some aspects of this imprinting (Adams et al., 2000; Luo et al., 1999). In addition, paternal *mea* silencing is reliant to some extent on the *ddm1* gene product. This is a member of the SWI/SNF family of DNA-dependent ATPase nucleosome remodelling factors- proteins which alter DNA-nucleosome interactions (Jeddeloh et al., 1999). Interestingly, although *ddm1* does not directly affect the plant methyltransferase enzymes, *ddm1* mutants show a 70% decrease in genome methylation compared to wild type plants (Vongs et al., 1993). Although *Arabidopsis* is clearly not the same as the mouse, by investigating the mammalian homologues of some of these key proteins we may begin to understand the epigenetic silencing processes at work in the germ line.

4. Preimplantation development

Preimplantation development is the second key stage of development involved in the maintenance of imprints, and is a complex time in terms of epigenetic changes. Fertilisation brings together two haploid genomes bearing distinct characteristics and epigenetic information. The sperm genome is transcriptionally inert and tightly compacted, with protamines replacing histones as the principal DNA packaging material. In terms of proteins and RNA, the male contributes relatively little to the zygote- a stock of maternal mRNAs and proteins supplied in the oocyte are required to sustain development until zygotic genes are transcribed. The parental genomes apparently remain as distinct nuclear entities for the first cell cycle in the form of the male and female pronuclei, and remain spatially segregated in the two- and four-cell embryo (Mayer et al., 2000b). The oocyte cytoplasm therefore has the capacity to profoundly influence the early stages of development and discriminate between the parental genomes, as can be seen in diverse organisms. For example, *Wolbachia* bacteria infecting many insect species can produce incompatibility between sperm and egg, leading to the breakdown of the paternal genome. This may even be responsible for driving species evolution (Bordenstein et al., 2001). In mammals, maternal modifier factors can affect gene expression by targeting transgenes for epigenetic modification in the zygote (Allen et al., 1990; Pickard et al., 2001), although it is not clear at the moment whether such activities may be responsible for driving evolution and speciation in mammals. However, the overbearing influence of the maternal contribution may have been an important factor in establishing epigenetic differences between the parental genomes during the evolution of imprinting.

4.1. Epigenetic changes in the early embryo

From fertilisation onwards a variety of complex epigenetic changes occur to the male and female genomes, including changes in DNA methylation and histone acetylation (Fig. 6). The sperm and oocyte chromatin must also be remodelled and organised into pronuclear structures, essentially reprogramming the gametes for development. Within a few hours of sperm entry, the sperm protamines which package the male DNA are removed

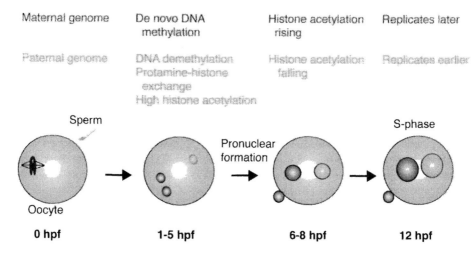

Fig. 6. Epigenetic events in early development. The maternal genome is represented in pink, while the paternal genome is blue. Timings are given as hours post-fertilisation (hpf). Between 1and 5 hpf a number of epigenetic events occur in the zygote. On the paternal genome, protamines are exchanged for histones, there is a rapid global DNA demethylation event, and high levels of histone acetylation can be detected. In contrast, the maternal genome undergoes some *de novo* DNA methylation. Later, between 6 and 8 hpf, levels of histone acetylation fall on the paternal genome, while rising on the maternal genome. S-phase (DNA replication) occurs around 12 hpf, with the paternal genome replicating prior to the maternal genome. (*For a coloured version of this figure, see plate section, page 280.*)

and replaced with histones (Nonchev and Tsanev, 1990). Shortly after fertilisation there is a dramatic and active demethylation of the paternal genome, as demonstrated by immuno-staining for methyl-cytosine and bisulphite sequencing analysis (Mayer et al., 2000a; Oswald et al., 2000; Santos et al., 2002). This paternal demethylation does not occur in animals that do not exhibit genomic imprinting, such as zebrafish (Macleod et al., 1999). Bisulphite sequencing analysis has also revealed an increase in *de novo* DNA methylation on the maternal genome at this time. Such a methylation process may be important for establishing correct maternal methylation imprints at certain loci (see later, El-Maarri et al., 2001; Reik and Walter, 2001). It is unknown how these two opposite methylation processes are regulated within the shared cytoplasm of the zygote.

Although *de novo* methyltransferases are well documented, little is known about the reverse process. One problem with achieving demethylation by directly removing the methyl group is the large kinetic barrier to this reaction. Recently, a putative *de novo* demethylase for the methyl CpG binding protein MBD2 was reported (Bhattacharya et al., 1999), although proof of its activity in this respect has not been confirmed by other labo-ratories (for example, Hendrich et al., 2001). The discovery of a ribozyme with demethy-lase activity was also discredited. An alternative mechanism for demethylation is the deamination of 5-methylcytosine, followed by mismatch repair. Another MBD protein, MBD4, has been proposed as a demethylating enzyme due to its T/G mismatch glycosy-lase activity (Hendrich et al., 1999). However, such a reaction requires a hemi-methylated substrate, so it is unlikely that this is the paternal genome demethylating activity in the zygote. It is also debatable as to whether the process of active demethylation actually exists (Bestor, 2000) in the absence of a characterised active demethylase enzyme there remains at least a possibility that it does not. As well as the active demethylation of the paternal genome at the zygotic stage there is also genome-wide demethylation at the morula stage of development (Kafri et al., 1993; Monk et al., 1987). It is not clear whether this demethylation is passive (i.e. as a result of failure to maintain methylation following DNA replication) or active. Some CpG sites at imprinted loci escape this global demethylation (e.g., Olek and Walter, 1997; Stoger et al., 1993; Tremblay et al., 1995) although it is unknown how these marks are preserved.

The somatic form of the putative maintenance DNA methyltransferase Dnmt1 is also absent from preimplantation embryos, although sex-specific promoters and additional 5' exons result in short forms which are specific for the male and female germ lines, and the early embryo. One of these variants, Dnmt1o, accumulates in the oocyte during the growth phase when methylation patterns and genomic imprints are established. This hints at a role for Dnmt1o as a *de novo* methyltransferase (Mertineit et al., 1998). Dnmt1o is locked at the cytoplasmic cortex in most stages of preimplantation embryogenesis, only entering and exiting the nucleus briefly at the eight-cell stage (Cardoso and Leonhardt, 1999; Carlson et al., 1992; Mertineit et al., 1998; Trasler et al., 1996) before being replaced by the somatic form after implantation. When the Dnmt1o-specific regions of the *Dnmt1* gene were deleted in mice, no effect is detected in animals homozygous for the deletion. However, when null mothers are crossed to wild-type males the resulting heterozygous foetuses die during late gestation. Interestingly, maternal imprints in the oocyte are not affected but methylation analysis of the foetuses showed a 50% postzygotic loss of normal methylation patterns at some imprinted loci, resulting in a loss of allele-specific expression. It can be concluded from this data that Dnmt1o is required for maintenance of methylation pattern of at least some imprinted genes during the process of global demethylation in pre-implantation embryos (Howell et al., 2001), although its role in establishing maternal imprints in the germ-line is disputable. By inference, there must also be methyltransferases functional both before and after the 8-cell stage in order to maintain essential methylation patterns.

4.2. Establishing and maintaining epigenetic asymmetry

It is not clear how the epigenetic asymmetry between the parental genomes in the zygote is defined and maintained. Various systems may be involved, including histone modifications, heterochromatic proteins and transcriptional activators. Histone modifications, including acetylation, methylation, phosphorylation and ubiquitination play a key role in transcriptional regulation (reviewed by Rice and Allis, 2001). Acetylation of lysine residues within the N-terminal tails of histones H3 and H4 principally contributes to gene activation while deacetylation is associated with gene silencing (Kuo and Allis, 1998). Differential histone acetylation of parental alleles has been detected at several imprinted loci, including *H19/Igf2* (Grandjean et al., 2001; Pedone et al., 1999), *U2af1-rs1* and *Snrpn* (Fulmer-Smentek and Francke, 2001; Gregory et al., 2001), and *Igf2r* (Hu et al., 2000). It is important to note that there are several lysine residues available for acetylation in H3 and H4, and modification at each of these may have a subtly different effect.

For example, modification of histone H3 but not H4 is associated with transcriptional effects at the *U2af-rs1* and *Snrpn* genes (Fulmer-Smentek and Francke, 2001; Gregory et al., 2001). Histone methylation has also recently received close scrutiny for a possible role in transcriptional control (Jenuwein, 2001). Histone H3 methylated at lysine 9 creates a binding site for the chromodomain of HP1 proteins (Bannister et al., 2001; Lachner et al., 2001; Nakayama et al., 2001), although there are other sites for histone methylation which are currently under investigation (Jenuwein and Allis, 2001). A recent report indicated the existence of allele-specific lysine 9 histone H3 methylation at the human PWS locus (Xin et al., 2001) suggesting that histone methylation may also play an important role in genomic imprinting. One might also predict that histone methylation and the subsequent recruitment of heterochromatic complexes, might be different between the parental genomes during gametogenesis and early development.

Investigation of histone H4 acetylation in the zygote by immunostaining reveals a rapid hyperacetylation of the paternal chromatin immediately after fertilisation, presumably correlating with the period of active demethylation and histone assembly (Adenot et al., 1997). No hyperacetylation can be detected on the maternal genome at these early stages, and this differential staining is maintained throughout G1 phase of the first cell cycle, equilibrating between the genomes by S/G2 phase independently of DNA replication. Given these findings, it is tempting to suggest that histone methylation differences also exist between the parental genomes in the zygote. One could visualise a model whereby recruitment of heterochromatic complexes to the maternal genome, whether mediated by histone methylation or not, may be responsible for protecting the maternal DNA from demethylation. The lack of heterochromatic complexes on the paternal genome therefore leaves it vulnerable to demethylation activity. There is currently also no explanation as to how *de novo* methylation is recruited to the maternal genome after fertilisation (Oswald et al., 2000). The discovery of factors and mechanisms by which these differences are established will have great significance for the establishment and reinforcement of parental imprints.

4.3. Transcriptional competency in the zygote

When small DNA fragments from the *Igf2r* locus imprinting box are injected into the maternal pronucleus, they become marked for subsequent DNA methylation during preimplantation embryogenesis. When injected into the paternal pronucleus, no such marking or methylation is seen (Birger et al., 1999). This suggests that an active mechanism is at work in the maternal chromatin, which is capable of targeting DNA methyltransferase activity to the appropriate sites in the genome. The paternal pronucleus also displays a much higher transcriptional competency than the maternal pronucleus in the zygote, as demonstrated by injection of reporter plasmids into either pronucleus (Aoki et al., 1997). This is prior to establishment of a more generally repressive chromatin state on the paternal genome, concurrent with the first S-phase (Rastelli et al., 2001), although the onset of zygotic gene activation actually requires this replication (Forlani et al., 1998). The repressive state can be reconstituted in the paternal pronucleus by the co-injection of unacetylated histones with the reporter plasmid (Rastelli et al., 2001). This reinforces the suggestion that histone modifications are of prime importance in enhancing and maintaining epigenetic states in the zygote. The further investigation of other histone modifications such as methylation and the attendant heterochromatic factors will yield important information as to how gene expression patterns and epigenetic states are established and maintained in the zygote.

We are only just beginning to understand the epigenetic modifications which occur during preimplantation development, and it is difficult to draw clear conclusions as to their precise importance and function. It is unknown how epigenetic information is preserved through the various methylation and demethylation processes which take place at this time, or how correct patterns of methylation are re-established following implantation. Little is also known about the role of endogenous *cis*-regulatory elements during very early development, although transgene injection studies have demonstrated a variety of phenomena. This include the importance of scaffold attachment regions for gene activity in the early embryo (Thompson et al., 1995), identification of sequences which can signal *de novo* methylation (Birger et al., 1999) and the importance of certain types of promoters and enhancers (Majumder and DePamphilis, 1994, 1995; Nothias et al., 1995).

5. Relevance of imprinting for stem cell and cloning technology

Headlines were made around the world in 1997, when the first successful mammalian clone was derived by nuclear transfer from an adult somatic cell into a donor oocyte (Wilmut et al., 1997). Since the creation of Dolly the cloned sheep, a number of other species have been cloned in this manner, including cows, pigs and mice. However, the failure rate in achieving live births in cloning experiments remains extremely high, although development to the blastocyst stage is relatively common, at least in the mouse (Wakayama et al., 1998; Wakayama and Yanagimachi, 1999). A major problem in these studies appears to be highly variable epigenetic reprogramming, for example at imprinted domains (Humpherys et al., 2001; Rideout et al., 2001). It is currently not clearly understood to what extent the somatic genome needs to be remodelled in order to direct development— certainly a totipotent state needs to be achieved, although one would imagine this ought not to extend to disruption of existing (correct) parental imprints. The role of epigenetic modifications in controlling totipotency is unknown, and an area for much future study.

Recently, both the scientific and popular press have focused much attention on the possible therapeutic applications of stem cell technology, ranging from overtly cautious to hysterical. Some of these therapies are based on the derivation of immortal human stem cells from a variety of tissues, including embryonic, with the aim of transplanting them into patients. Given the power of mouse ES cells to be differentiated into a number of cell types and contribute to all tissues of the body, attention has turned to the possibility of using human ES cells in therapy. However, human ES cells do not appear to have exactly the same nature as their mouse counterparts—indeed there are several properties of human ES cells which cannot be confirmed experimentally, such as germline transmission. A major problem with transferring any foreign cells into humans is immune rejection. Some authorities suggest this problem might be overcome by so-called therapeutic cloning. This would involve creating cloned blastocysts, using somatic nuclei from the patient transferred into enucleated donor eggs. ES cells could then be derived from these blastocysts, differentiated into the required tissue type and transplanted back into the patient with no fear of an immune response. However, the significant epigenetic defects at imprinted loci in cloned mouse embryos (Humpherys et al., 2001; Rideout et al., 2001) casts doubt on the long term prospects for such therapies. It remains to be seen whether transplanted stem cells with defective imprints are capable of making a useful contribution to an organism. On a more positive note, a recent study has revealed stem cells in human skin explants. If these cells could be successfully reprogrammed to a "naïve" state, they could replace the need for embryo-derived stem cells. Again, the status of imprints in such cells would need to be determined.

It is clear that in the past 15 years the field of imprinting has grown exponentially, both in terms of understanding and complexity. In addition, our knowledge of epigenetic regulatory mechanisms has greatly increased and touches on many areas of biology. Far from being a peculiar piece of embryonic phenomenology, imprinting has now emerged as a major area of research. Epigenetics and imprinting will therefore be of key importance for future studies of reproductive biology, including stem cell and nuclear transfer experiments.

Acknowledgements

K.L.A. gratefully acknowledges support from the Elmore Bursary, Gonville and Caius College, Cambridge and the Isaac Newton Trust. S.E. gratefully acknowledges support from Boehringer Ingelheim Fonds. Work in the laboratory of M.A.S. is funded by the Wellcome Trust and the BBSRC.

Note added in proof

Since this chapter was written, a study has been published characterising the distribution of histone H3 lysine 9 methylation and HP1 proteins in the mouse zygote (Arney et al., 2002. Int. J. Dev. Biol. 46, 317–320).

References

Aapola, U., Kawasaki, K., Scott, H.S., Ollila, J., Vihinen, M., Heino, M., Shintani, A., Minoshima, S., Krohn, K., Antonarakis, S.E., Shimizu, N., Kudoh, J., Peterson, P. 2000. Isolation and initial characterization of a novel zinc finger gene, DNMT3L, on 21q22.3, related to the cytosine-5-methyltransferase 3 gene family. Genomics 65, 293–298.

Aapola, U., Lyle, R., Krohn, K., Antonarakis, S.E., Peterson, P. 2001. Isolation and initial characterization of the mouse Dnmt3l gene. Cytogenet. Cell. Genet. 92, 122–126.

Adams, S., Vinkenoog, R., Spielman, M., Dickinson, H.G., Scott, R.J. 2000. Parent-of-origin effects on seed development in Arabidopsis thaliana require DNA methylation. Development 127, 2493–2502.

Adenot, P.G., Mercier, Y., Renard, J.P., Thompson, E.M. 1997. Differential H4 acetylation of paternal and maternal chromatin precedes DNA replication and differential transcriptional activity in pronuclei of 1-cell mouse embryos. Development 124, 4615–4625.

Ainscough, J.F., Koide, T., Tada, M., Barton, S., Surani, M.A. 1997. Imprinting of *Igf2* and H19 from a 130 kb YAC transgene. Development 124, 3621–3632.

Ainscough, J. F.-X., John, R.M., Barton, S.C., Surani, M.A. 2000. A skeletal muscle-specific mouse *Igf2* repressor lies 40 kb downstream of the gene. Development 127, 3923–3930.

Allen, N.D., Norris, M.L., Surani, M.A. 1990. Epigenetic control of transgene expression and imprinting by genotype-specific modifiers. Cell 61, 853-861.

Aoki, F., Worrad, D.M., Schultz, R.M. 1997. Regulation of transcriptional activity during the first and second cell cycles in the preimplantation mouse embryo. Dev. Biol. 181, 296-307.

Arima, T., Drewell, R.A., Arney. K.L., Inoue, J., Makita, Y., Hata, A., Oshimura, M., Wake, N., Surani, M.A. 2001. A conserved imprinting control region at the HYMAI/ZAC domain is implicated in transient neonatal diabetes mellitus. Hum. Mol. Genet. 10, 1475-1483.

Ashe, H.L., Monks, J., Wijgerde, M., Fraser, P., Proudfoot, N.J. 1997. Intergenic transcription and transinduction of the human beta-globin locus. Genes. Dev. 11, 2494-2509.

Ballestar, E., Wolffe, A.P. 2001. Methyl-CpG-binding proteins. Targeting specific gene repression. Eur. J. Biochem. 268, 1-6.

Bannister, A.J., Zegerman, P., Partridge, J.F., Miska, E.A., Thomas, J.O., Allshire, R.C., Kouzarides, T. 2001. Selective recognition of methylated lysine 9 on histone H3 by the HP1 chromo domain. Nature 410, 120-124.

Bao, S., Obata, Y., Carroll, J., Domeki, I., Kono, T. 2000. Epigenetic modifications necessary for normal development are established during oocyte growth in mice. Biol. Reprod. 62, 616-621.

Barlow, D.P. 1993. Methylation and imprinting: from host defense to gene regulation? Science 260, 309-310.

Barlow, D.P. 1997. Competition—a common motif for the imprinting mechanism? EMBO J. 16, 6899-6905.

Beard, C., Li, E., Jaenisch, R. 1995. Loss of methylation activates Xist in somatic but not in embryonic cells. Genes. Dev. 9, 2325-2334.

Bell, A.C., Felsenfeld, G. 2000. Methylation of a CTCF-dependent boundary controls imprinted expression of the *Igf2* gene. Nature 405, 482-485.

Bestor, T.H. 2000. The DNA methyltransferases of mammals. Hum. Mol. Genet. 9, 2395-2402.

Bhattacharya, S.K., Ramchandani, S., Cervoni, N., Szyf, M. 1999. A mammalian protein with specific demethylase activity for mCpG DNA. Nature 397, 579-583.

Bickmore, W.A., Carothers, A.D. 1995. Factors affecting the timing and imprinting of replication on a mammalian chromosome. J. Cell. Sci. 108 (Pt 8), 2801-2809.

Birger, Y., Shemer, R., Perk, J., Razin, A. 1999. The imprinting box of the mouse Igf2r gene. Nature 397, 84-88.

Bordenstein, S.R., O'Hara, F.P., Werren, J.H. 2001. Wolbachia-induced incompatibility precedes other hybrid incompatibilities in Nasonia. Nature 409, 707-710.

Brandeis, M., Kafri, T., Ariel, M., Chaillet, J.R., McCarrey, J., Razin, A., Cedar, H. 1993. The ontogeny of allele-specific methylation associated with imprinted genes in the mouse. EMBO J. 12, 3669-3677.

Brockdorff, N., Duthie, S. M. 1998. X chromosome inactivation and the Xist gene. Cell. Mol. Life Sci. 54, 104-112.

Buiting, K., Lich, C., Cottrell, S., Barnicoat, A., Horsthemke, B. 1999. A 5-kb imprinting center deletion in a family with Angelman syndrome reduces the shortest region of deletion overlap to 880 bp. Hum. Genet. 105, 665-666.

Buiting, K., Saitoh, S., Gross, S., Dittrich, B., Schwartz, S., Nicholls, R.D., Horsthemke, B. 1995. Inherited microdeletions in the Angelman and Prader-Willi syndromes define an imprinting centre on human chromosome 15. Nat. Genet. 9, 395-400.

Cardoso, M.C., Leonhardt, H. 1999. DNA methyltransferase is actively retained in the cytoplasm during early development. J. Cell. Biol. 147, 25-32.

Carlson, L.L., Page, A.W., Bestor, T.H. 1992. Properties and localization of DNA methyltransferase in preimplantation mouse embryos: implications for genomic imprinting. Genes. Dev. 6, 2536-2541.

Cattanach, B.M., Beechey, C.V. 1990. Autosomal and X-chromosome imprinting. Dev. Suppl. 63-72.

Cattanach, B.M., Kirk, M. 1985. Differential activity of maternally and paternally derived chromosome regions in mice. Nature 315, 496-498.

Chen, R.Z., Pettersson, U., Beard, C., Jackson-Grusby, L., Jaenisch, R. 1998. DNA hypomethylation leads to elevated mutation rates. Nature 395, 89-93.

Davis, T.L., Yang, G.J., McCarrey, J.R., Bartolomei, M.S. 2000. The H19 methylation imprint is erased and re-established differentially on the parental alleles during male germ cell development. Hum. Mol. Genet. 9, 2885-2894.

Dorsett, D. 1999. Distant liaisons: long-range enhancer-promoter interactions in Drosophila. Curr. Opin. Genet. Dev. 9, 505-514.

Drewell, R.A., Arney, K.L., Arima, T., Barton, S.C., Brenton, J.D., Surani, M.A. 2002. Novel conserved elements upstream of the H19 gene are transcribed and act as mesoderm enhancers. Development 129, 1205-1213.

Drewell, R.A., Brenton, J.D., Ainscough, J. F.-X., Barton, S.C., Hilton, K.J., Arney, K.L., Dandolo, L., Surani, M.A. 2000. Deletion of a silencer element disrupts H19 imprinting independently of a DNA methylation epigenetic switch. Development 127, 3419–3428.

El-Maarri, O., Buiting, K., Peery, E.G., Kroisel, P.M., Balaban, B., Wagner, K., Urman, B., Heyd, J., Lich, C., Brannan, C.I., Walter, J., Horsthemke, B. 2001. Maternal methylation imprints on human chromosome 15 are established during or after fertilization. Nat. Genet. 27, 341–344.

Feil, R., Walter, J., Allen, N.D., Reik, W. 1994. Developmental control of allelic methylation in the imprinted mouse Igf2 and H19 genes. Development 120, 2933–2943.

Ferguson-Smith, A.C., Surani, M.A. 2001. Imprinting and the epigenetic asymmetry between parental genomes. Science 293, 1086–1089.

Ferreira, J., Carmo-Fonseca, M. 1997. Genome replication in early mouse embryos follows a defined temporal and spatial order. J. Cell Sci. 110 (Pt 7), 889–897.

Forlani, S., Bonnerot, C., Capgras, S., Nicolas, J.F. 1998. Relief of a repressed gene expression state in the mouse 1-cell embryo requires DNA replication. Development 125, 3153–3166.

Fulmer-Smentek, S.B., Francke, U. 2001. Association of acetylated histones with paternally expressed genes in the Prader—Willi deletion region. Hum. Mol. Genet. 10, 645–652.

Fundele, R.H., Surani, M.A., Allen, N.D. 1997. Consequences of genomic imprinting for fetal development. In: "Genomic Imprinting" (W. Reik and M.A. Surani, Eds.), 98–117. IRL Press, Oxford.

Georgiades, P., Watkins, M., Burton, G.J., Ferguson-Smith, A.C. 2001. Roles for genomic imprinting and the zygotic genome in placental development. Proc. Natl. Acad. Sci. USA 98, 4522–4527.

Grandjean, V., O'Neill, L., Sado, T., Turner, B., Ferguson-Smith, A. 2001. Relationship between DNA methylation, histone H4 acetylation and gene expression in the mouse imprinted Igf2-H19 domain. FEBS Lett 488, 165–169.

Gregory, R.I., Randall, T.E., Johnson, C.A., Khosla, S., Hatada, I., O'Neill, L.P., Turner, B.M., Feil, R. 2001. DNA methylation is linked to deacetylation of histone H3, but not H4, on the imprinted genes Snrpn and U2af1-rs1. Mol. Cell. Biol. 21, 5426–5436.

Grossniklaus, U., Vielle-Calzada, J.P., Hoeppner, M.A., Gagliano, W.B. 1998. Maternal control of embryogenesis by MEDEA, a polycomb group gene in Arabidopsis. Science 280, 446–450.

Gunaratne, P.H., Nakao, M., Ledbetter, D.H., Sutcliffe, J.S., Chinault, A.C. 1995. Tissue-specific and allele-specific replication timing control in the imprinted human Prader-Willi syndrome region. Genes. Dev. 9, 808–20.

Haig, D., Graham, C. 1991. Genomic imprinting and the strange case of the insulin-like growth factor II receptor. Cell 64, 1045–1046.

Hajkova, P., Erhardt, S., Lane, N., Haaf, T., El-Maarri, O., Reik, W., Walter, J., Surani, M.A. 2002. Epigenetic reprogramming in mouse primordial germ cells. Mech. Dev., in press.

Hark, A.T., Schoenherr, C.J., Katz, D.J., Ingram, R.S., Levorse, J.M., Tilghman, S.M. 2000. CTCF mediates methylation-sensitive enhancer-blocking activity at the H19/Igf2 locus. Nature 405, 486–489.

Hark, A.T., Tilghman, S.M. 1998. Chromatin conformation of the H19 epigenetic mark. Hum. Mol. Genet. 7, 1979–1085.

Hayashizaki, Y., Shibata, H., Hirotsune, S., Sugino, H., Okazaki, Y., Sasaki, N., Hirose, K., Imoto, H., Okuizumi, H., Muramatsu, M. 1994. Identification of an imprinted U2af binding protein related sequence on mouse chromosome 11 using the RLGS method. Nat. Genet. 6, 33–40.

Heard, E., Clerc, P., Avner, P. 1997. X-chromosome inactivation in mammals. Annu. Rev. Genet. 31, 571–610.

Hendrich, B., Guy, J., Ramsahoye, B., Wilson, V.A., Bird, A. 2001. Closely related proteins MBD2 and MBD3 play distinctive but interacting roles in mouse development. Genes. Dev. 15, 710–723.

Hendrich, B., Hardeland, U., Ng, H.H., Jiricny, J., Bird, A. 1999. The thymine glycosylase MBD4 can bind to the product of deamination at methylated CpG sites. Nature 401, 301–304.

Holdeman, R., Nehrt, S., Strome, S. 1998. MES-2, a maternal protein essential for viability of the germline in Caenorhabditis elegans, is homologous to a Drosophila Polycomb group protein. Development 125, 2457–2467.

Holmgren, C., Kanduri, C., Dell, G., Ward, A., Mukhopadhya, R., Kanduri, M., Lobanenkov, V., Ohlsson, R. 2001. CpG methylation regulates the Igf2/H19 insulator. Curr. Biol. 11, 1128–1130.

Howell, C.Y., Bestor, T.H., Ding, F., Latham, K.E., Mertineit, C., Trasler, J.M., Chaillet, J.R. 2001. Genomic imprinting disrupted by a maternal effect mutation in the Dnmt1 gene. Cell 104, 829–838.

Hsieh, C.L. 1999. Evidence that protein binding specifies sites of DNA demethylation. Mol. Cell Biol. 19, 46–56.

Hu, J.F., Pham, J., Dey, I., Li, T., Vu, T.H., Hoffman, A.R. 2000. Allele-specific histone acetylation accompanies genomic imprinting of the insulin-like growth factor II receptor gene. Endocrinology 141, 4428–4435.

Humpherys, D., Eggan, K., Akutsu, H., Hochedlinger, K., Rideout, W.M., Biniszkiewicz, D., Yanagimachi, R., Jaenisch, R. 2001. Epigenetic instability in ES cells and cloned mice. Science 293, 95-97.

Hurst, L.D. 1997. Evolutionary theories of genomic imprinting. In: *Genomic Imprinting* (W. Reik and M.A. Surani, Eds.), Oxford: IRL Press, pp. 211–237.

Hurst, L.D., McVean, G.T. 1998. Do we understand the evolution of genomic imprinting? Curr. Opin. Genet. Dev. 8, 701–708.

Ishihara, K., Hatano, N., Furumi, H., Kato, R., Iwaki, T., Miura, K., Jinno, Y., Sasaki, H. 2000. Comparative genomic sequencing identifies novel tissue-specific enhancers and sequence elements for methylation-sensitive factors implicated in *Igf2/H19* imprinting. Genome. Res. 10, 664–671.

Jackson-Grusby, L., Beard, C., Possemato, R., Tudor, M., Fambrough, D., Csankovszki, G., Dausman, J., Lee, P., Wilson, C., Lander, E., Jaenisch, R. 2001. Loss of genomic methylation causes p53-dependent apoptosis and epigenetic deregulation. Nat. Genet. 27, 31–39.

Jeddeloh, J.A., Stokes, T.L., Richards, E.J. 1999. Maintenance of genomic methylation requires a SWI2/SNF2-like protein. Nat. Genet. 22, 94–97.

Jenuwein, T. 2001. Re-SET-ting heterochromatin by histone methyltransferases. Trends Cell. Biol. 11, 266–73.

Jenuwein, T., Allis, C.D. 2001. Translating the histone code. Science 293, 1074–1080.

John, R.M., Aparicio, S.A., Ainscough, J.F., Arney, K.L., Khosla, S., Hawker, K., Hilton, K.J., Barton, S.C., Surani, M.A. 2001. Imprinted expression of neuronatin from modified BAC transgenes reveals regulation by distinct and distant enhancers. Dev. Biol. 236, 387–399.

John, R.M., Surani, M.A. 2000. Genomic imprinting, mammalian evolution, and the mystery of egg-laying mammals. Cell 101, 585–588.

Jones, C.A., Ng, J., Peterson, A.J., Morgan, K., Simon, J., Jones, R.S. 1998. The Drosophila esc and E(z) proteins are direct partners in polycomb group-mediated repression. Mol. Cell. Biol. 18, 2825–2834.

Kafri, T., Ariel, M., Brandeis, M., Shemer, R., Urven, L., McCarrey, J., Cedar, H., Razin, A. 1992. Developmental pattern of gene-specific DNA methylation in the mouse embryo and germ line. Genes Dev. 6, 705–714.

Kafri, T., Gao, X., Razin, A. 1993. Mechanistic aspects of genome-wide demethylation in the preimplantation mouse embryo. Proc. Natl. Acad. Sci. USA 90, 10558–10562.

Kanduri, C., Holmgren, C., Pilartz, M., Franklin, G., Kanduri, M., Liu, L., Ginjala, V., Ulleras, E., Mattsson, R., Ohlsson, R. 2000a. The 5′ flank of mouse H19 in an unusual chromatin conformation unidirectionally blocks enhancer-promoter communication. Curr. Biol. 10, 449–457.

Kanduri, C., Pant, V., Loukinov, D., Pugacheva, E., Qi, C.F., Wolffe, A., Ohlsson, R., Lobanenkov, V.V. 2000b. Functional association of CTCF with the insulator upstream of the H19 gene is parent of origin-specific and methylation-sensitive. Curr. Biol. 10, 853–856.

Kearns, M., Preis, J., McDonald, M., Morris, C., Whitelaw, E. 2000. Complex patterns of inheritance of an imprinted murine transgene suggest incomplete germline erasure. Nucleic Acids Res. 28, 3301–3309.

Kehle, J., Beuchle, D., Treuheit, S., Christen, B., Kennison, J.A., Bienz, M., Muller, J. 1998. dMi-2, a hunchback-interacting protein that functions in polycomb repression. Science 282, 1897–1900.

Kelly, W.G., Fire, A. 1998. Chromatin silencing and the maintenance of a functional germline in Caenorhabditis elegans. Development 125, 2451–2456.

Khosla, S., Aitchison, A., Gregory, R., Allen, N.D., Feil, R. 1999. Parental allele-specific chromatin configuration in a boundary-imprinting-control element upstream of the mouse H19 gene. Mol. Cell Biol. 19, 2556–66.

Killian, J.K., Byrd, J.C., Jirtle, J.V., Munday, B.L., Stoskopf, M.K., MacDonald, R.G., Jirtle, R.L. 2000. M6P/IGF2R imprinting evolution in mammals. Mol. Cell. 5, 707–716.

Kinoshita, T., Yadegari, R., Harada, J.J., Goldberg, R.B., Fischer, R.L. 1999. Imprinting of the MEDEA polycomb gene in the Arabidopsis endosperm. Plant Cell 11, 1945–1152.

Kitsberg, D., Selig, S., Brandeis, M., Simon, I., Keshet, I., Driscoll, D.J., Nicholls, R.D., Cedar, H. 1993. Allele-specific replication timing of imprinted gene regions. Nature 364, 459–463.

Koide, T., Ainscough, J., Wijgerde, M., Surani, M.A. 1994. Comparative analysis of Igf-2/H19 imprinted domain: identification of a highly conserved intergenic DNase I hypersensitive region. Genomics 24, 1–8.

Korf, I., Fan, Y., Strome, S. 1998. The Polycomb group in Caenorhabditis elegans and maternal control of germline development. Development 125, 2469–2478.

Kuo, M.H., Allis, C.D. 1998. Roles of histone acetyltransferases and deacetylases in gene regulation. Bioessays 20, 615–626.

Lachner, M., O'Carroll, D., Rea, S., Mechtler, K., Jenuwein, T. 2001. Methylation of histone H3 lysine 9 creates a binding site for HP1 proteins. Nature 410, 116–120.

Lawson, K.A., Dunn, N.R., Roelen, B.A., Zeinstra, L.M., Davis, A.M., Wright, C.V., Korving, J.P., Hogan, B.L. 1999. Bmp4 is required for the generation of primordial germ cells in the mouse embryo. Genes Dev. 13, 424–436.

Lefebvre, L., Viville, S., Barton, S.C., Ishino, F., Keverne, E.B., Surani, M.A. 1998. Abnormal maternal behaviour and growth retardation associated with loss of the imprinted gene Mest. Nat. Genet. 20, 163–9.

Leighton, P., Saam, J., Ingram, R., Stewart, C., Tilghman, S. 1995. An enhancer deletion affects both H19 and Igf2 expression. Genes Dev. 9, 2079–2089.

Li, E., Beard, C., Jaenisch, R. 1993. Role for DNA methylation in genomic imprinting. Nature 366, 362–365.

Li, E., Bestor, T.H., Jaenisch, R. 1992. Targeted mutation of the DNA methyltransferase gene results in embryonic lethality. Cell 69, 915–926.

Li, L., Keverne, E.B., Aparicio, S.A., Ishino, F., Barton, S.C., Surani, M.A. 1999a. Regulation of maternal behavior and offspring growth by paternally expressed Peg3. Science 284, 330–333.

Li, Q., Harju, S., Peterson, K.R. 1999b. Locus control regions: coming of age at a decade plus. Trends Genet. 15, 403–408.

Luo, M., Bilodeau, P., Koltunow, A., Dennis, E.S., Peacock, W.J., Chaudhury, A.M. 1999. Genes controlling fertilization-independent seed development in Arabidopsis thaliana. Proc. Natl. Acad. Sci. USA 96, 296–301.

Macleod, D., Clark, V.H., Bird, A. 1999. Absence of genome-wide changes in DNA methylation during development of the zebrafish. Nat. Genet. 23, 139–140.

Majumder, S., DePamphilis, M.L. 1994. TATA-dependent enhancer stimulation of promoter activity in mice is developmentally acquired. Mol. Cell. Biol. 14, 4258–4268.

Majumder, S., DePamphilis, M.L. 1995. A unique role for enhancers is revealed during early mouse development. Bioessays 17, 879–889.

Matsui, Y. 1998. Developmental fates of the mouse germ cell line. Int. J. Dev. Biol. 42, 1037–1042.

Mayer, W., Niveleau, A., Walter, J., Fundele, R., Haaf, T. 2000a. Demethylation of the zygotic paternal genome. Nature 403, 501–502.

Mayer, W., Smith, A., Fundele, R., Haaf, T. 2000b. Spatial separation of parental genomes in preimplantation mouse embryos. J. Cell. Biol. 148, 629–634.

McGrath, J., Solter, D. 1984. Completion of mouse embryogenesis requires both the maternal and paternal genomes. Cell 37, 179–183.

Mertineit, C., Yoder, J.A., Taketo, T., Laird, D.W., Trasler, J.M., Bestor, T.H. 1998. Sex-specific exons control DNA methyltransferase in mammalian germ cells. Development 125, 889–897.

Mitsuya, K., Meguro, M., Lee, M.P., Katoh, M., Schulz, T.C., Kugoh, H., Yoshida, M.A., Niikawa, N., Feinberg, A.P., Oshimura, M. 1999. LIT1, an imprinted antisense RNA in the human KvLQT1 locus identified by screening for differentially expressed transcripts using monochromosomal hybrids. Hum. Mol. Genet. 8, 1209–1217.

Monk, M., Boubelik, M., Lehnert, S. 1987. Temporal and regional changes in DNA methylation in the embryonic, extraembryonic and germ cell lineages during mouse embryo development. Development 99, 371–382.

Moore, T., Constancia, M., Zubair, M., Bailleul, B., Feil, R., Sasaki, H., Reik, W. 1997. Multiple imprinted sense and antisense transcripts, differential methylation and tandem repeats in a putative imprinting control region upstream of mouse Igf2. Proc. Natl. Acad. Sci. USA 94, 12509–12514.

Morgan, H.D., Sutherland, H.G., Martin, D.I., Whitelaw, E. 1999. Epigenetic inheritance at the agouti locus in the mouse. Nat. Genet. 23, 314–318.

Nabetani, A., Hatada, I., Morisaki, H., Oshimura, M., Mukai, T. 1997. Mouse U2af1-rs1 is a neomorphic imprinted gene. Mol. Cell. Biol. 17, 789–798.

Nakatsuji, N., Chuma, S. 2001. Differentiation of mouse primordial germ cells into female or male germ cells. Int. J. Dev. Biol. 45, 541–548.

Nakayama, J., Rice, J.C., Strahl, B.D., Allis, C.D., Grewal, S.I. 2001. Role of histone H3 lysine 9 methylation in epigenetic control of heterochromatin assembly. Science 292, 110–113.

Naumova, A.K., Greenwood, C.M., Morgan, K. 2001. Imprinting and deviation from Mendelian transmission ratios. Genome 44, 311–320.

Neumann, B., Kubicka, P., Barlow, D.P. 1995. Characteristics of imprinted genes. Nat. Genet. 9, 12–3.

Nonchev, S., Tsanev, R. 1990. Protamine-histone replacement and DNA replication in the male mouse pronucleus. Mol. Reprod. Dev. 25, 72–76.

Nothias, J.Y., Majumder, S., Kaneko, K.J., DePamphilis, M.L. 1995. Regulation of gene expression at the beginning of mammalian development. J. Biol. Chem. 270, 22077–22080.

O'Carroll, D., Erhardt, S., Pagani, M., Barton, S.C., Surani, M.A., Jenuwein, T. 2001. The polycomb-group gene Ezh2 is required for early mouse development. Mol. Cell Biol. 21, 4330–4336.

O'Neill, M.J., Ingram, R.S., Vrana, P.B., Tilghman, S.M. 2000. Allelic expression of IGF2 in marsupials and birds. Dev. Genes. Evol. 210, 18–20.

Obata, Y., Kaneko-Ishino, T., Koide, T., Takai, Y., Ueda, T., Domeki, I., Shiroishi, T., Ishino, F., Kono, T. 1998. Disruption of primary imprinting during oocyte growth leads to the modified expression of imprinted genes during embryogenesis. Development 125, 1553–1560.

Ohad, N., Yadegari, R., Margossian, L., Hannon, M., Michaeli, D., Harada, J.J., Goldberg, R.B., Fischer, R.L. 1999. Mutations in FIE, a WD polycomb group gene, allow endosperm development without fertilization. Plant. Cell. 11, 407–416.

Ohta, T., Buiting, K., Kokkonen, H., McCandless, S., Heeger, S., Leisti, H., Driscoll, D.J., Cassidy, S.B., Horsthemke, B., Nicholls, R.D. 1999. Molecular mechanism of angelman syndrome in two large families involves an imprinting mutation. Am. J. Hum. Genet. 64, 385–396.

Okano, M., Bell, D.W., Haber, D.A., Li, E. 1999. DNA methyltransferases Dnmt3a and Dnmt3b are essential for de novo methylation and mammalian development. Cell 99, 247–257.

Olek, A., Walter, J. 1997. The pre-implantation ontogeny of the H19 methylation imprint. Nat. Genet. 17, 275–6.

Oswald, J., Engemann, S., Lane, N., Mayer, W., Olek, A., Fundele, R., Dean, W., Reik, W., Walter, J. 2000. Active demethylation of the paternal genome in the mouse zygote. Curr. Biol. 10, 475–478.

Paldi, A., Gyapay, G., Jami, J. 1995. Imprinted chromosomal regions of the human genome display sex-specific meiotic recombination frequencies. Curr. Biol. 5, 1030–1035.

Paulsen, M., El-Maarri, O., Engemann, S., Strodicke, M., Franck, O., Davies, K., Reinhardt, R., Reik, W., Walter, J. 2000. Sequence conservation and variability of imprinting in the Beckwith-Wiedemann syndrome gene cluster in human and mouse. Hum. Mol. Genet. 9, 1829–1841.

Pedone, P.V., Pikaart, M.J., Cerrato, F., Vernucci, M., Ungaro, P., Bruni, C.B., Riccio, A. 1999. Role of histone acetylation and DNA methylation in the maintenance of the imprinted expression of the H19 and Igf2 genes. FEBS Lett. 458, 45–50.

Pesce, M., Scholer, H.R. 2000. Oct-4: control of totipotency and germline determination. Mol. Reprod. Dev. 55, 452–457.

Pesce, M., Scholer, H.R. 2001. Oct-4: gatekeeper in the beginnings of mammalian development. Stem. Cells 19, 271–278.

Peters, J., Wroe, S.F., Wells, C.A., Miller, H.J., Bodle, D., Beechey, C.V., Williamson, C.M., Kelsey, G. 1999. A cluster of oppositely imprinted transcripts at the Gnas locus in the distal imprinting region of mouse chromosome 2. Proc. Natl. Acad. Sci. USA 96, 3830–3835.

Pickard, B., Dean, W., Engemann, S., Bergmann, K., Fuermann, M., Jung, M., Reis, A., Allen, N., Reik, W., Walter, J. 2001. Epigenetic targeting in the mouse zygote marks DNA for later methylation: a mechanism for maternal effects in development. Mech. Dev. 103, 35–47.

Pirrotta, V. 1998. Polycombing the genome: PcG, trxG, and chromatin silencing. Cell 93, 333–336.

Rastelli, L., Robinson, K., Xu, Y., Majumder, S. 2001. Reconstitution of enhancer function in paternal pronuclei of one-cell mouse embryos. Mol. Cell. Biol. 21, 5531–5540.

Reik, W., Dean, W., Walter, J. 2001. Epigenetic reprogramming in mammalian development. Science 293, 1089–1093.

Reik, W., Walter, J. 1998. Imprinting mechanisms in mammals. Curr. Opin. Genet. Dev. 8, 154–64.

Reik, W., Walter, J. 2001. Evolution of imprinting mechanisms: the battle of the sexes begins in the zygote. Nat. Genet. 27, 255–256.

Rice, J.C., Allis, C.D. 2001. Histone methylation versus histone acetylation: new insights into epigenetic regulation. Curr. Opin. Cell. Biol. 13, 263–273.

Rideout, W.M., Eggan, K., Jaenisch, R. 2001. Nuclear cloning and epigenetic reprogramming of the genome. Science 293, 1093–1098.

Ringrose, L., Paro, R. 2001. Remembering silence. BioEssays 23, 566–570.

Robinson, W.P., Lalande, M. 1995. Sex-specific meiotic recombination in the Prader-Willi/Angelman syndrome imprinted region. Hum. Mol. Genet. 4, 801–806.

Rossant, J., Spence, A. 1998. Chimeras and mosaics in mouse mutant analysis. Trends Genet. 14, 358–363.

Santos, F., Hendrich, B., Reik, W., Dean, W. 2002. Dynamic reprogramming of DNA methylation in the early mouse embryo, Dev. Biol. 241, 172–182.

Sasaki, H., Jones, P.A., Chaillet, J.R., Ferguson-Smith, A.C., Barton, S.C., Reik, W., Surani, M.A. 1992. Parental imprinting: potentially active chromatin of the repressed maternal allele of the mouse insulin-like growth factor II (Igf2) gene. Genes. Dev. 6, 1843–1856.

Schmidt, J.V., Levorse, J.M., Tilghman, S.M. 1999. Enhancer competition between H19 and Igf2 does not mediate their imprinting. Proc. Natl. Acad. Sci. USA 96, 9733–9738.

Schmidt, J.V., Matteson, P.G., Jones, B.K., Guan, X.J., Tilghman, S.M. 2000. The Dlk1 and Gtl2 genes are linked and reciprocally imprinted. Genes. Dev. 14, 1997–2002.

Schumacher, A., Faust, C., Magnuson, T. 1996. Positional cloning of a global regulator of anterior-posterior patterning in mice. Nature 384, 648.

Sewalt, R.G., van der Vlag, J., Gunster, M.J., Hamer, K.M., den Blaauwen, J.L., Satijn, D.P., Hendrix, T., van Driel, R., Otte, A.P. 1998. Characterization of interactions between the mammalian polycomb-group proteins Enx1/EZH2 and EED suggests the existence of different mammalian polycomb-group protein complexes. Mol. Cell. Biol. 18, 3586–3595.

Seydoux, G., Strome, S. 1999. Launching the germline in Caenorhabditis elegans: regulation of gene expression in early germ cells. Development 126, 3275–3283.

Shao, Z., Raible, F., Mollaaghababa, R., Guyon, J.R., Wu, C.T., Bender, W., Kingston, R.E. 1999. Stabilization of chromatin structure by PRC1, a Polycomb. complex. Cell 98, 37–46.

Shemer, R., Hershko, A.Y., Perk, J., Mostoslavsky, R., Tsuberi, B., Cedar, H., Buiting, K., Razin, A. 2000. The imprinting box of the Prader-Willi/Angelman syndrome domain. Nat. Genet. 26, 440–443.

Simon, I., Tenzen, T., Reubinoff, B.E., Hillman, D., McCarrey, J.R., Cedar, H. 1999. Asynchronous replication of imprinted genes is established in the gametes and maintained during development. Nature 401, 929–932.

Sleutels, F., Barlow, D.P. 2001. Investigation of elements sufficient to imprint the mouse Air promoter. Mol. Cell Biol. 21, 5008–5017.

Stoger, R., Kubicka, P., Liu, C.G., Kafri, T., Razin, A., Cedar, H., Barlow, D.P. 1993. Maternal-specific methylation of the imprinted mouse Igf2r locus identifies the expressed locus as carrying the imprinting signal. Cell 73, 61–71.

Struhl, G. 1983. Role of the esc+ gene product in ensuring the selective expression of segment-specific homeotic genes in Drosophila. J. Embryol. Exp. Morphol. 76, 297–331.

Surani, M.A., Barton, S.C., Norris, M.L. 1984. Development of reconstituted mouse eggs suggests imprinting of the genome during gametogenesis. Nature 308, 548–550.

Szabo, P.E., Mann, J.R. 1995. Biallelic expression of imprinted genes in the mouse germ line: implications for erasure, establishment, and mechanisms of genomic imprinting. Genes. Dev. 9, 1857–1868.

Szabo, P.E., Pfeifer, G.P., Mann, J.R. 1998. Characterization of novel parent-specific epigenetic modifications upstream of the imprinted mouse H19 gene. Mol. Cell. Biol. 18, 6767–6776.

Tada, M., Tada, T., Lefebvre, L., Barton, S.C., Surani, M.A. 1997. Embryonic germ cells induce epigenetic reprogramming of somatic nucleus in hybrid cells. EMBO J. 16, 6510–6520.

Tada, M., Takahama, Y., Abe, K., Nakatsuji, N., Tada, T. 2001. Nuclear reprogramming of somatic cells by in vitro hybridization with ES cells. Curr. Biol. 11, 1553–1558.

Tada, T., Tada, M., Hilton, K., Barton, S.C., Sado, T., Takagi, N., Surani, M.A. 1998. Epigenotype switching of imprintable loci in embryonic germ cells. Dev. Genes. Evol. 207, 551–561.

Taddei, A., Roche, D., Sibarita, J.B., Turner, B.M., Almouzni, G. 1999. Duplication and maintenance of heterochromatin domains. J. Cell Biol. 147, 1153–1166.

Takada, S., Tevendale, M., Baker, J., Georgiades, P., Campbell, E., Freeman, T., Johnson, M.H., Paulsen, M., Ferguson-Smith, A.C. 2000. Delta-like and gtl2 are reciprocally expressed, differentially methylated linked imprinted genes on mouse chromosome 12. Curr. Biol. 10, 1135–1138.

Takagi, N., Sasaki, M. 1975. Preferential inactivation of the paternally derived X chromosome in the extraembryonic membranes of the mouse. Nature 256, 640–642.

Tam, P.P., Zhou, S.X. 1996. The allocation of epiblast cells to ectodermal and germ-line lineages is influenced by the position of the cells in the gastrulating mouse embryo. Dev. Biol. 178, 124–132.

Tanaka, M., Puchyr, M., Gertsenstein, M., Harpal, K., Jaenisch, R., Rossant, J., Nagy, A. 1999. Parental origin-specific expression of Mash2 is established at the time of implantation with its imprinting mechanism highly resistant to genome-wide demethylation. Mech. Dev. 87, 129–142.

Tanaka, S.S., Toyooka, Y., Akasu, R., Katoh-Fukui, Y., Nakahara, Y., Suzuki, R., Yokoyama, M., Noce, T. 2000. The mouse homolog of Drosophila Vasa is required for the development of male germ cells. Genes. Dev. 14, 841–853.

Thompson, E.M., Legouy, E., Christians, E., Renard, J.P. 1995. Progressive maturation of chromatin structure regulates HSP70.1 gene expression in the preimplantation mouse embryo. Development 121, 3425–3437.

Toyooka, Y., Tsunekawa, N., Takahashi, Y., Matsui, Y., Satoh, M., Noce, T. 2000. Expression and intracellular localization of mouse Vasa-homologue protein during germ cell development. Mech. Dev. 93, 139–149.

Trasler, J.M., Trasler, D.G., Bestor, T.H., Li, E., Ghibu, F. 1996. DNA methyltransferase in normal and Dnmtn/Dnmtn mouse embryos. Dev. Dyn. 206, 239–247.

Tremblay, K.D., Saam, J.R., Ingram, R.S., Tilghman, S.M., Bartolomei, M.S. 1995. A paternal-specific methylation imprint marks the alleles of the mouse H19 gene. Nat. Genet. 9, 407–413.

Tucker, K.L., Beard, C., Dausmann, J., Jackson-Grusby, L., Laird, P.W., Lei, H., Li, E., Jaenisch, R. 1996. Germline passage is required for establishment of methylation and expression patterns of imprinted but not of non-imprinted genes. Genes. Dev. 10, 1008–1020.

Ueda, T., Abe, K., Miura, A., Yuzuriha, M., Zubair, M., Noguchi, M., Niwa, K., Kawase, Y., Kono, T., Matsuda, Y., Fujimoto, H., Shibata, H., Hayashizaki, Y., Sasaki, H. 2000. The paternal methylation imprint of the mouse H19 locus is acquired in the gonocyte stage during foetal testis development. Genes Cells 5, 649–59.

van der Vlag, J., Otte, A.P. 1999. Transcriptional repression mediated by the human polycomb-group protein EED involves histone deacetylation. Nat. Genet. 23, 474–478.

van Lohuizen, M., Tijms, M., Voncken, J.W., Schumacher, A., Magnuson, T., Wientjens, E. 1998. Interaction of mouse polycomb-group (Pc-G) proteins Enx1 and Enx2 with Eed: indication for separate Pc-G complexes. Mol. Cell. Biol. 18, 3572–3579.

Vielle-Calzada, J.P., Thomas, J., Spillane, C., Coluccio, A., Hoeppner, M.A., Grossniklaus, U. 1999. Maintenance of genomic imprinting at the Arabidopsis medea locus requires zygotic DDM1 activity. Genes. Dev. 13, 2971–2982.

Vongs, A., Kakutani, T., Martienssen, R.A., Richards, E.J. 1993. Arabidopsis thaliana DNA methylation mutants. Science 260, 1926–1928.

Wade, P.A., Gegonne, A., Jones, P.L., Ballestar, E., Aubry, F., Wolffe, A.P. 1999. Mi-2 complex couples DNA methylation to chromatin remodelling and histone deacetylation. Nat. Genet. 23, 62–66.

Wakayama, T., Perry, A.C., Zuccotti, M., Johnson, K.R., Yanagimachi, R. 1998. Full-term development of mice from enucleated oocytes injected with cumulus cell nuclei. Nature 394, 369–374.

Wakayama, T., Yanagimachi, R. 1999. Cloning of male mice from adult tail-tip cells. Nat. Genet. 22, 127–128.

Wang, J., Mager, J., Chen, Y., Schneider, E., Cross, J.C., Nagy, A., Magnuson, T. 2001. Imprinted X inactivation maintained by a mouse Polycomb group gene. Nat. Genet. 28, 371–375.

Whitelaw, E., Martin, D.I. 2001. Retrotransposons as epigenetic mediators of phenotypic variation in mammals. Nat. Genet. 27, 361–365.

Wilmut, I., Schnieke, A.E., McWhir, J., Kind, A.J., Campbell, K.H. 1997. Viable offspring derived from fetal and adult mammalian cells. Nature 385, 810–813.

Wroe, S.F., Kelsey, G., Skinner, J.A., Bodle, D., Ball, S.T., Beechey, C.V., Peters, J., Williamson, C.M. 2000. An imprinted transcript, antisense to Nesp, adds complexity to the cluster of imprinted genes at the mouse Gnas locus. Proc. Natl. Acad. Sci. USA 97, 3342–3346.

Wylie, A.A., Murphy, S.K., Orton, T.C., Jirtle, R.L. 2000. Novel imprinted DLK1/GTL2 domain on human chromosome 14 contains motifs that mimic those implicated in IGF2/H19 regulation. Genome. Res. 10, 1711–1718.

Wylie, C. 1999. Germ cells. Cell 96, 165–174.

Xin, Z., Allis, C.D., Wagstaff, J. 2001. Parent-Specific Complementary Patterns of Histone H3 Lysine 9 and H3 Lysine 4 Methylation at the Prader-Willi Syndrome Imprinting Center. Am. J. Hum. Genet. 69, 1389–1394.

Xu, G.L., Bestor, T.H., Bourc'his, D., Hsieh, C.L., Tommerup, N., Bugge, M., Hulten, M., Qu, X., Russo, J.J., Viegas-Pequignot, E. 1999. Chromosome instability and immunodeficiency syndrome caused by mutations in a DNA methyltransferase gene. Nature 402, 187–191.

Yates, P.A., Burman, R.W., Mummaneni, P., Krussel, S., Turker, M.S. 1999. Tandem B1 elements located in a mouse methylation center provide a target for de novo DNA methylation. J. Biol. Chem. 274, 36357–36361.

Ying, Y., Liu, X.M., Marble, A., Lawson, K.A., Zhao, G.Q. 2000. Requirement of Bmp8b for the generation of primordial germ cells in the mouse. Mol. Endocrinol. 14, 1053–1063.

Yoshimizu, T., Obinata, M., Matsui, Y. 2001. Stage-specific tissue and cell interactions play key roles in mouse germ cell specification. Development 128, 481–490.

Zernicka-Goetz, M. 1998. Fertile offspring derived from mammalian eggs lacking either animal or vegetal poles. Development 125, 4803–4808.

Zhang, Y., Reinberg, D. 2001. Transcription regulation by hpistone methylation: interplay between different covalent modifications of the core histone tails. Genes. Dev. 15, 2343–2360.

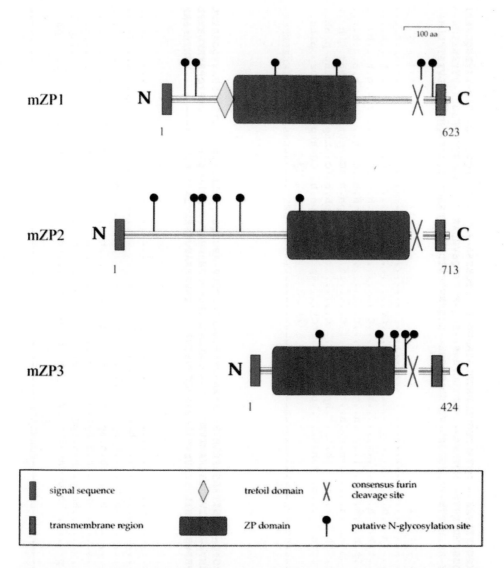

Fig. 2. (Jovine et al., see page 37, this volume.) Schematic representation of the overall architecture of mouse ZP glycoproteins, ZP1, ZP2, and ZP3. The polypeptide sequence of each ZP glycoprotein is depicted as a purple bar, drawn to scale, with the N- and C-termini indicated. Domains were identified with *SMART* (Schultz et al., 2000; http://smart.embl-heidelberg.de/) and signal peptides and transmembrane regions with *SignalP* (Nielsen et al., 1997; http://www.cbs.dtu.dk/services/SignalP/) and *PHDhtm* (Rost et al., 1996; http://maple.bioc.columbia.edu/pp/), respectively. Only putative N-linked glycosylation sites conforming to the strict pattern Asn-[^Pro]-Ser/Thr-[^Pro], where [^Pro] can be any amino acid other than Pro, are shown.

```
ZP1_MUMU          QCFKSGYFTLVMS---QETALTHGVLLDNVHLAYAPNGCP--PTQKTSA-FVVFHVPLTLCGTAIQVVGEQ-LIYENQLVSDI--
ZP2_MUMU          LCAQDGFMDFEVYS--HQTKPA--LMLDTLLVGNS--SCQ-PIFKVQSVGLARFHIPLNGCGTRQKFEGDK-VIYENEIHA--
ZP3_MUMU          ECLE-AELVVTVS-RDLFGTGKL-VQPGDLTLGSE--GCQPRVSVDTD--VVRFNAQLHECSSRVQMTKD-ALVYSTFLLH--

ClustalX          *  :. : .:       :  .    : . ::    *        .     .  *:   *: .: :::*:.*:.  :
Consensus 100%    .C...................................s.................C.......................h..
Consensus  95%    .C....h.h............................C.......h....t.C...................h..h....
Consensus  90%    .C...t.h.h..............tth..........C.......h.h..t.Cs..................h.h....
Consensus  85%    .C...t.h.h...........h..tth..........C.......hth.httCssh.............h.h....
Consensus  80%    .C...sth.l.h......h.t..h.tth.........C.......t........hph.httCsssht.....hhhp.tlh..
Consensus  75%    pCtt.sth.l..p...sh.t..l.hpsh..........tC.........ttt....hph.hstCsshp...tsthhpsplhh..
Consensus  70%    pCtt.sthl..p..hshst..l.hpshhht......tCt....tpst..h.hphshstCGshp...tsthhapsplhh..
Conserved Cys     ①                                    ②                            ③
Jpred2            ------EEEEE--------E------------------EEE------EEEE----EEEEEEEE-


ZP1_MUMU          DVQKGPQ---GSITRDSAFRLHVRCIFNAS-DFLPIQASIFSPQPP-APVTQSGPLRLELRIAT--DKT--FSSYYQGSDYPL
ZP2_MUMU          -LMENP--PSNIVFRNSEFRMTVRCYYIR--DSMLLNAHVKGHPSPEAFVKP---GPLVLVLQTYP----DQSYQRPYRKDEYPL
ZP3_MUMU          --DPRPVS-GLSILRTNRVEVPIECRYPRQGNVSSHPIQP-TWVPFRATVSSEEKLAFSLRLMEEN------WNTEKSAP

ClustalX          *           : * . .::..*. :           .           * *.  * . *:
Consensus 100%    .....................s..............................................
Consensus  95%    ..................h.C.h...........h........h.hth.............
Consensus  90%    .....h.t....h.hpC.h...........h........h.htlh................
Consensus  85%    .....h.p...h.h.hpC.a....th....h........h...h.hplht...............
Consensus  80%    ...s....hhpt.thth.hpC.a.t.th....ht....h........h.hphplht...t....
Consensus  75%    ...s....hhpp.thth.hpC.attt.ht.t.ht...h......sht...thphplhp.......p.s..
Consensus  70%    ...ps....hhpptshthtlpC.Yttt.ph.shthps.shs....shtt.t..thphplhp.......ttphs.
Conserved Cys                        ④
Jpred2            ------EEEE---EEEEEEE-----------------------------------EEEEEE--------
```

266

```
ZP1_MUMU    VRLLREPVYEVRLL--QRTDPSLVLLLHQCWATPT--TSPFEQ-PQWPILSD-GCPFKGD-NYRTQVVAADKEAL
ZP2_MUMU    VRYLRQPIYMEVKVLSRNDPN--IKLVLDDCWATSS--EDPAS-APQWQIVMD-GCEYELDN-YRTTFHPAG-SSA
ZP3_MUMU    TFHLGEVAHLQAEVQTGSHLP--LQLFVDHCVATPSPLPDPNSS-PYHFIVDFHGGLVDGLSESFSAFQVP-----

ClustalX        .    *    :::::...   ::   *.:...* **.:   .*    *    *:   **
Consensus 100%  ......h.h.............h..............h....h......
Consensus 95%   .h.htt.l.h.hth..........h.h.th.h......t.........hh...uC...
Consensus 90%   .h.ltp.l.h.hth..........h.hhhtpChst...t......ph.hl.t.GC...
Consensus 85%   .h.ltp.l.h.hth..........h.hhlppChsp.t...s....ph.hl.t.GC..t...s.h.....
Consensus 80%   hh.ltp.lhhthph...ttt...h.hhlppChspss...s....ph.ll.p.GC.hp...s.h.....
Consensus 75%   hh.ltp.lhhthph...ttt...hthhlppChsssss...s....ph.lltp.GC.hpt...sth.....
Consensus 70%   hh.lsp.lhhphph...ttt...htlhlppChssss...s.....ph.llpp.GC.hst...hhsphh.....
Conserved Cys
Jpred2          EEE---EEEEEEEE----------------------EEEEE

                                                        ⑤                          ⑥


ZP1_MUMU    PFWSHYQRFTITTFMLLDSSSQNALRGQVYFFCSASACHPLG--SDT---CSTTCDSG
ZP2_MUMU    AHSGHYQRFDVKTFAFVS--EARGLSSLIYFHCSALICNQVS--LDSP-LCSVTCPAS
ZP3_MUMU    RPRPETLQFTVDVFHFAN----SSRNTLYITCHLKVAPANQIP-DKLN--KACSFN

ClustalX        .    :*:.*:.   .    .    :*:   *.       :*  .
Consensus 100%  ...............................h.............
Consensus 95%   ...............a.h..........l.h.sth.hh...........s...
Consensus 90%   ...t.....h.Fth...........l.h.Cph.hs..........C...
Consensus 85%   ...t.....h.Fta.t..........lhh.Cph.hs..........C...
Consensus 80%   ...tthhth.h.Fta.t.........tt.lhhpCph.hs..........tC...
Consensus 75%   ...tthhthph.hFpF.s........ts.lhhpCphths..........tCs...
Consensus 70%   ...tphhphphpFpFss.......ps.lahpCplphC.tt..........tCs...
Conserved Cys
Jpred2          ------EEEEEE------------------EEEEEEEEEE

                                                    ⑦                    ⑧
```

Consensus keys
----

a : aromatic (F, Y, W, H)

l : aliphatic (I, V, L)

h : hydrophobic (F, Y, W, H, I, V, L, A, G, M, C, K, R, T)

p : polar (H, K, R, D, E, Q, N, S, T, C)

u : tiny (G, A, S)

s : small (G, A, S, V, T, D, N, P, C)

t : turnlike (G, A, S, H, K, R, D, E, Q, N, S, T, C)

. : any (G, A, V, I, L, M, F, Y, W, H, C, P, K, R, D, E, Q, N, S, T)

Jpred2 keys

E : sheet

- : loop

Fig. 3. (Jovine et al., see pages 38–40, this volume.) Consensus analysis and secondary structure prediction of ZP domain sequences. The polypeptide sequences of the ZP domains of mouse ZP glycoproteins, aligned using a non-redundant 70% threshold ZP domain sequence collection (see text), are superimposed on the consensus patterns for the domain, calculated at different thresholds using the program *Consensus* (http://www.bork.embl-heidelberg.de/Alignment/consensus/html). Also shown are the *ClustalX* (Thompson et al., 1997) output for the alignment of the three ZP glycoproteins, a *Jpred2* (Cuff and Barton 1999) secondary structure prediction based on the full sequence collection, and the positions of the conserved 8 Cys residues. Identical amino acids in both the alignment and consensus patterns are indicated in red and marked with "*" in the *ClustalX* output. Conserved positions are encoded and color-coded using the *Consensus* keys reported below the alignment, and marked with either ":" or "." in the *ClustalX* output according to their degree of conservation. The *Jpred2* output is also encoded using the keys specified.

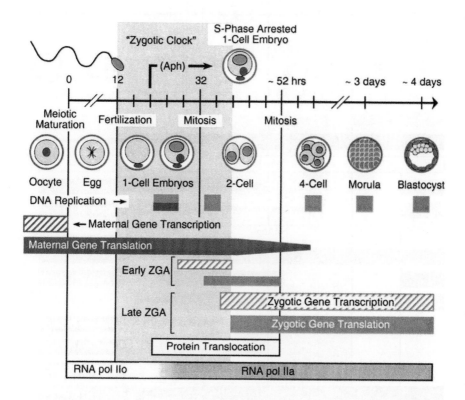

Fig. 1. (DePamphilis et al., see page 57, this volume.) Maternal to zygotic gene transition in the mouse. Maternal events are indicated in *red*, paternal events in *blue*, and zygotic events in *green*. Open bars apply to both. Embryonic stem cells ("inner cell mass") are indicated in *yellow*; trophectodermal cells in *orange*. Periods of transcription are indicated by hatched bars; translation by solid bars.

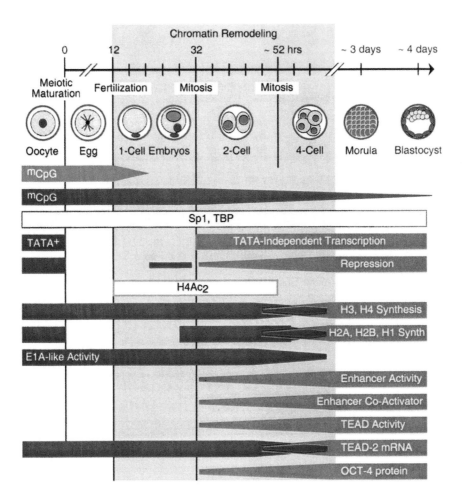

Fig. 2. (DePamphilis et al., see page 63, this volume.) Developmental changes affecting regulation of gene expression at the beginning of mouse development. These include changes in DNA methylation of the paternal and maternal genomes, changes in histone synthesis and modification that lead to chromatin-mediated repression, acquisition of chromatin-mediated repression, and acquisition of enhancer function. Some transcription factors such as Sp1 and TBP are ubiquitous. TATA-box function appears to be restricted to differentiated cells. Color coding and bar coding are the same as in Fig. 1.

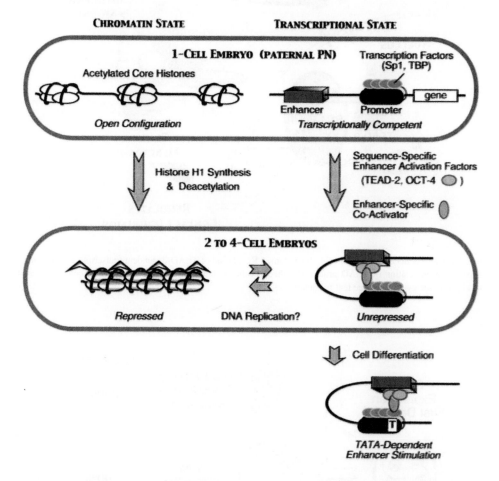

Fig. 3. (DePamphilis et al., see page 67, this volume.) Relationships between chromatin structure, promoter/ enhancer activity and DNA replication during the maternal to zygotic gene transition in the mouse. Chromatin in the paternal pronucleus is in an open configuration that does not suppress promoter activity. Formation of a 2-cell embryo is accompanied by changes in chromatin structure that repress promoter activity. This repression can be relieved by enhancer activity which requires sequence specific-enhancer binding proteins such as TEAD-2 or OCT-4, and an as yet unidentified enhancer specific coactivator, all of which first become available during ZGA. DNA replication may facilitate activation of some genes. Cell differentiation is accompanied by the need for a TATA-box element in the promoter in order for enhancers to function.

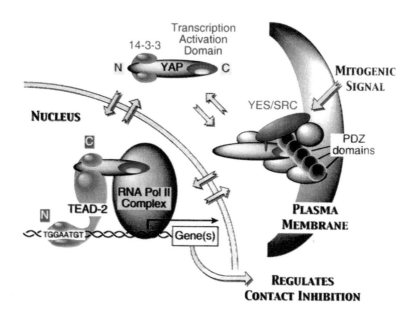

Fig. 4. (DePamphilis et al., see page 74, this volume.) Regulation of TEAD-dependent transcription in mammals depends on association of TEAD protein (localized in the nucleus) with YAP65, a transcriptional coactivator localized in the cytoplasm. See text for details.

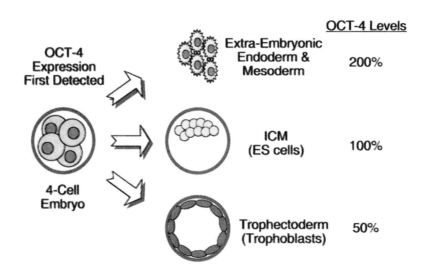

Fig. 5. (DePamphilis et al., see page 76, this volume.) Regulation of mouse ES cell development depends on the level of OCT-4 expression. Normal pre-implantation levels stabilize totipotent ES cells. Higher levels cause differentiation into endoderm and mesoderm, while lower levels induce differentiation into trophectoderm. See text for details.

A. Early blastula

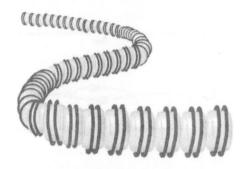

Fig. 1. (Veenstra, see page 88, this volume.) Model of transcriptional regulation during early embryonic development in *Xenopus*.

A. The embryonic genome is transcriptionally quiescent before the MBT, due to a repressive chromatin structure (symbolized by a high density of nucleosomes in this panel), and constraints on the transcription machinery. Constraints on the transcription machinery include rate-limitation of transcription initiation due to low levels of TATA binding protein (TBP), cytoplasmic retention of transcriptional regulators, and other constraints on the transcriptional activation.

B. In late blastula embryos, chromatin is less repressive toward transcription (symbolized by lower density of nucleosomes) and many genes are transcribed, which for a subset of genes is facilitated by the developmentally regulated translation of TBP mRNA. A number of genes are transcribed at low levels by the general transcription machinery, with constraints on activator function still in place. The general transcription machinery is depicted at "open" spots of the chromatin, with the transcription start site depicted with an arrow.

C. During gastrulation chromatin becomes more repressive towards transcription due to incorporation of linker histone H1 into chromatin and a more prominent role of histone deacetylases. A more prominent role for targeted, gene-selective activation and repression events is observed, symbolized by the presence of additional proteins (circle and oval) in the vicinity of the general transcription machinery. As a consequence, some genes are induced to high levels whereas some of the genes that were transcribed initially are repressed by the time of gastrulation, symbolized in this panel by disappearance of one transcription complex (compare with panel B).

B. Late blastula

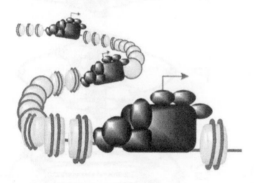

C. Gastrula

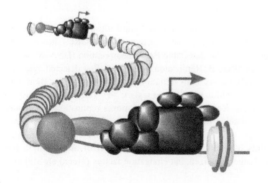

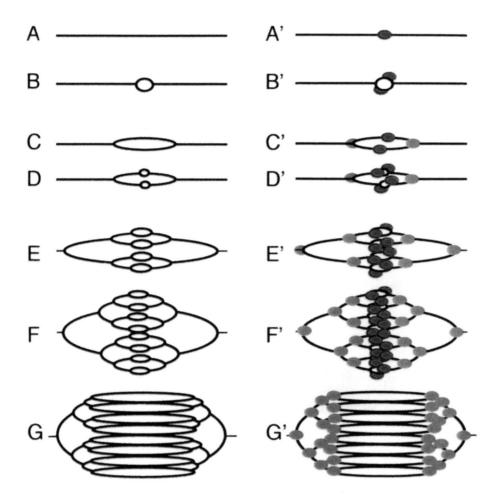

Fig. 3. (Bosco and Orr-Weaver, see page 126, this volume.) Schematic and cytological illustration of gene ampli-fication. A. A linear chromosome before replication initiates loads initiator proteins, such as ORC (shown in red, A′), and is first detected in stage 10A. At this stage there is no detectable BrdU incorporation (DNA synthesis) and it is not clear whether initiation events are taking place. B. Replication initiation is first detected cytologi-cally at stage 10B and initiator proteins are still bound (B′). C–F. Many initiation events occur within a short win-dow of time, forming an "onion skin" structure of bubbles within bubbles (see text). C′–F′. Initiator proteins (red) are still bound to chromatin as intiation events continue. Replication elongation moves replication forks away from the origin and replication factors involved in elongation (e.g. PCNA and MCMs, depicted in green) travel with the replication forks. G. Late stage 10B and stages 11–13 continue to synthesize DNA as replication forks proceed away from the origin, but initiation events have ceased. The maximum level of amplification has occurred at the origin. G′. Initiator proteins are no longer localized, but elongation factors (green) are still pre-sent at replication forks.

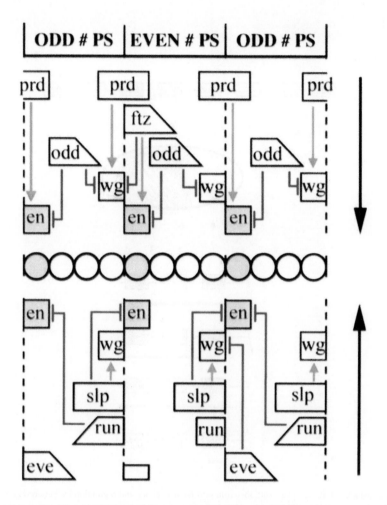

Fig. 9. (Nasiadka et al., see page 177, this volume.) Initial regulation of *engrailed* (*en*) and *wingless* (*wg*). Expression of *en* and *wg* begins at the end of cellular blastoderm (stage 5) as a result of complex gene-regulatory interactions mediated by pair-rule gene products. A schematic representation of three consecutive parasegments is shown. The circles in the center represent a row of cells along the anterior–posterior axis. These are placed in the center of the figure for clarity with gene interactions diagramed both above and below. Stripes of gene expression are depicted as boxes or trapezoids. Sloped sides indicate decreasing levels of expression. Positive gene-regulatory interactions are marked in green and negative interactions in red. Stripes of *en* and *wg* initiate in single rows on either side of the parasegmental border.

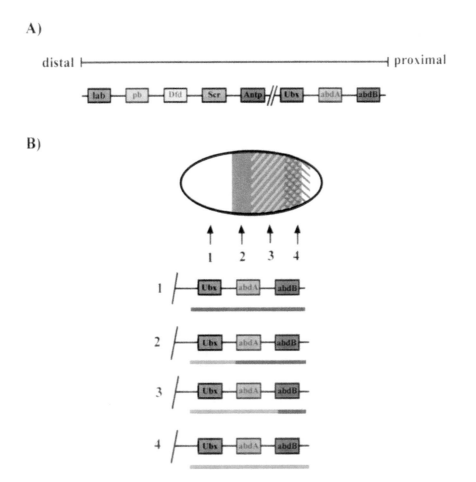

Fig. 10. (Nasiadka et al., see page 180, this volume.) *Hom-C* genes and regulation by PcG/trxG complexes. (A) The eight genes of the *Hom-C* are displayed with their positions along the chromosome (proximal to distal) indicated. In *Drosophila*, the ANT-C (left) and BX-C (right) complexes are not contiguous (break indicated by //). (B) Postulated control of the BX-C genes by the PcG/trxG. In the anterior region where genes of the BX-C are inactive (1), PcG complexes create a repressive chromatin domain along most of the BX-C (denoted by red line). In more posterior regions, where genes of the BX-C become activated (2 and 3), this repressive complex is replaced by transcriptionally active trxG complexes (green line) such that the boundary between PcG and trxG complexes shifts in a proximal to distal direction. In more primitive insects, this movement of the PcG/trxG boundary may be processive. *lab, labial; pb, proboscipedia; Dfd, Deformed; Scr, Sex combs reduced; Antp, Antennapedia; Ubx, Ultrabithorax; abdA, abdominalA; AbdB, AbdominalB.*

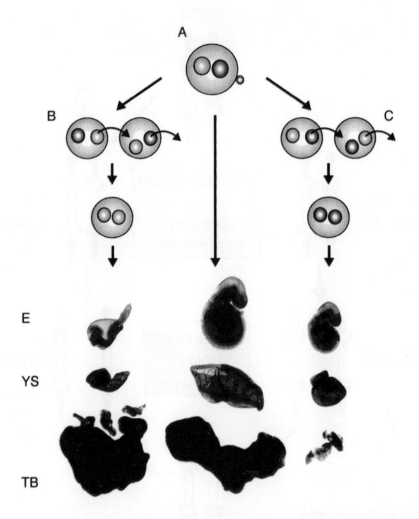

Fig. 1. (Arney et al., see page 235, this volume.) Both maternal and paternal contributions are required for normal mammalian development. A. A normal fertilised mouse zygote contains both a paternal (blue) and maternal (pink) genome, in the form of pronuclei. The normal embryo (E) at 10.5 days post coitum (dpc), yolk sac (YS) and trophoblast (TB) are shown in the centre panel. B. If pronuclear transfer is used to generate embryos containing two paternal genomes (androgenetic embryos), an abnormally small and retarded embryo is seen at 10.5 dpc. Note the deficient yolk sac and excessive amounts of trophoblast tissue. C. Similarly, two maternal genomes lead to abnormal development (gynogenetic or parthenogenetic embryos). In this case, the embryo appears normal, but small, yet is extremely lacking in trophoblastic tissue at 10.5 dpc. Neither the androgenetic nor parthenogenetic embryos can sustain development much further than this timepoint and are subsequently resorbed by the mother.

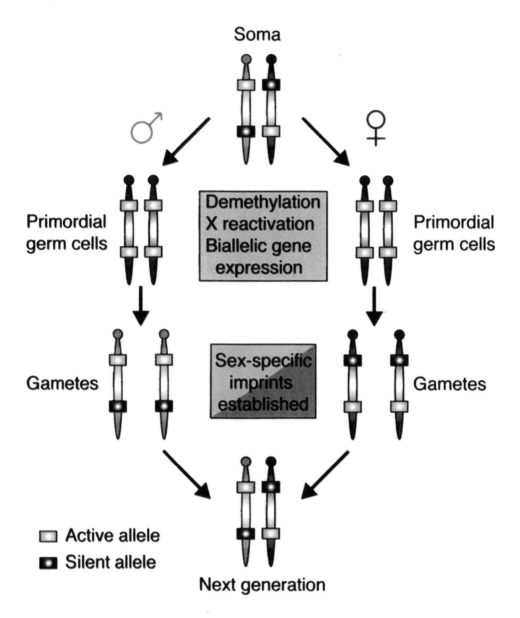

Soma

Primordial germ cells

Demethylation
X reactivation
Biallelic gene
expression

Primordial germ cells

Gametes

Sex-specific
imprints
established

Gametes

☐ Active allele
▣ Silent allele

Next generation

Fig. 2. (Arney et al., see page 236, this volume.) The imprinting cycle. The somatic cells of a mammal bear both maternal and paternal imprints, interpreted as monoallelic gene expression. In the primordial germ cells of the developing animal, these imprinting marks are completely erased, leading to demethylation in a global and locus-specific manner, biallelic expression of imprinted genes and reactivation of the silent X-chromosome in females. During subsequent game to genesis, sex-specific imprints are re-established on the chromosomes. Fertilisation results in a new organism with a correct complement of parental imprints in the next generation.

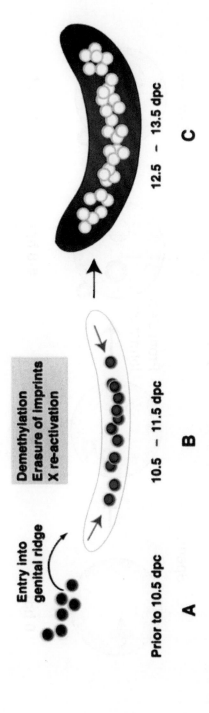

Fig. 5. (Arney et al., see page 247, this volume.) Demethylation of PGCs in vivo. A. Migrating PGCs have a somatic methylation pattern (indicated by red nuclei) at the time they colonise the sexually bipotential developing genital ridges (10.5–11.5 days post coitum). B. Reprogramming of the PGC genome may occur in response to signals from the surrounding gonadal cells (blue arrows). This reprogramming is seen as demethylation of the genome, erasure of parental imprints and reactivation of the silent X chromosome in female cells. C. By 12.5 dpc the demethylation of single copy loci is complete (indicated by pink nuclei). Note that the gonads are sexually dimorphic at this point.

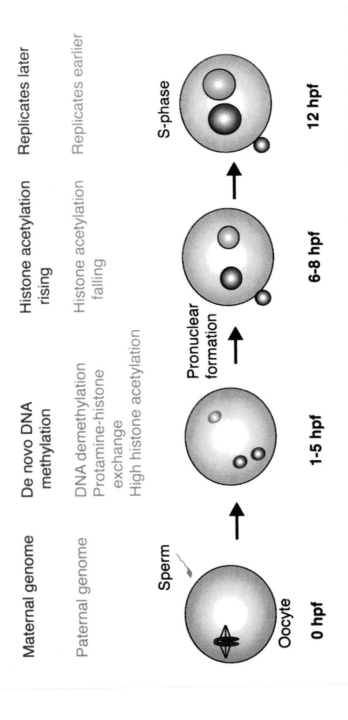

Fig. 6. (Amey et al., see page 252, this volume.) Epigenetic events in early development. The maternal genome is represented in pink, while the paternal genome is blue. Timings are given as hours post-fertilisation (hpf). Between 1and 5 hpf a number of epigenetic events occur in the zygote. On the paternal genome, protamines are exchanged for histones, there is a rapid global DNA demethylation event, and high levels of histone acetylation can be detected. In contrast, the maternal genome undergoes some *de novo* DNA methylation. Later, between 6 and 8 hpf, levels of histone acetylation fall on the paternal genome, while rising on the maternal genome. S-phase (DNA replication) occurs around 12 hpf, with the paternal genome replicating prior to the maternal genome.

Printed and bound by CPI Group (UK) Ltd, Croydon, CR0 4YY

08/05/2025

01864966-0009